The Final Theory
Rethinking Our Scientific Legacy

Mark McCutcheon

Universal Publishers
Boca Raton • 2004

The Final Theory: Rethinking Our Scientific Legacy

Universal Publishers/uPUBLISH.com
Boca Raton • 2004 • 1st revised edition

ISBN: 1-58112-601-8

www.universal-publishers.com

Library of Congress Cataloging-in-Publication Data

McCutcheon, Mark, 1965-
 The final theory : rethinking our scientific legacy / Mark McCutcheon.
  p. cm.
 ISBN 1-58112-601-8 (alk. paper)
  1. Quantum theory. 2. Special relativity (Physics) 3. General relativity (Physics) 4.
Expanding universe. I. Title.

QC174.13.M33 2004
530.14'2--dc22
                                        2004051668

To my parents for their support, my father for his considered feedback throughout and many long hours of editing, and friends who offered their time and comments along the way

Introduction

We are all born into this universe and live out our lives within its laws and principles. From the inescapable law of gravity extending across the universe to the fundamental principles behind the tiniest atoms, our lives are immersed in the laws of nature. As intelligent beings it is only natural for us to wonder about the world around us, and as children of this universe it seems reasonable that we should be able to arrive at an understanding of it all – that this understanding is very much our birthright. In fact, to many it may seem as if we have already arrived at this understanding, with only a few loose ends remaining. Isaac Newton gave us an understanding of gravity as an attracting force in nature, and from there many others have contributed to our understanding of light, electricity, magnetism, atomic structure, etc. This process has finally brought us to a point where science today contains theories that cover every known observation, collectively known as *Standard Theory*. This age of understanding has made it possible to invent radio, television, and computers, even allowing us to build spacecraft that have visited distant planets. Although scientists continue to pursue deeper questions, it may seem that Standard Theory provides us with a fairly comprehensive scientific understanding of our universe. But is this really the case?

How much do we *truly* understand about gravity, for example? Do we know the physical reasons why gravity attracts objects together instead of repelling them away from one another? Newton gave us a compelling *description* of this observation as an apparent attracting force, but provided no *explanation* for the existence and nature of this force itself. Does it really make sense that a force holds objects to the surface of planets, and moons in orbit, all with no known power source? Do we know if it is possible to create some type of anti-gravity device, what principles might underlie such a device, or for that matter, even what principles underlie gravity itself? And despite Newton's concept of gravity, Albert Einstein found it necessary to continue searching for answers, arriving at a very different description of gravity, while scientists continue to search for still other explanations. Why is it that we have two explanations for the same effect in our science today, and continue to search for still others – and do any of them truly answer our most basic questions about gravity?

Do we *truly* understand light? For centuries a debate raged back and forth as to whether light was composed of waves or particles. Today we have settled on a belief that somehow light is *both* a wave *and* a particle (the photon) – sometimes behaving as one and sometimes as the other, depending on the situation or experiment. Even today this remains a very mysterious and poorly understood characteristic of light as part of a theory known as *Quantum Mechanics* – a theory whose very creators and practitioners readily describe as bizarre and mysterious.

Do we *truly* understand magnetism? We know that two magnets will repel each other if both of their north poles or south poles face each other, but can we truly explain this? If we try to hold these two magnets together against this repelling force our muscles will tire as we continuously expend energy, but the repelling force from within the magnet does not. Is it reasonable that an apparently *endless* force from within magnets will continuously battle any external power source in this manner, eventually exhausting all external power sources without an equivalent weakening itself? In fact, there is *no identifiable power source at all* within these magnets to support this endless force from within. Do we even know what magnetic fields are, or have we simply discovered how to create them and learned to model their behavior with equations? That is, are we confusing practical know-how and abstract models with true knowledge and understanding?

A closer look shows that solid answers to these and many other questions about everyday occurrences are not to be found in today's Standard Theory. Science has managed to *model* our observations rather well, but many of these models lack a clear physical explanation. Newton worked out a *model* of gravity as an attracting force but couldn't tell us *why* it should attract and *how* matter does this endlessly simply by existing; in fact, we still lack these answers three hundred years later. We have equations that *model* magnetic fields, and theories that describe their obvious observed behaviors, but we have little clear physical explanation for *why* they behave as they do, leaving mysteries such as the apparently endless energy emanating from within a simple permanent magnet. In fact, many scientists recognize that we still lack a deep understanding of our universe, which is why there are ongoing efforts to further our knowledge using high-energy particle accelerators and powerful space telescopes. The hope is that these investigations will lead to a key breakthrough in understanding – perhaps through the discovery of a currently unknown fundamental subatomic particle or principle, or possibly via some new type of energy or cosmological phenomenon detected in the heavens. It is expected that if such a key fundamental discovery is made, it will have a ripple effect that runs through the patchwork of often poorly understood theories in our Standard Theory today, ideally transforming them into a single clear theory that simplifies and truly explains everything. This much-hoped-for theory is known by physicists as the *Theory Of Everything* – and is considered the ultimate goal of much fundamental research in physics today.

A key expectation of the Theory Of Everything is not only that it will finally explain all of physics – gravity, light, magnetism, etc. – with a clarity and simplicity that is unknown today, but that it will do so via *one single unifying principle* in nature that has so far eluded us. Once found, this theory is expected to provide a clarity and understanding akin to turning on a light to see the contents of a room at a glance, where current theory is like a flashlight in the dark, giving only disconnected glimpses here and there. A less comprehensive form of this theory, known as the *Unified Field Theory*, would explain and unify everything *except* gravity, since it is thought that gravity may have a very different nature than the other fields and forces once we come to truly understand them all. Both theories are sought after by physicists around the world today, with the ultimate goal being the arrival at an understanding that explains all the forces of nature *including* gravity – i.e. the all-encompassing Theory Of Everything.

Although this fairly formal definition of the Theory Of Everything has only taken shape within the last century, it has actually been the ultimate goal of science ever since the earliest times; even medieval alchemists were, in their own way, searching for this ultimate understanding of the physical world. Some of Newton's many contributions to science were his descriptions of gravity, light, and the mechanics of moving objects, while Einstein provided quite different descriptions of these phenomena, with additional ideas about energy, mass, space and time. Both of these scientists were essentially in pursuit of the Theory Of Everything, whether or not their efforts were formally presented as such, as are many scientists who pursue basic research in an attempt to discover fundamental truths about our universe.

So far, our efforts have not yielded the Theory Of Everything, but rather a "theory of everything" known as Standard Theory. Although it isn't typically represented this way, Standard Theory is indeed a "theory of everything" since it attempts to explain every known observation and phenomenon. It has evolved from many hypotheses presented over the centuries, with the most successful ones incorporated as sub-theories *within* Standard Theory. Even such radical and mysterious theories as *Quantum Mechanics* and *Special Relativity* are not considered part of some other "theory of everything" but part of Standard Theory today. Therefore, Standard Theory is not only a "theory of everything," but it is also the *only* one so far. In order for a new theory to truly form the basis of another "theory of everything" it would have to be based on a principle that lies entirely outside of known physics – *and* provide a sweeping rewrite of everything in Standard Theory based entirely on this new principle. Figure 1-1 shows the patchwork of theories within Standard Theory today that have resulted from our "flashlight-in-the-dark" approach to science over the past few centuries, as well as the single illuminating perspective of the Theory Of Everything that is expected once the correct underlying principle is discovered.

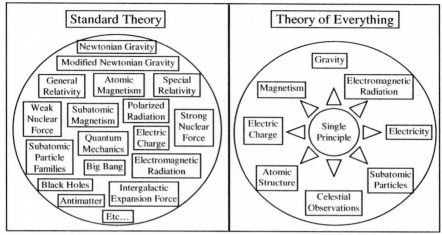

Fig. 1-1 Patchwork of Theories Today vs. Theory Of Everything

The chapters to follow present just such a new principle in physics, showing that all matter may well possess this important new property that has so far been overlooked or misunderstood, and developing this principle into a second "theory of everything" for us to consider. This new theory begins with a clear physical explanation for gravity that resolves the many questions and mysteries surrounding it today, such as why it behaves as an apparent attracting force and how it functions without a power source. Planetary orbits, ocean tides, and all other known gravitational observations are entirely explained by this new theory without relying on our current theories of gravity. New insights and possibilities are also suggested by this new theory that are unknown today and would not be predicted by our current gravitational theories.

This same new principle further explains the structure of the atom, as well as the nature of the individual electrons, protons, and neutrons composing atoms, with a physical simplicity and clarity that is unknown today. This new perspective on atomic structure shows how the gravity of objects can be directly related to the electricity and magnetism produced by the flow of electrons in wires, since this new principle underlies both atoms and electrons. The apparently endless energy within magnets mentioned earlier is also explained by this new principle, and a clear physical reason is given for why electricity and magnetism are so closely related. This principle also suggests an explanation of electron orbits within atoms that resolves this still mysterious aspect of atomic theory in our science today.

This same new principle is further shown to explain the nature of light, suggesting a resolution to the age-old question of whether light is a particle or a wave ... or indeed something else entirely. Since the mysterious wave-particle beliefs about light in Standard Theory support a sizable portion of the theory of *Quantum Mechanics*, resolving this issue has serious implications for *Quantum Theory*. In fact, our current quantum mechanical descriptions of atomic structure, light, and energy are shown to be unnecessary once the new unifying principle is considered. This should be expected of any alternate "theory of everything" since, by definition, it would have to be entirely separate and self-sustaining without relying on any of the patchwork of theories that compose Standard Theory today – of which *Quantum Mechanics* is one. As might be further expected then, Einstein's *Special Relativity Theory* is also shown to have serious problems, and is also replaced by this new principle. This means we can now replace the complexities and mysteries of *Quantum Mechanics* and *Special Relativity* with one simple principle that runs throughout our science, dispelling some long-standing mysterious beliefs such as the speed-of-light limit that we accept as true today. All of the well-known thought experiments and real-world experiments that are used to support these mysterious theories and beliefs are re-examined and shown to have serious flaws, misunderstandings, or even clear errors upon closer examination.

Finally, the same simple principle is shown to explain the many mysterious phenomena and particles that have emerged from high-energy particle accelerator experiments in recent decades, such as *virtual particles* and *antimatter*, removing the mystique that surrounds them today. This new explanation of subatomic particle experiments also suggests a new interpretation for the increasing number of new particle types that are being discovered in ever more powerful particle accelerators. It also provides a new perspective on Einstein's idea that matter and energy can be converted back and forth (according to his famous equation, $E=mc^2$). Rather than this mysterious conversion of matter into energy in the explosion of an atomic bomb, or energy into matter when subatomic particles apparently materialize out of pure energy in particle accelerators, this new unifying principle provides a clear, demystifying explanation for both effects. This principle also speaks to many of our celestial observations, suggesting simple alternate explanations for observations leading to today's more mysterious theories about Black Holes, the "Big Bang" creation event, and the apparently accelerating expansion of our universe.

The alternate explanations presented throughout this book do not constitute a string of proposed new theories *within* Standard Theory, but belong to a new and *entirely alternate theory* – an alternate "theory of everything." This parallel explanation of our universe provides answers to the many questions and mysteries in our science today with a clarity that allows even non-scientists to truly comprehend our universe – and does so via one simple unifying principle that is consistent with all known experiments and observations. It is worth noting that this last point is a claim that cannot be made even of Standard Theory today. That is, as shown in each of the following chapters, within many of our everyday experiences lie unanswered questions, unexplained mysteries, and even clear violations of our most elementary laws of physics when explained with Standard Theory. Therefore, as it stands today, our current body of scientific knowledge is not merely lacking some answers, but is actually a *fatally flawed* "theory of everything." While it is possible that our ongoing search for answers will be able to resolve these flaws and turn Standard Theory into the much-sought-after Theory Of Everything, it is equally possible that the answers can only be found in an entirely new "theory of everything." It is suggested that the new theory presented in the following chapters does not merely provide an entirely alternate way of viewing our universe, but that it is the only one to meet the criteria of the Theory Of Everything for which science has been searching for centuries. However, this will be up to the scientific community, as well as each individual reader, to decide for themselves. We now begin the journey toward discovery and understanding of this new principle with an exploration of *gravity*.

First ... A Note on Format

Although this book is intended for both scientists and non-scientists alike, it does represent a sweeping re-think of our complete body of scientific knowledge today. Therefore, in order to help organize the discussions, as well as to quickly identify key points and their significance, summary boxes or icons will accompany key sections or phrases as follows:

NOTE
    Highlights a key point in a discussion.

WATCH FOR...
    • Lists key points in the discussion to follow.
     •

NEW IDEA
    Introduces a new idea for consideration.

LAW
    Reminder of a current law of physics in Standard Theory.

VIOLATION
    Indicates a physical law violation in a current scientific belief.

MYSTERY
    Indicates an unexplained mystery in a current scientific belief.

ERROR
    Indicates a logic or math error in a current scientific belief.

EXPERIMENT

    Presents a thought experiment or real-world experiment.

OPTIONAL MATH

    Indicates that math follows, but is optional reading which is explained in either the preceding or following section.

- 1 -

Investigating

Gravity

The Theory of Gravity

Gravity as One of Four Basic Forces in Nature

Gravity is one of the most fundamental and familiar forces of nature. As such, before discussing gravity in particular, it is important to clarify what the forces of nature are considered to be and how they relate both to Standard Theory and to our ultimate quest for understanding. Although Standard Theory is a composite of many sub-theories, some of which were listed earlier in Figure 1-1, most scientists believe the search for the Theory Of Everything is a quest to understand and unify what are currently considered to be the four separate fundamental forces of nature:

- *Gravity* – the familiar attraction between all matter, first described by Isaac Newton.

- *Electromagnetism* – the closely related phenomena of electricity and magnetism, as well as electromagnetic radiation such as radio waves and light.

- *Strong Nuclear Force* – a powerful, short-range force thought to be holding atomic nuclei together. Atomic nuclei have many positively charged protons in close proximity, which should strongly repel each other and cause the nucleus to fly apart according to the theory of *Electric Charge*. Therefore, the concept of an attracting *Strong Nuclear Force* between protons in the nucleus was introduced to explain how the nucleus is held together in apparent violation of *Electric Charge Theory*.

- *Weak Nuclear Force* – another nuclear force, considered to be much weaker than the *Strong Nuclear Force*. Phenomena such as the random decay of populations of subatomic particles (i.e. radioactivity) were difficult to explain until the concept of this additional nuclear force was introduced.

It is currently believed that these are the four fundamental forces in nature, and that, in essence, they are merely different manifestations of one single underlying force or principle that has so far eluded science. To discover this underlying force or principle would be to arrive at the Theory Of Everything since, at a glance, it would show the single underlying cause for every observation, belief, and theory in science today. Such a unified understanding is expected to transform the patchwork of separate abstract theories in Standard Theory into a much simpler, coherent whole that shows a true *physical* explanation for everything, sparking a scientific revolution.

The new theory discussed throughout these chapters suggests that while this vision is the proper intuition, there are several reasons why success has eluded us so far. First, since we obviously lack the deeper understanding that we are seeking, we cannot be certain we have properly identified the fundamental forces of nature. If, for example, our theory of *Electric Charge* is an imperfect model of the true underlying principle behind many of our observations, then our current model of proton behavior as positively charged particles that always repel each other may not be an accurate description of the nucleus of an atom. Instead, it may be perfectly natural for protons to cluster together when in the nucleus of an atom, according to an undiscovered principle in nature that may have been misunderstood and represented as a "positive electric charge" upon protons. In that case, the concept of a *"Strong Nuclear Force"* keeping the nucleus from flying apart would be a completely unnecessary fabrication, and our attempts to find a unifying theory would be based in part on a force that doesn't even exist. Our current goal of unifying these four forces may be based on such flawed assumptions from the start.

Secondly, much of our current and largely mathematical approach to finding a unifying theory may be straying from the original spirit and purpose of the quest. The goal of a new and deep physical understanding of our universe may be in danger of merely becoming an exercise in mathematical manipulation of our current equations. Since arrival at this deep physical understanding is expected to yield a common mathematical framework for all the forces of nature, it is often assumed that if we simply pursue this mathematical end result directly –

using our current models – we will achieve this deeper understanding. However, this approach may be unsound since it assumes we have correctly identified the fundamental forces of nature and simply need to rearrange our mathematical models. Yet, if this turns out to be an incorrect assumption, then such an approach would only achieve a largely meaningless mathematical link between flawed models of the physical world. This approach also risks trivializing our search for deeper *physical* understanding into an attempt to achieve a mere *mathematical* goal, bringing no deeper meaning. We may expect mathematically unified models to emerge once we achieve a deep physical understanding of our universe, but this does not necessarily mean this deep physical understanding will emerge by mathematically unifying our current models. It is possible that this approach may provide some useful insights, but it may also result in little more than contrived mathematical relationships between essentially the same equations modeling the same limited physical understanding we have today.

For the reasons mentioned above, the discussions of this new "theory of everything" in the coming chapters do not strictly follow the format of a mathematical unification of the "four fundamental forces" in nature. In fact, there is very little math and only loose references to these forces amidst a broad and rich discussion of science in clear physical and common-sense terms. The discussions do, however, begin with the first of these forces – *gravity* – showing the numerous problems with our current gravitational beliefs, and leading to an introduction of the new unifying principle behind a new theory of gravity that resolves these problems. Once this new principle is established, it does indeed ripple through the rest of Standard Theory in the chapters that follow, not only redefining our concept of the "four fundamental forces," but redefining the complete patchwork of theories in science today in clear physical terms.

The Trouble with Gravity

Newton's Theory of Gravity is undoubtedly one of the most universally recognized and accepted theories in all of science. It has become so deeply ingrained in our thinking and our science over the centuries that this theory has largely become synonymous with the very phenomenon of gravity itself. It is almost inconceivable today to separate our everyday experience of gravity from Newton's proposal of an attracting force emanating from all matter; yet, as shown in the following discussions, Newton's theory actually contains many unexplained mysteries and scientifically impossible claims. Such problems should prevent any new theory from becoming widely accepted as fact, leaving it only with the status of a proposal or *hypothesis*; however, the compelling nature of Newton's proposal combined with the lack of a more viable theory has meant that it has largely escaped such scrutiny.

WATCH FOR...

- Newton's theory of gravity does not explain *why* objects attract one another; it simply models this observation.
- There is no known power source supporting the gravitational field that Newton claims to be emanating from our planet and from all objects.
- Despite the ongoing energy expended by Earth's gravity to hold objects down and the moon in orbit, this energy never diminishes in strength or drains a power source – in violation of one of our most fundamental laws of physics: the *Law of Conservation of Energy*.
- These mysteries and violations are overlooked today because of a flawed explanation that arises from the improper use of an equation known as the *Work Function*.
- Every effect explained by Newton's theory of gravity today is accurately modeled by *non-gravitational* equations that existed even before Newton.
- Newton's gravitational force is actually an entirely redundant and superfluous concept providing no additional usefulness and having no proven existence in nature or scientific support.

Newton's Error – Violations of the Laws of Physics

Gravity is one of the most familiar and important phenomena in nature. Although it has always been known that *something* obviously causes objects to fall, it wasn't until Isaac Newton (1642-1727) that we had a clear model of this *something* as an attracting force emanating from all matter in a manner that is precisely describable via an equation. Newton also claimed that this very same attracting force was responsible for the orbits observed in the heavens, making our universe as comprehensible and predictable as a clockwork mechanism for the first time in history. This was such a monumental achievement in Newton's day that it set the stage for other models of forces described by equations in similar fashion ever since.

Although today we commonly speak of such forces, it is often overlooked that modern science still has little or no solid physical explanation for many of them. The legacy of theories and equations that compose our body of scientific knowledge today works rather well, making it easy to forget that these are largely *abstract models* – not solid physical explanations. Newton was the first in a long line of scientists to produce explanatory models for various classes of phenomena, which can be very compelling and useful but cannot be fully explained in physically meaningful and scientifically viable ways even today.

In fact, there was a strong undercurrent of resistance to Newton's gravitational force concept when it was introduced, since it seemed to represent an almost magical force at a time when solid rational thought was finally beginning to prevail over the mysticism and superstition of ages past. Today, largely as a result of the scientific acceptance of Newtonian gravity, we have grown accustomed to the idea of unexplained forces reaching across empty space to affect objects at a distance in some equally unexplained manner. We have even grown accustomed to the fact that many of these forces (gravity, magnetism, electric charge, etc.) have no known power source. However, in Newton's time such concepts were only known in stories of myth and magic. To philosophers such as René Descartes (1596-1650), it had been a long journey for society to shake off the mysticism of the past and finally enter a welcome era of solid rational thought and debate.

In fact, Descartes himself had an earlier and widely accepted *physical* theory of orbits that claimed the planets were dragged along by an invisible material, known as the ether, which presumably swirled around the sun. Although this theory had its own problems, in this era of rationality many considered Newton's idea of a completely unexplained force acting across empty space to be an unwelcome return to the magical thinking of the past. Newton realized this fundamental problem with his theory of a gravitational force, and never claimed to be able to explain it. However, the compelling and rational nature of his accompanying mathematical model soon solidified the force of gravity as a physical reality and a scientific fact that continued to grow in acceptance for centuries, being the predominant theory even today.

It is important to note, however, that although it is generally recognized that Newton's gravitational force lacks a proper physical explanation, the much larger issue – that it *violates the laws of physics* – has gone almost entirely unnoticed. This point will be clearly illustrated, beginning with a reminder of one of the most fundamental and unbreakable laws of physics – *The Law of Conservation Of Energy*.

LAW

 The Law of Conservation Of Energy

Energy can neither be created nor destroyed, but merely changes from one form to another.

This is one of the most fundamental and unbreakable laws of physics, serving as a test for the scientific validity of any proposed theory or invention. If a proposed theory or device either uses or produces energy it must draw on an existing power source to do so, merely transforming energy from one form to another in the process. For example, the stored chemical energy in gasoline changes to kinetic energy as it is "used up" to accelerate a vehicle. In accordance with the *Law of Conservation Of Energy*, the chemical energy in the gasoline does not actually vanish, but is converted into another form of energy – the kinetic energy of the vehicle's motion. Similarly, the kinetic energy of the vehicle did not simply appear out of nowhere, but was converted from an existing chemical energy source – the gasoline. Although we commonly refer to power sources being *drained*, what we actually mean by this is that the energy from a given power source is converted into another form of energy elsewhere. This is the law that tells us perpetual motion machines are impossible since they are considered to be devices capable of producing or expending energy continually without draining a power source. There is no such thing as "energy for free" in our science. Free energy devices violate our most elementary laws of physics.

Also noteworthy, once it was realized that energy (denoted by the symbol *E*) and matter (denoted by *m* for mass) can change form back and forth, modeled by Einstein's famous equation $E=mc^2$, the *Law of Conservation Of Energy* included *matter* as one of the energy forms. The explosion of an atomic bomb, for example, does not actually *create* the enormous amount of energy in its explosion, but is considered to *release* it by converting its original core of matter into energy. Therefore, in all things the *Law of Conservation Of Energy* must be upheld.

VIOLATION

 Newton's Gravitational Force Violates the *Law of Conservation Of Energy*

There is nothing in Newton's gravitational theory stating that the force of gravity weakens as it expends energy. The mass of the moon exceeds one percent of the Earth's mass and would fly past the Earth and off into space if not forcefully constrained by gravity to circle the Earth, according to Newton's theory. Yet this tremendous continual effort expended by Earth's gravitational field is not considered to diminish the strength of this field at all – millennium after millennium.

Returning to the vehicle analogy, when a car increases its speed it is said to accelerate, which is only possible by drawing on a power source, converting its energy into the car's increased speed or kinetic energy. Turning the vehicle in a circle is another form of speed change or acceleration, involving a constant, forced change from its natural straight-line direction of travel. This continuously forced circular direction change is known as *centripetal acceleration*, and also requires energy to maintain this constant diversion from the natural straight-line path of objects. Likewise, the natural forward momentum of the moon would carry it away from our planet and off into space in a straight line if gravity were not forcefully pulling it into a circular orbit moment by moment. Yet this tremendous energy expenditure is not balanced by a conversion of energy from any known power source. This is a *creation* of energy from nothing – energy for free – rather than a *conversion* of energy from one form (a power source) to another (circular centripetal acceleration). This situation is a clear violation the *Law of Conservation Of Energy*.

Gravity also forcefully holds down all objects on the surface of our planet, which would drift off into space otherwise. In fact, the pull of gravity helps to hold our very planet together, creating tremendous crushing forces within the center of the Earth. This has been going on for well over 4 billion years, yet no known power source is being drawn upon to support this tremendous ongoing energy expenditure.

This mystery is further deepened when we consider that not only is there no *drainage* of energy from a power source to support the effort expended by the gravitational force, but in fact there is *no power source at all*. A gravitational force is considered to emanate from within each atom of matter, adding up to the tremendous overall gravity of the Earth, yet we still have no explanation for its endless power source despite having created detailed atomic theories – and even having split the atom. This is a textbook case of an impossible free energy device.

This discussion naturally raises the question of why such a fundamental violation of our laws of physics doesn't generate intense scientific concern, curiosity, and investigation. Why is Newtonian gravitational theory simply accepted and its mysteries left uninvestigated? This question brings a curious mixture of responses. One answer is that science *has* responded to these concerns by accepting a very different explanation of gravity proposed by Albert Einstein (1879-1955) known as *General Relativity Theory*, which will be explored further in later discussions. However, Einstein's theory offers no solutions to these problems either. In fact, these violations are not generally acknowledged as the reasons for accepting Einstein's alternate theory of gravity, nor are these violations even generally acknowledged at all today.

Perhaps more curious is the fact that even though *General Relativity Theory* is generally accepted in academic circles as the proper description of gravity, it is not widely taught or used by engineers and physicists – usually being reserved for optional or advanced study, and mostly for rare and exotic applications. Most university science and engineering graduates know little or nothing about Einstein's theory of gravity despite the fact that it is presumably the true explanation of this phenomenon, and it is not generally used in our space programs. Newton's concept of gravity is by far the main gravitational theory used in space missions today, despite the fact that there was apparently good reason to accept Einstein's quite different theory of gravity into our science. All of this further deepens the mystery surrounding gravitational theory today, so let's take a closer look at these issues starting with the currently unrecognized law violations in Newtonian theory.

The serious law violations and mysteries found in Newtonian gravitational theory have just been clearly pointed out in reference to one of our most fundamental laws of physics, yet science does not generally recognize these violations. How can this be? Why might those who are the most highly educated in physics be the least likely to acknowledge these mysteries and violations? The answer is that when Newton's theory of gravity is taught, it is usually accompanied by further instruction on how to resolve these mysteries and violations by referring to an equation called the *Work Function*. Although it will be shown shortly that this is a fatally flawed explanation attempt that gives a false sense of closure on these issues, this fact is overlooked by our educational institutions today since there is no other explanation for Newtonian gravity. Therefore, all properly educated scientists have firmly learned the standard (though erroneous) logical techniques that have been taught for generations to provide ready answers for the mysteries and violations of Newtonian gravity. This leads to the curious fact that, on the one hand, science found it necessary to search for and accept such alternate gravitational theories as Einstein's *General Relativity Theory*, while on the other hand, Newtonian gravity is still widely accepted by scientists. This makes the *Work Function* an important pivotal element in this whole mystery, and therefore worthy of a closer look.

ERROR

The Work Function – A Flawed Explanation

Physical labor typically involves moving heavy objects or material from one place to another. The heavier the object and the further it is moved, the more energy must be expended in the process. The *Work Function* is merely an attempt to describe this fact using a simple equation – originally designed to help engineer mechanical devices that use energy to do work, such as steam engines that burn fuel to move trains. This equation is written as $W = F\,d$, which is read as *work* (*W*) equals *force* (*F*) times *distance* (*d*). That is, the more force required to move an object, and the further the object is moved by that force, the more work is done in performing this task.

The *Work Function* can be a very useful tool in analyzing and quantifying the amount of work done by a given process or machine, and has served engineers well for over a century. However, serious problems arise when its use is extended beyond its design intent. Its original purpose was as an engineering tool to compute how much *work* is done when a force moves an object across a distance, which also corresponds to how much *energy* was expended, since an equivalent amount of fuel must be used in the process. This all seems quite reasonable; however, over the years the *Work Function* has undergone a subtle and surprisingly deceptive transformation into a "work detector," whose result is taken as the final word on how much *energy* was used in any given process. This is such a subtle yet powerfully deceptive transformation that it needs to be clarified with an example:

Consider the situation where an object is simply too heavy to move, despite all efforts to push it. There is no question that one could expend a tremendous amount of effort and energy attempting to move the object, yet never actually manage to move it an inch. However, applying the *Work Function* as a "work detector," it calculates that zero work was done. A tremendous amount of force was applied to the object, but the object was nevertheless moved *zero* distance, and since *work* equals *force* times *distance*, the *Work Function* calculates that zero work was done. If this were further taken to mean no *energy* was expended, we would have a worker who is exhausted from attempting to move such a heavy object, yet who is considered to have expended *no energy*. Of course, this is obviously a serious misapplication of the *Work Function* that brings nonsensical results, yet this is precisely the logic used to justify the gravitational force, as we will see shortly. The *Work Function* is only designed to help organize and quantify situations where a force clearly moves an object through a distance, but is not meant to function as a generic "work detector" that further tells us whether any energy was expended by an arbitrary event.

Now, to complete the improper transformation of the original *Work Function* from a simple engineering tool to a generic "work detector," it has evolved from its original form of $W = F\,d$ to its current form $W = F\,d\,cos(\theta)$. The additional term here, *cos(θ)*, is the cosine function, which transforms any angle from 0 to 360 degrees into a value that lies between -1 and 1. Therefore, the original result from the *Work Function* calculation is now multiplied by a value between -1 and 1 that corresponds to the angle (θ) between the direction the object is pushed and the direction it actually ends up moving. If the object simply moves in the direction it is pushed, which is the usual case, this zero-degree angle between force and movement results in the work calculation being multiplied by 1, since *cos(0) = 1*. This means nothing changes from the original *Work Function* when force and movement are in the same direction. However, if the object somehow managed to move completely sideways despite a forward push being applied to it, this 90-degree angle between force and movement means the resulting work calculation must be multiplied by 0, since *cos(90) = 0*. Therefore, the work done in this scenario would be calculated as zero. This *modified Work Function*, $W = F\,d\,cos(\theta)$, is said to calculate the amount of *useful* work, since only the amount of work done in the direction of the force is considered to be desired and therefore useful work.

This is how the *Work Function* is taught today, which now sets the stage to explain why the previously mentioned violations of the laws of physics by Newton's gravitational force cause no particular concern for most scientists. First, the issue of objects being held to the planet's surface by a force that has no known power source is easily dismissed by noting that an object held down by the gravitational force does not move. If the object doesn't move, there is no work done according to the *Work*

Function, and therefore no energy is expended and no energy *source* is required to explain how things are forcefully held down by gravity. The serious law violation that results from gravity forcefully holding objects to the planet's surface with no known power source suddenly vanishes. This is the same flawed logic used earlier, which left our worker exhausted after trying unsuccessfully to move a heavy object despite having apparently expended no energy. Yet, of course, both the worker and gravity must expend energy in these examples.

In similar fashion, the *modified Work Function* is used to justify the tremendous energy required to hold our moon in orbit, again with no known power source. Since the moon is actually traveling *past* the Earth in a straight line but is continuously constrained in its orbit by the gravitational force pulling it *down* toward the planet, this is considered to be a situation much like an object that slides sideways when a force pushes forward. The angle between the direction of the moon's travel *past* the Earth and the direction of gravity pulling *down* is the same 90-degree angle as in the earlier example of the sideways-sliding object, meaning the *Work Function* must be multiplied by 0. This gives the result that the gravitational force does zero useful work and thus expends no energy in constantly constraining our moon from flying off into space, removing the need to look for a power source. Once again, a serious violation of the laws of physics suddenly vanishes. Yet, a person who must constantly struggle to constrain a heavy, speeding rock into traveling in a circle on the end of a rope might disagree with this zero-work, zero-energy conclusion for orbits.

Finally, there is the situation where objects fall straight down. Surely the *Work Function* would have to give a non-zero result here since the direction of movement is in the same direction as the downward pull of gravity. Indeed, the *Work Function* does calculate a positive amount of work, which should mean energy has been expended by the gravitational force, requiring an energy source be identified within the Earth that is drained by an equivalent amount if this event is to remain within our laws of physics. Since there is no such energy source known to science, we must either admit that Newtonian gravity cannot be scientifically explained, or arrive at some further justification. Indeed, an additional logical abstraction *has* been invented for this type of situation to avoid the search for a power source, which runs along the following lines:

In order for an object to drop from a given height, work had to be done earlier *against* the pull of gravity to lift it to that height in the first place. Since this upward lifting could be considered *negative* work from the perspective of the downward-pulling gravitational force, the positive work done by gravity when the object falls could be considered to cancel with this earlier negative work. This zero overall work then corresponds to zero net energy expenditure, and thus we are once again saved from looking for the energy source for gravity. Of course, this abstract exercise overlooks the *physical* reality that the falling object must still somehow drain gravity's unknown energy source, and no known theory states how lifting the object earlier would have *charged* this power source in order to compensate for this later energy drain. Further, this explanation implies the *existence* of such a mysterious and currently unknown power source, which is the very issue it was invented to avoid. So the "energy balance" in this logic is a meaningless abstraction that merely diverts attention from the physical law violation that gravity somehow pulls objects to the ground while expending no energy.

Once again, the reason this logical conundrum has arisen in our science is due to the deceptively subtle, yet powerful difference between using the *Work Function* to describe clear situations where a force moves an object through a distance, and using it as a generic "work detector" in all situations. In fact, in the case of Newtonian gravity, not only has the *Work Function* been misused as a "work detector" but also as a "force authenticator." That is, not only is it used to alleviate concerns about law violations by calculating that the gravitational force does no work and expends no energy, but it is put to this use in order to help justify or authenticate the very *existence* of the gravitational force. After all, any theory involving a force that violates our most fundamental laws of physics is unacceptable as anything other than a purely abstract model of a still unexplained physical process. It cannot literally be taken as the proper physical explanation since this is precisely why our laws of physics exist – as a litmus test or sanity test for such proposed new ideas. The *Work Function* is simply intended to describe the work done by *known* forces as they move objects, but here it is being used in an attempt to *authenticate the existence* of the previously *unknown* force introduced by Newton – a force that is otherwise *scientifically unexplainable*. This misapplication of the *Work Function* essentially creates a loophole in the *Law of Conservation Of Energy*, corrupting the original purpose of *both* of these concepts.

This *Work Function* discussion shows the type of logic that keeps most physicists from acknowledging that Newton's gravitational force violates the *Law of Conservation Of Energy*. However, once the flawed *Work Function* explanation is exposed and removed, there are simply no excuses remaining for this unexplained force. The rationalists who followed Descartes had good reason to see Newton's gravitational force as a return to the magical thinking of the past. Perhaps in Newton's day it was reasonable to expect that future generations of scientists would find a scientifically viable explanation or even a true power source for the gravitational force. However, three centuries later we have found no answers, instead opting to turn a blind eye to its violations of our laws of physics by installing a flawed logical justification for this force into our science. Regardless of its original purpose, the *Work Function* has now been incorporated into our science in such a manner that most scientists clearly believe a zero-value result from its calculation always means there has been no expenditure of energy. This has led to the logical oversight that gravity need not expend energy to hold objects to the planet, since there is no motion involved, nor to constrain the moon from speeding away, since the pull of gravity is perpendicular to the moon's orbit.

This state of affairs exists because we very much *want* to believe in this force. For centuries it has been the only reasonable explanation we have had, and in fact, it is still the only compelling and intuitive physical explanation for falling objects and orbiting moons even today. The official position in science today does state that another viable explanation exists in Einstein's *General Relativity Theory* of a "warped space-time continuum," but this does not address our everyday experiences and seems far off the mark compared with Newton's intuitive gravitational force. And indeed, as shown in the following chapter where the new principle is introduced, gravity *can* be explained in a simple, intuitive, and scientifically viable manner – but without appealing to either an unexplained force or an abstract and largely incomprehensible "warping of space-time."

So far, we have seen a number of questions, mysteries, and even violations of physical laws surrounding the concept of a gravitational force. We have no answer for why it attracts rather than repels objects, we know of no power source within matter that would produce this force, and it expends energy without diminishing in strength or draining a power source – an "energy-for-free" scenario that violates the *Law of Conservation Of Energy*. In addition, there is yet another troublesome issue with Newtonian gravity to consider – the issue of its speed of travel through space. We begin with a reminder of our currently accepted universal speed limit, the speed of light.

LAW

 ### The Speed-of-Light Limit

Neither matter nor energy can travel through space faster than the speed of light.

This is a currently accepted law in our science today, stating that the speed of light in the vacuum of empty space represents an absolute upper speed limit on all objects and also on the speed of propagation of all fields and all forms of energy through space. According to this law, nothing known to man can travel faster than light. This is an idea that Einstein proposed as part of his *Special Theory of Relativity*, and which currently stands as an unbreakable law of nature.

VIOLATION

 ### Newton's Gravitational Force Exceeds the Speed of Light

Newtonian gravitational theory comes with no speed limit. A common example of this is to imagine our sun suddenly vanishing. While it would still appear as if the sun were present for roughly eight minutes as the last rays of light eventually made their way to Earth at light-speed, the gravitational field of the sun would vanish immediately along with the sun. The Earth would not experience eight additional minutes of the sun's gravity constraining it in orbit, but would *immediately* leave its orbit about the sun and begin to drift off into space. This is because the loss of gravity from the sun would be immediately felt at any distance throughout the solar system, and indeed throughout the universe according to Newtonian theory. This faster-than-light transmission of the gravitational force through space – and indeed even *instantaneous* transmission across *any* distance in our universe – is a great, unexplained mystery in our science today.

This is *one* violation in Newtonian gravitational theory for which a logical justification has *not* been found that allows it to be dismissed or overlooked. That is, unlike the law-violating behaviors mentioned earlier that were justified with a misapplication of the *Work Function*, this speed-of-light violation remains in plain view. However, although this violation lacks a logical justification, a resolution can be found in Einstein's *General Relativity Theory*, since one of the key differences with this alternate theory of gravity is that the element of *time* is built into its equations. This provides a description of gravity that allows it to take time to travel or propagate through space, proposing a solution to this issue. However, this is only a *proposed* solution since the actual speed of gravity is unknown – no definitive tests have been done to determine it.

In fact, the issue of the speed of gravity is still a very contentious one in our science, and there is often sizable disagreement on how to even go about measuring it. Some scientists do claim to have already measured it, and that, according to their method of determining it, gravity travels at the speed of light. Yet, other scientists attempting to independently verify any such claims to date typically report that the speed-of-light result was built into the very design of the experiment and the assumptions made by the researchers. In other words, the procedures and analyses were not those of a truly open experiment, but more self-contained contrivances that guaranteed agreement with Einstein before even being carried out, invalidating the experiment and leaving the actual speed of gravity still an open question. This is the likely reason why such presumably momentous claims as determining the speed of gravity – something mankind has wondered about for centuries – have so far quickly vanished from the headlines with little impact.

So, we have the choice of Newton's simple and intuitive theory, which violates the speed-of-light limit, or Einstein's complex and mysterious theory, which offers an unproven solution to this violation. As a result of this type of interplay between these two theories, we are left with an odd combination of *both* theories in our science today. Neither theory truly stands alone today as the singular, correct description of gravity, as both theories tend to complement each other's weaknesses. It is this type of interplay between them that leaves us with two very different explanations for gravity in our science today, even though common sense tells us there can be only one clear physical explanation underlying any observation. Clearly one of these theories must be fatally flawed, or *both* theories are merely useful interim models that have captured one aspect or another of the true and as-yet-undiscovered physical explanation for gravity. It is precisely this as-yet-undiscovered explanation that is proposed in the next chapter, offering a resolution to this odd state of affairs in our science today.

The Origin of Newton's Gravitational Force

The discussions so far have largely taken for granted that we are all very familiar with the Newtonian explanation of gravity as an attracting force that somehow emanates from matter; as such, the details and origin of this theory have not yet been addressed. If we could examine the progression of ideas that led to Newton's theory of gravity, perhaps we could identify once and for all either where the overlooked power source may be for this force, or alternatively, how this *fictitious* force came to be invented.

The first publication of Newton's *Law of Universal Gravitation* appeared in his famous work, widely known as *"The Principia"* today, published in 1687. In this publication Newton describes his proposed new force, showing how it explains our observations of falling objects and orbiting bodies, and even providing a simple and intuitive mathematical formula for calculating the strength of this gravitational force between any two objects. To arrive at this equation Newton would have had to follow the clues available to him at the time, both from his own experience and education as well as from the available astronomical data of his day. Let's now follow the type of thought processes that would have led to Newton's formal theory of a gravitational force.

At the time, a formal mathematical description of the orbits of moons and planets was already in existence – provided by Johannes Kepler (1571-1630) – based on the astronomical data of the day. In fact, Kepler's three laws of planetary motion are very accurate and useful indeed, still remaining as some of the most important tools used in our space programs. Yet, despite this great achievement by Kepler, these laws only provided a *mathematical* description of planetary motion without explaining *why* and *how* this motion occurs. In essence, Kepler's Laws described only the *geometry* of planetary motion, but not the *physical* reason for this geometry.

Prior to Newton's *Law of Universal Gravitation* there were suspicions that some type of attracting force might be at work, but no one had managed to arrive at a solid theory or justification for such a force. Newton's well-developed theory of a gravitational force finally managed to achieve this convincingly, bridging the gap between Kepler's purely geometric laws of planetary motion and the strong suspicion that some type of attracting force in nature may underlie them. Newton's *Law of Universal Gravitation* is now presented, followed by a consideration of its origins to see what insights can be gained into the source of the familiar, yet still quite mysterious gravitational force that we believe in today.

LAW

 Newton's Law of Universal Gravitation

There is an attracting force in nature emanating from all objects, pulling them toward one another with a strength that increases with their masses and decreases with the distance between them squared.

According to this claim made by Newton, now considered a law of nature, the greater an object's mass the greater its gravitational field strength, and this gravitational field diminishes rapidly in strength the further it extends out into space away from the object. Specifically, the strength of this gravitational force between any two objects is calculated by multiplying their masses together then dividing by the square of the distance between their centers. Finally, this result is multiplied by a constant, known as the *gravitational constant*, to present it in standard units of force. The resulting equation of the strength of the gravitational force, F, between two objects is written as:

$$F = \frac{G \cdot (m_1 m_2)}{R^2}$$

where m_1 and m_2 are the masses of the two objects.

R is the distance (radius) between their centers.

G is a constant, called the gravitational constant.

This equation is known as the *Law of Universal Gravitation*. Yet this represented much more than just another equation when Newton introduced it. It ushered a completely new force of nature into our awareness and our science. It was not merely an abstract model of observations, but a statement of an *actual* force in nature emanating from objects – varying in strength with their mass, which we can lift, and their distance, which we can measure. This is a concept that we are now taught as children and have grown accustomed to, but it would have been truly revolutionary when it was first introduced in Newton's day. Some had suspected that something of this nature might exist to explain falling objects and orbiting bodies, but Newton was the first to actually *show* that this force apparently did exist, and to describe it in very definite, concrete terms. Further, it is fairly straightforward to derive today's *Newtonian Orbit Equation* from Newton's *Law of Universal Gravitation*, as will be shown shortly, which very accurately predicts the motions of the planets and plays a central role in our space programs even today. All of this made Newton's theory of gravity a revolutionary discovery, as well as apparently irrefutable proof of the existence of such a force in nature.

But where did this revelation come from? Somehow we went from a vague suspicion that an attracting force might be operating in the world around us, to a definite statement of its existence, its source in all material objects, and its precise behavior captured in an equation. How does something like this happen? The following investigation into this issue will help to clear up this mystery, showing that Newton's gravitational theory is actually a completely superfluous and unnecessary invention that is based on a logically and scientifically flawed assumption. As a result of this invention, a crucially important equation for the orbits of planets was overlooked, then recast in Newtonian gravitational terms and presented as an entirely new equation – the *Newtonian Orbit Equation* that is currently in use today. The story of how this occurred and its enormous implications follows, showing surprising revelations about Newtonian gravitational theory.

An Alternate Origin

Although Newton provided a mathematical derivation for his law of gravity based on Kepler's laws of planetary motion, the somewhat different derivation below provides a clearer picture of the origin of the gravitational force in our science, addressing the issues that still remain a mystery even today.

WATCH FOR...

- Kepler developed three *purely geometric* equations of planetary motion involving no gravitational force, which described the heavens extremely well prior to Newton, and still do even today·

- A fourth *purely geometric* orbit equation of great importance is easily identifiable in the astronomical data available at the time, yet no formal record of this *Geometric Orbit Equation* exists.

- Newton's gravitational force equation can be easily arrived at by equating the *Geometric Orbit Equation* to the equation for a rock swung by a string, thereby *inventing* Newton's force by making the same rock-and-string assumption made by Newton.

- This assumed equality between swinging rocks and orbiting planets is seriously flawed, leading to the unexplainable mysteries and violations still present in Newtonian gravitational theory today.

- The *Newtonian Orbit Equation* widely used today is derived from Newton's gravitational theory; however, this only *appears* to give an entirely new and important orbit equation, but is actually merely a disguised return to the original *Geometric Orbit Equation* that pre-dated Newton.

- In actuality, Newton's whole theory of gravity is a pure invention with no scientific support, based on the pre-existing *Geometric Orbit Equation* combined with a flawed rock-and-string equality to orbits.

The Orbit Equation Actually Existed Prior to Newton

The analysis of the origin of Newton's proposed gravitational force begins with Kepler's three laws of planetary motion. Unlike Newton's *Law of Universal Gravitation* and the Newtonian orbit equation that follows from it, Kepler's laws are *purely geometric* descriptions of planetary motion based on observations of the heavens. They were arrived at prior to Newton's theory of gravity, and make no reference to a gravitational force. These laws are as follows:

LAW

 Kepler's Laws of Planetary Motion

- *Kepler's First Law* states that the planets move in oval-shaped ellipses around the sun, with the sun at one end of the ellipse.

- *Kepler's Second Law* states that as a planet proceeds in its elliptical orbit, an imaginary line joining the sun and the planet would always sweep out the same area in a given time period regardless of where the planet is along its elliptical path.

- *Kepler's Third Law* provides an equation that calculates the average distance of a planet from the sun simply by measuring the time it takes to make a complete orbit.

These three laws are very accurate, reliable, and central to our space programs today. However, an additional and very important geometric relationship regarding orbits can be readily seen in the astronomical data that would have been available to Kepler and Newton, yet it is missing from both *Kepler's Laws* and Newton's gravitational theory. In fact, there is no formal record of it at all in our scientific history. This purely geometric relationship is so "Kepler-ian" in nature that it is tempting to call it *Kepler's Fourth Law*, but since this would obviously be inappropriate, we'll call it the *Geometric Orbit Equation*:

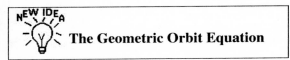 **The Geometric Orbit Equation**

The *Geometric Orbit Equation* is a previously unrecognized, purely geometric equation embodying a relationship in the standard astronomical data showing that the orbital radius of any planet in our solar system (i.e. its distance from the sun) multiplied by the square of its velocity always gives the same constant value. This would be written as:

$$v^2 R = K,$$

where K is a constant with the unchanging value of 1.325×10^{20} $[m^3/s^2]$
R is the orbital radius of the planet (distance from the sun)
v is the velocity of the planet

This relationship can be readily deduced from any standard table of planetary data that can be found in most introductory physics textbooks. The constant, K, is the same for all planets orbiting the sun, but differs for other orbital systems. For instance, the value of K for objects orbiting the *Earth* rather than the sun can be readily calculated as 3.7×10^{14} by referring to these same tables of planetary data. This value of K for our Earth-based orbital system would apply to the orbit of the moon, for instance, as well as the orbits of the various satellites and spacecraft about our planet.

This geometric orbit equation allows the distance of orbiting objects to be calculated if their speed is known. Perhaps more importantly, it allows for the planning or alteration of satellite and spacecraft orbits by indicating the speed required to achieve a given orbit, and the required speed change to transfer from one orbital trajectory to another. This type of calculation would underlie everything from fuel requirement planning for space shuttle missions to orbital insertion of satellites around Mars. Notably, the *Geometric Orbit Equation* pre-dates Newton and achieves these results in a purely geometric fashion, as its name implies, *without any reference to masses or gravitational forces.*

The *Geometric Orbit Equation* is the type of important astronomical observation that we might expect to be noticed and identified in the time of Kepler and Newton. Although there is no clear record of this occurring, the existence of this earlier geometric relationship provides an intriguing alternate derivation for Newton's gravitational force and the final form of his *Law of Universal Gravitation.* To see this, we turn to the common analogy for planetary orbits taught in all elementary physics courses – the presumably equivalent scenario of a rock swung in a circle at the end of a string, as assumed by Newton.

The Rock-And-String Assumption

The idea of the moon being forcefully constrained by gravity to circle the Earth seems very reasonable at first, since we are all familiar with the seemingly similar concept of swinging a rock on the end of a string, causing it to "orbit" about us. Of course, this is not truly an orbit since it involves a physical length of string with clear physical tension throughout it as our muscles strain to keep the rock from flying off. This leads to the mysterious concept that the orbit of our moon involves a mysterious attracting force acting across space in a manner that is still unexplained by science, apparently forcefully keeping the moon from flying off without drawing on any power source. However, since this is the equivalence made by Newton and widely accepted today, we will follow this same assumed rock-and-string equivalence in this alternate derivation of Newton's gravitational force.

Once this assumption is made, it may then seem reasonable to equate the force required to constrain the rock in a circular path about us with the gravitational force said to constrain the moon in its orbit about the Earth. The *Centripetal Force Equation* for calculating the force, F, required to constrain a rock swung by a string is well known, as it was in Newton's day:

Centripetal Force Equation ("rock-and-string")

$F = mv^2/R$ where m is the mass of the rock

v is the velocity of the rock

R is the radius of swing (string length)

Equating this with the scenario of gravitational orbits gives the picture of equivalence between all elements involved, as shown in Figure 1-2.

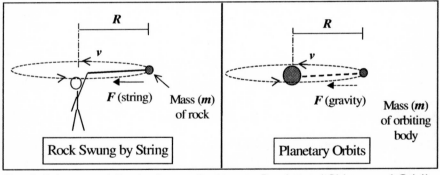

Fig. 1-2 Assumed Equivalence between Rock-and-String and Orbits

At this point, we have an equation for orbits (the *Geometric Orbit Equation*), an equation for a rock swung by a string (the *Centripetal Force Equation*), and an assumed equivalence between them. So then, it should be valid to combine these two separate equations to create one single equation that embodies this equivalence. This can be done by first rearranging the *Geometric Orbit Equation* in terms of its velocity parameter ($v = \sqrt{K/R}$), then substituting this velocity expression into the *Centripetal Force Equation*, resulting in the equation:

Hypothetical Gravitational Force Equation

$F = \dfrac{mK}{R^2}$ where m is the mass of the orbiting body

K is the constant from the *Geometric Orbit Equation*

R is the orbital radius, also from the *Geometric Orbit Equation*

This new equation is a hybrid of the *Geometric Orbit Equation* and the *Centripetal Force Equation*, obtained by making the *completely arbitrary assumption* that swinging rocks are *physically* equivalent to orbiting objects – and not simply similar in *appearance*. This would mean that there must somehow be an actual physical force pulling on objects to constrain them in orbit, just as there is a physical tension force in the rock-and-string equivalent as shown in Figure 1-2. As we will see soon, this new equation forms the foundation of Newton's *Law of Universal Gravitation*, and the force, F, is the first-ever occurrence of a *hypothetical* "gravitational force."

NOTE

 This new hybrid equation marks the first appearance of an attracting gravitational force in our science.

As noted above, this new hybrid equation is no mere mathematical exercise, but the literal *creation point* for the supposed "gravitational force," and the first point where a force of any kind appears in relation to orbits. Prior to this a description of orbits would have been possible, provided by the *Geometric Orbit Equation*, but in completely geometric fashion involving only velocity and distance, with no mention of an attracting force emanating from the mass of the orbiting body. Now we have an equation that implies a gravitational force may be at work, which is somehow directly related to the mass of the orbiting body, m, and diminishes with the square of its orbital radius, R.

While this would be an exciting result for a scientist in Newton's day when this issue was a deep mystery and a very hot topic in science, we must keep in mind that this is still an unsupported hypothesis in the derivation so far. We went from a fully functional, purely geometric orbit equation to an equation implying that forces and masses are involved in orbits merely by making a few simple assumptions and mathematical manipulations. This hypothetical force is still just as mysterious as it always was in scientific circles, with no scientific explanation for why it should spring forth from matter and pull on other objects. However, this new equation does give *form* to this proposed force. Instead of being just a vague suspicion, now it has an equation describing it, an identifiable material source (presumably the mass, *m*, of the orbiting object), and the characteristic that it diminishes in strength with the square of the distance between the object and the orbited body. Whether or not this is based on pure assumption, it is certainly a very compelling result.

To review, at this point we have a hybrid equation involving mass and a force, resulting from the assumption that a rock swung forcefully by a string is equivalent to the otherwise purely geometric orbits in the heavens. This hypothetical gravitational force equation has the form:

$$F = mK/R^2$$ – Hypothetical Gravitational Force Equation
(shown earlier)

This equation claims that there is an attracting force holding objects in orbit, whose strength varies directly with the mass of the orbiting object, diminishes with distance squared, and is also dependent on a mysterious constant, *K*, that differs from one orbital system to another. But what could this constant refer to?

Since this new, hypothesized gravitational force presumably emanates from the orbiting object, *m*, it stands to reason that it should also emanate from the object that is being orbited; therefore, we would expect the mass of the *orbited* body to appear in this equation as well. So then, if we assume that the constant, *K*, is actually the mass of the orbited body, we have a viable explanation. It seems quite reasonable that this constant that differs between orbital systems may well be the mass of the orbited body, which is also a constant that differs between orbital systems. So then, replacing K by this second mass, m_2, now gives our hypothetical gravitational force equation the form:

$$F = m_1 m_2 / R^2$$ – Hypothetical Gravitational Force Equation
with K replaced by m_2

The only remaining step is to make sure the results from this calculation are expressed in the units of force, and are reasonable values. Currently this equation multiplies two masses and divides by a distance squared, giving the units of $[\text{kg}^2/\text{m}^2]$ – that is, kilograms squared per meter squared. These are not the proper units for a force, and the values that result when using reasonable estimates for the mass of the Earth or the sun as the larger mass, m_2, are also millions of times too large to be sensible. However, this problem is easily solved by multiplying our equation by a value that reduces the results to within a reasonable range and alters the units into those of a force. This simply involves the arbitrary introduction of a *constant of proportionality* that has these qualities. However, if we now assume that our hypothetical gravitational force equation truly describes an actual attracting force in nature, then this arbitrarily invented constant of proportionality would have to be a true *natural constant*. Although all of this is still only an assumption, if true, this constant would become what is known as the *gravitational constant*, G, today, giving the final form:

$$F = G(m_1 m_2) / R^2$$ – Newton's *Law of Universal Gravitation*

NOTE

 This is precisely the form of Newton's *Law of Universal Gravitation* shown earlier and presented in his *Principia*.

As noted above, this final result is precisely the equation for the gravitational force that Newton presented in his *Principia* in 1687. Although this alternate derivation differs somewhat from that provided by Newton, it shows that the origin for his gravitational force can be clearly found in the *Geometric Orbit Equation*. Given this, we can now evaluate where our current belief in this force comes from, and the firmness of the foundation for this belief. We now know, for example, that there was no advanced knowledge or understanding of a hidden power source that led Newton to this belief. Instead, it is simply based on the assumption that the scenario of a rock swung by a string is the literal physical equivalent to that of objects in orbit. Yet the rock-and-string scenario *does* have an identifiable power source – our muscles, while the gravitational force maintaining orbits does not. Also, the rock-and-string scenario *does* have a physical explanation for the attracting force constraining the rock – the tension in the string, while Newton's proposed gravitational force has no clear physical explanation. In short, the assumption that these two scenarios are equivalent is based more on their similarities in *appearance* as systems involving circling objects than on any verified *physical* equivalence.

Further, there are other physical systems that may have even more similarities to orbiting objects than a rock swung by a string; consider a rock swung by a *spring*, for example. One of the problems with the rock-and-string equivalence assumption is that the rock can be swung faster and faster while remaining the same distance away at the end of the string – the tension in the string simply increases. If this were a true physical equivalence to orbits then gravity would have to increase its attracting force to constrain a faster moving object at the same orbital distance. However, this does not happen, either in theory or in practice. Instead, orbiting objects that are given more forward thrust move further out into space, much the way the rock would if it were swung faster at the end of a stretchable *spring* instead of a rigid string.

So, as long as we're making arbitrary intuitive guesses at familiar mechanisms that might possibly be a literal *physical* equivalent to orbiting objects, we would have to seriously consider abandoning the rock-and-string idea for that of a rock-and-*spring*. This is not to say that orbits are the physical equivalent of a rock-and-spring either – this model also has its limitations and problems, and is just as arbitrarily chosen since we are merely going on superficial similarities in appearance. Still, as an educated guess it is perhaps more functionally similar to orbits than the rock-and-string scenario upon which today's gravitational theory is built, exposing the weak and arbitrary foundation of Newtonian gravitational theory.

Interestingly, if we used the rock-and-spring model, we would end up with an entirely different version of Newton's *Law of Universal Gravitation* since the centripetal force equation for the rock-and-spring is different than for the rock-and-string. That is, this difference in the centripetal force equation for circling rocks using springs means that when we substitute the velocity from the *Geometric Orbit Equation* into the *Centripetal Force Equation* as we did before, the resulting expression for the gravitational force must also differ. Yet this resulting spring-based gravitational force equation would still give us a numeric value for the gravitational force, just as Newton's current equation does. And although this numeric value is not directly measurable – even from Newton's current equation – it gives the *appearance* of an actual force in nature; one whose strength we can even *calculate*, using the concrete attributes of mass and distance.

NOTE

 Therefore, the familiar form of Newton's *Law of Universal Gravitation* is not a true law of nature, but merely a *flawed invention* based on superficial similarities in appearance between orbits and the very different scenario of a rock-and-string.

The preceding alternate origin for Newton's gravitational force shows that the introduction of an attracting gravitational force in orbits was *completely arbitrary and unnecessary,* considering the contributions by the already existing body of purely geometric equations, i.e. Kepler's three laws plus the *Geometric Orbit Equation.* But this is a fact that could not have been realized without this alternate derivation since *the Geometric Orbit Equation is unknown to science,* at least in the formal manner presented in this discussion. Instead, we have the *Newtonian Orbit Equation* today, derived from Newton's *Law of Universal Gravitation.* Since this *Newtonian* orbit equation is central to our science of astronomy and our space programs, Newton's theory of gravity is considered to be of immense importance as the origin of this equation. However, it is now possible to show that the *Newtonian Orbit Equation* is simply the pre-existing *Geometric Orbit Equation* in disguise. To see this, let's take a closer look at the origin of the *Newtonian Orbit Equation* in use today.

The Invention of the Newtonian Orbit Equation

Throughout the following discussion it is important to keep in mind that the progression from the *Geometric Orbit Equation* to Newton's *Law of Universal Gravitation* that was just shown is unknown to science, just as the formal *Geometric Orbit Equation* itself is unknown. Therefore, the following derivation of today's *Newtonian* orbit equation from Newton's *Law of Universal Gravitation* is currently believed to be the sole origin and form of the orbit equation in our science. The fully equivalent, pre-existing, and in fact, more proper *Geometric Orbit Equation* is unknown today, as is the flawed foundation of Newton's *Law of Universal Gravitation* itself. This gives the appearance that the existence of today's *Newtonian* orbit equation, as well as its tremendous contributions to astronomy and our space programs, is owed entirely to Newtonian gravitational theory. In actuality, however, this homage that is commonly paid to Newtonian theory is quite unfounded, as will now be shown.

The standard derivation of the *Newtonian Orbit Equation* in use today begins with the assumption that the rock-and-string scenario is equivalent to orbiting bodies in the heavens – a centuries-old assumption that is simply accepted unquestioningly today. Therefore, since Newton's gravitational force and the rock-and-string centripetal force shown earlier are considered equivalent physical concepts today, the derivation of the *Newtonian Orbit Equation* starts by simply equating these two forces:

Newton's Equation $\rightarrow$ $GmM/_{R^2} = mv^2/_R$ $\leftarrow$ Rock-and-String Equation

Here, the two masses, m_1 and m_2, in Newton's equation are named m and M to signify the smaller mass, m, of the orbiting object and the typically much larger mass, M, of the orbited body. The above equality immediately simplifies to the familiar form of the *Newtonian Orbit Equation* that exists in our science today:

$$v^2R = GM \qquad -\textit{Newtonian Orbit Equation}$$

Note that, although this *appears* to be a completely new and important equation derived from Newton's law of gravity, in actuality it is merely a reversal of the steps performed earlier in the derivation of Newton's *Law of Universal Gravitation* from the original *Geometric Orbit Equation*. That is, where we started with the *Geometric Orbit Equation* and arrived at Newton's *Law of Universal Gravitation* by making the (flawed) rock-and-string assumption, we now have simply used this same flawed assumption to work backwards from Newton's equation to the original *Geometric Orbit Equation* again. The *Newtonian Orbit Equation* above looks a bit different from the *Geometric Orbit Equation*, but as we'll soon see, this is only a cosmetic difference in appearance.

Today this fact is not recognized since Newton's derivation for his *Law of Universal Gravitation* does not show its origin in the *Geometric Orbit Equation*. Therefore, it *appears* as if the orbit equation we use today is a perfectly valid Newtonian result derived solely from "solid gravitational theory." Today, this mere reversal from Newton's gravitational force equation to a disguised version of the *Geometric Orbit Equation* is unknown, lending unwarranted credibility both to Newton's gravitational theory and to the assumed physical equivalence of the rock-and-string analogy. In actuality, the flawed rock-and-string analogy was used to *invent* Newton's equally flawed equation of a gravitational force in the first place, then used again to undo this logic, merely arriving at a slightly disguised version of the only correct equation in this whole process – the original *Geometric Orbit Equation*.

A review of the earlier derivation for Newton's gravitational equation shows that the constant, *K*, was essentially arbitrarily replaced with the two multiplied constants, *GM*. Recall that this occurred after assuming that *K* must refer to the mass of the orbited body, then realizing that the "natural constant," *G*, had to be introduced to alter the size and units of the final result. But this switch from *K* to *GM* earlier was merely based on an arbitrary and unsupported assumption; as such, it is not only valid but also more correct to return to the original constant, **K**. Therefore, if we simply continue with the step-reversals that were started above and that led from Newton's gravitational equation to the *Newtonian Orbit Equation*:

$$v^2R = GM \qquad - \textit{Newtonian Orbit Equation}$$

the next step in the reversal is to replace **GM** with **K**, giving the original *Geometric Orbit Equation*:

$$v^2R = K \qquad - \textit{Geometric Orbit Equation}$$

This means the *Newtonian Orbit Equation* used today, based on the Newtonian theory of gravity, provides exactly the same function as the *Geometric Orbit Equation*, which can be derived purely from astronomical observations without appealing to a gravitational force at all. *Indeed, they are the same equation.* In fact, this explains why the *geometric* orbit equation is unknown today – we already believe we have the proper *gravitational* version, including its reference to mass, **M**, and the "gravitational constant of nature," **G**. Given this, there is no need to even take notice of the obvious, simple, and entirely equivalent geometric form that pre-dates our familiar orbit equation today. Yet, it is this very fact – that a simple and *fully functional* geometric form already exists – which is of such great significance, especially since we also widely use Kepler's three laws in our science and space programs, which also have nothing to do with a gravitational force. This means that even when we use our *Newtonian* orbit equation, we are actually unknowingly using the *geometric* orbit equation, and so, all of astronomy as well as our space programs are actually based *solely on geometry* – and *not on Newton's gravitational force at all*. The apparently insignificant fact that a simple *geometric* orbit equation can be easily identified which parallels our gravitational version is actually not so insignificant at all, but of *great significance indeed.*

NOTE

 Though not recognized today, Newton's gravitational force is a completely superfluous and redundant abstraction, both in theory and in practice.

The above statement may seem premature since the *Newtonian* orbit equation involves the mass of the orbited body, *M*, while the *geometric* orbit equation has only an arbitrary constant, *K*. It might seem that, if nothing else, Newton's gravitational theory shows that this constant actually refers to the mass of the orbited body, which could prove to be a very useful realization. In fact, one very important result from today's *Newtonian Orbit Equation* is that it apparently allows us to calculate the mass of distant bodies, such as the planets in our solar system. That is, if we know the speed, *v*, with which an object is orbiting and the radius of its orbit, *R*, we can use the *Newtonian Orbit Equation* to calculate the mass, *M*, of the larger body it is orbiting. This would tell us the mass of a distant planet simply by observing the motion of its moons, for example, which is precisely how we have arrived at the values we believe to be the masses of the planets today.

Yet, if we used the *Geometric Orbit Equation*, knowing the speed and orbital radius of orbiting objects would only allow us to calculate the constant, *K*, for that orbital system rather than the mass of the body they are orbiting. Knowing the value of this constant for a particular orbital system is still very useful for calculating the speed or orbital radius of other orbiting objects in that system, but it would not tell us the mass of the orbited body. Therefore, it would appear that if we had never known of Newton's gravitational theory we would not have been able to determine the masses of the moons, planets, and sun of our solar system – at least not by using Kepler's three laws and a purely geometric orbit equation. And so, it might *appear* that Newton's gravitational theory somehow provides a deeper physical meaning and insight into nature. However, the following discussion shows that this is not the case at all, and that it is merely an illusion that Newton's gravitational theory provides any additional insight or utility beyond what was already possible prior to its introduction.

Newtonian Theory Does Not Give Mass-At-A-Distance

Newton's theory of gravity claims that a gravitational force emanates from planets (and all objects) to act across space and out to remote distances, allowing a planet's mass to be determined remotely since its mass is claimed to be directly related to the strength of this force. In particular, referring to the *Newtonian Orbit Equation*, $v^2R = GM$, it would *appear* that we only need to note the velocity and orbital radius of an object in order to determine the mass of the body it is orbiting. However, the following discussion shows that it is only an illusion that mass can be directly determined at a distance in this manner.

WATCH FOR...

- The orbit equation expresses a relationship between the speed and the orbital distance of an orbiting object; in this respect, both the geometric and Newtonian versions function equally.

- The known masses of moons and planets are merely approximations based on an unsupported assumption that is built into Newtonian theory – they are not the literal, accurate masses we believe them to be.

- The above-mentioned assumption is that mass is directly related to orbits – an assumption that is neither scientifically proven nor entirely correct as it turns out, giving arbitrary, inaccurate mass values.

- We are still able to use these inaccurate mass values in other calculations of orbital velocity and distance since these mass values are typically not used alone, but as part of the expression *GM*, which is entirely equivalent to using the original constant, *K*, in the original *Geometric Orbit Equation*.

We first begin by noting that whether we use the geometric or the Newtonian form of the orbit equation, the function of the orbit equation is to describe the relationship between the velocity and the orbital radius of an orbiting object. This role is equally fulfilled by *either* orbit equation since the *Newtonian* "gravitational" version is merely the original *geometric* equation with an arbitrary cosmetic change in the appearance of its constant, *K*. That is, we can arbitrarily change the *symbol* of the constant *K* in the *geometric* orbit equation into the two multiplied constants *GM* if we wish, creating the *appearance* of a new "gravitational" orbital equation but not actually altering the *function* of the original equation at all. The orbit equation still provides the same relationship between velocity and orbital radius as always, regardless of this cosmetic change.

However, since the value of K is easily determined by remote observation of orbiting objects, then arbitrarily changing K to GM would allow us to calculate M (since G is a known constant value), creating the *illusion* that we can remotely determine the mass of the orbited body. The *possibility* that K may actually be a direct reference to the mass of the orbited body is merely an interesting *conjecture* of Newtonian theory, but one that is both scientifically unproven and also *irrelevant to our orbital calculations*. This is an important point to note, since today we are under the illusion that we use the masses of moons and planets in the orbital calculations of our space missions. In actuality, we typically do not use these supposed masses alone, but as part of the expression GM. And as we now know, this expression is nothing other than the original constant, K, in the original *Geometric Orbit Equation*. The Newtonian exercise of redefining K as GM, solving for M, then using M in the expression GM is merely a winding path of logic disguising the fact that we are still simply using the original constant, K. The implied existence of a "gravitational force" in this circular Newtonian logic, as well as the supposed remotely-determined mass, are only conjectures at best – and at worst, *pure fictions*.

It is a powerful illusion that our current Newtonian orbit equation, $v^2 R = GM$, is the true original orbit equation, and that it contains an actual physical mass. This illusion arises because its purely geometric origins are well hidden under a compelling gravitational overlay. All of the previous discussions comparing Newtonian theory with the original *Geometric Orbit Equation* are impossible today, since this equation is not formally known in our science; its existence and significance have been buried for centuries beneath our unwavering and largely unquestioned Newtonian beliefs. We simply accept the mass of the sun listed in our textbooks, overlooking the fact that it was arrived at by plugging the known velocities and orbital radii of the planets into our current *Newtonian* orbit equation, which actually calculates K but disguises it as GM. We unknowingly accept that this hidden redefinition from K to GM is correct, arbitrarily turning a *purely geometric* constant calculated from *purely geometric* observations of our planets, into the solid mass of the sun. Without benefit of the analysis given in the previous discussions, we could not even know that we are making such an unsupported and arbitrary assumption. We *believe* in Newtonian gravity ... we *believe* today's orbit equation is solely a product of Newtonian theory ... we *believe* the mass in today's orbit equation describes a real mass ... and we are fundamentally unable to contemplate the geometric origins of it all since they are firmly buried beneath these beliefs and illusions.

But then, it is natural to wonder if there remains any significance to the values listed as masses in our textbooks. Even though we may have arrived at these values by making the unsupported assumption that *K* is actually *GM*, it still seems reasonable that *K* must correspond to *some* material aspect of the orbited body. And further, the value of *K* does vary between different orbital systems in a manner that seems to reasonably reflect the expected mass differences between the central orbited bodies in these separate orbital systems. So, what are we to make of this situation?

This issue of mass will be more fully understood once the new principle in nature is introduced in the next chapter; however, for now it can be said that today's mass values represent *approximate* masses – essentially reasonable educated guesses. This is because the observed gravitational *effect* that we call orbits (which does not involve a confirmed gravitational *force* unless proven scientifically viable) does indeed turn out to be related to the mass of the orbited body – though not *directly* related as assumed today. Therefore, our assumption that it is valid to arbitrarily replace the constant, *K*, in the orbit equation with the expression *GM*, involving the mass of the orbited body, is somewhat justified but inaccurate. That is, despite the fact that Newton's model of a gravitational force emanating from matter cannot describe the true physical reality – for all the reasons mentioned so far – it still is undeniable that our massive planets and sun *somehow* cause our observations of falling objects and orbiting bodies. So then, since we know that one of the main defining qualities of our sun and planets is their mass, it would be expected that mass would be involved in our observations of the heavens – if not directly then at least indirectly. And as we will see in the next chapter, mass is only *indirectly* involved.

As an example of how mass might be *indirectly* involved in observations, just for illustration purposes lets consider a hypothetical scenario where all bodies in the heavens have an attracting magnetic field, but where we also have not discovered magnetism yet. In this case, we might tend to think that the mass of an object somehow *directly* causes the attraction that we observe in orbits, which would mean that an object with double the observed attraction must have double the mass. However, unknown to us, the doubled attraction would actually be due to double the *magnetic field*, which may or may not correspond to double the mass depending on whether magnetic field strength is correlated with mass in a direct one-to-one relationship. If two objects with the same mass but different material composition could have different magnetic field strengths, then this direct relationship would not hold. An observation of double the orbital attraction may be caused by a planet with only 30% more mass than another (though mass of a different material), yet our assumption of a direct relationship between orbital observations and mass would cause us to incorrectly list that planet as having *double* the mass.

This is similar to today's belief that mass is directly related to orbital observations. This direct mass relationship supposedly occurs via Newton's mysterious "gravitational force" – a force that has never been felt or detected remotely, but whose strength is said to directly mirror any changes in mass. So, if our Newtonian calculations tell us that an orbital observation corresponds to double the gravitational pull, we note the orbited body to have double the mass. However, the new principle in the next chapter shows that orbits are not caused by a "gravitational force," and that, although the actual cause *is* related to mass, the relationship is not strictly a direct one-to-one correspondence. It is a reasonable assumption that a larger planet with a greater *effective* gravitational influence on orbiting objects would also have a correspondingly greater mass, but this assumption cannot be verified with certainty from a distance. It would be necessary to physically analyze the material composition of the planet to know for sure. This is analogous to the hypothetical magnetic field scenario, where a stronger influence on orbits (a greater magnetic field in this case) would seem to imply a correspondingly greater planetary mass, but could simply be due to a different magnetic material regardless of mass.

It is for this reason that the accepted masses today of the sun, planets, and moons of our solar system were stated earlier to be only approximations – not true mass measurements. Some of these values may be very close to the actual mass of the body, while others may be far off the mark. This has not been a problem for most standard orbital calculations since, as mentioned earlier, we typically use these mass values in the expression GM, which simply returns us to the constant K in the original *Geometric Orbit Equation*, and makes the actual individual mass value irrelevant. However, it is important to understand this mass issue for other reasons. For example, planetary geologists cannot gather a proper understanding of planetary formation, composition, and geology if the assumed mass is far from the actual mass of the planet. Also, theoretical fusion reaction calculations for our sun include mass in their calculations, and it may well be crucial to have the correct mass value for our sun in order to properly understand the physics of fusion itself.

Despite all of the preceding discussions suggesting that orbits are not ruled by Newton's mass-based gravitational force, there can still be some compelling illusions that appear to support Newton's theory. One such example from our space programs is the need to include the mass of our spacecraft in all trajectory calculations – even down to the diminishing weight of the fuel as it is expended or the additional weight of any rock samples that may be carried back to Earth from a distant moon or planet. If the mass of our spacecraft is an important consideration in the accuracy of our current trajectory calculations, doesn't the success of most missions validate our Newtonian calculations and beliefs?

The answer is that the mass of the spacecraft is only important to the *inertial* calculations of the mission – not the *orbital* calculations. Inertial calculations involve any attempt to forcefully alter the trajectory of the spacecraft using a fuel burn. Just as the mass of a football player is of crucial importance to any player attempting a tackle, the precise mass of the spacecraft is of crucial importance to know how much fuel to burn for a given maneuver. A more massive spacecraft requires a longer or more powerful fuel burn, just as a heavier football player is harder to tackle. This is merely a classical Newtonian *inertial* calculation (not a *gravitational* one), given by Newton's equation $F = ma$ (*force* equals *mass* times *acceleration*). The fact that such mass-based *inertial* calculations are crucial to any space mission lends unwarranted credibility to the illusion that mass is further useful and necessary in our current *Newtonian* "gravitational" orbit calculations. Orbits (which form the basis of all spacecraft trajectories) are still completely described by the *purely geometric* equations of Kepler and the *Geometric Orbit Equation*, which do not involve mass or force.

Does the Evidence Support a Gravitational Force?

Despite the fact that Newton's concept of a gravitational force violates our laws of physics and is unnecessary to describe orbits and spacecraft trajectories, it is still credited with explaining many other facets of life on Earth. For example, the reason objects have weight here on Earth is supposedly because a gravitational force emanates from our planet and pulls them down, forcefully and continuously holding them in place in proportion to their mass and giving them their mass-dependent weight. Even though we have no scientifically viable explanation for this constant pulling force, it would certainly *appear* as if such a force existed, nonetheless.

Yet, we have always known that *something* creates this effect, even before Newton arrived on the scene, but it wasn't necessarily considered to be an attracting gravitational force from within the planet. It could have been due to the Earth's magnetic field, or some type of downward repelling force from the stars in the heavens above, or any manner of other ideas. The weight of objects was simply an experience that was undeniable and common sense – no one expected objects to fall *up* when they were dropped – but the underlying cause could have been almost anything; it was simply unknown. We design spring-loaded measuring scales that we deliberately calibrate to properly weigh objects, but this is merely a device that takes advantage of this obvious *weight effect* all around us. Our mechanical scales are not actually based on a *gravitational force* principle, but rather, on a *spring* principle that takes advantage of whatever is causing the *weight effect* around us.

Even the science of calculating how a projectile, such as a cannonball, flies through the air is not actually based on Newton's gravitational force, though this is commonly thought to be the case today. The work of Galileo Galilei (1564-1642) provided a very useful constant-acceleration equation for falling bodies or flying cannonballs, but a quick look at this equation shows no particular reference to a gravitational force:

$$d = \tfrac{1}{2}at^2 \qquad - Constant\text{-}Acceleration\ Equation$$

This equation essentially states that the vertical distance, *d*, that an object falls as it is either dropped or shot through the air is determined by a constant downward acceleration upon it, *a*, multiplied by the square of the time, *t*, that it takes to hit the ground. It is worth noting that this equation is a *purely geometric equation* involving *no physical masses or forces*, merely embodying the obvious fact that objects in free-fall experience a constant downward *acceleration effect*. It does not state the *cause* of this effect any more than the cause for the weight of objects was universally settled upon prior to Newton. This observable and measurable downward *acceleration effect* on Earth is the same for all objects *no matter how massive they are*, and can easily be measured to be 9.8 m/s^2 and substituted directly into the above equation to give:

$$d = \tfrac{1}{2}(9.8)t^2$$

We typically use the symbol, *g*, for this measured *constant-acceleration effect* upon earthbound objects, giving us:

$$d = \tfrac{1}{2}gt^2$$

The symbol, *g*, is taken to mean the acceleration due to gravity (9.8 m/s^2), in reference to Newton's proposed gravitational force; but that interpretation, of course, is only an assumption.

Equal Acceleration Regardless of Mass

As mentioned above, whatever the cause may be for the *acceleration effect* of falling objects, it manages to accelerate all objects with equal ease at the same rate and with no noticeable stresses upon them. This is true whether they are as light as a golf ball or as massive as an ocean-liner. If a force were at work here, it would have to be quite a *mysterious and unprecedented force indeed* to achieve such a feat.

 The "Gravity Shield" Mystery

Another ongoing mystery surrounding gravity is the idea of a "gravity shield." After all, by using various materials we are able to insulate against electricity, electric fields, magnetic fields, light, radio waves, and radioactivity, so why not the gravitational field as well? Since science has never had a clear understanding of gravity, it has been impossible to either conceive of or rule out the possibility of developing some material or device to shield us from gravity. Such an invention would allow an object to levitate in mid-air simply by inserting this gravity shield between the object and the ground. If the attracting force of gravity cannot reach up past the gravity shield, then any objects above the shield should float and not be pulled downward. Such ideas have surfaced repeatedly over the years (and continue still), being shrouded in secrecy and mystery, and drawing short-lived interest and funding until ultimately fizzling out.

The preceding discussions have shown that, while Newton's proposed gravitational force is a very compelling and intuitive idea, it is *rife* with problems. As a *model* of the true, and as-yet-unknown, underlying cause for many observations it has proven very useful – which is the purpose of any model or equation – but things become very problematic and mysterious when the model is taken as the literal reality. And in fact, as was also shown, Newton's model is not even strictly necessary, as everything from falling apples to orbiting moons can be dealt with equally well with purely geometric equations. This model is part of our scientific legacy from centuries past, and as such, it sits largely unquestioned in our science today despite the fact that it clearly is not a scientifically viable theory.

We have tried applying logical patches, such as the misapplied *Work Function*, and even invented entirely new theories, such as *General Relativity Theory* – but to no avail. We have been unable to find true scientific justification for Newton's gravitational force, yet we also have been unable to develop a truly viable theory to completely replace it. As a result, Newtonian gravitational theory remains our main and most compelling explanation for falling objects and orbiting bodies, while also clearly being a fatally flawed theory in our science.

The reason Newton's gravitational explanation was so revolutionary when it was proposed is that it was thought to have finally provided a *physical* understanding of the underlying cause for these observations – something mankind had wondered about through the ages. However, if a gravitational force is *not* a viable scientific explanation for the underlying cause, then what is? An answer to this question that provides a clear physical explanation for gravity and resolves all of the mysteries and violations mentioned so far is provided in the following chapter, where a new principle in nature is presented – *one that has been overlooked so far in our science.*

- 2 -

Encountering

the

New Principle

A New Theory of Gravity

As the exploration of gravity in the previous chapter shows, while Newton's model of an attracting gravitational force emanating from matter is very compelling and intuitive, serious problems arise when this model is taken to describe a *literal* force in nature. It would have to forcefully confine objects to the surface of planets and powerfully constrain bodies in orbit millennium after millennium, while never diminishing in strength or even drawing on a power source – all clear violations of the *Law of Conservation Of Energy*. It would also have to act instantaneously across any distance throughout the known universe – a violation of the speed-of-light limit. It would further have to rapidly accelerate all objects with equal ease, no matter how large and massive they may be – an unprecedented feat for any other force known to man. Three centuries after its introduction into our science we still cannot explain why the gravitational force would emanate from matter, how it might reach across empty space, and why it should pull on other objects. Although we are familiar with the *effects* of gravity and have *models* of these observed effects – compliments of Newton, and more recently, Einstein – we do not truly *know* gravity at all; its *physical nature* is completely foreign and mysterious to us even today.

Further, there is no particular reason to conclude that we live in a universe ruled by a gravitational force, as thought today, or even to conclude that such a force exists at all in nature. The motion of every object in the heavens can be described and calculated by our existing suite of purely geometric equations, which contain no forces and no masses. Newton's universe of an attracting gravitational force emanating from every body in the heavens according to their mass is merely an arbitrary overlay atop a universe that proceeds according to predictable geometric patterns, whether we use Newton's gravitational overlay or not. Here on Earth we experience a *weight effect* and a *downward acceleration effect* upon objects, both of which have been instinctively recognized and exploited by every living being on the planet since time immemorial. Whether we explain these effects in terms of a gravitational force emanating from within our planet or not, they remain just as they are and have always been. We measure and calculate these effects using purely empirical or geometric methods that also require no particular appeal to Newton's gravitational force.

Yet, it is not enough for us to know *of* planetary motion – we want to know *why*. It is not enough for us to know *of* the weight and acceleration effects here on Earth – we also want to know *why*. In fact, if we truly want to advance technologically and intellectually as a species, we *need* to know why. But there is only limited value in arriving at passable explanations; with enough resourcefulness we can invent any number of them – Descartes proposed one, Newton proposed another, and Einstein proposed yet another – yet none have answered our ultimate questions. In fact, it could even be said that our current gravitational theories deepen the mystery further, by introducing a scientifically unexplainable force in Newton's theory, or a mysterious warping of a four-dimensional space-time continuum in Einstein's theory.

There have been accurate predictions made with both Newton and Einstein's theories, though this should not be surprising from models that have proven useful enough to make their way into our science. Yet, although any useful model captures at least a few features of the underlying reality with a reasonable amount of accuracy and utility, mere models eventually break down at some point, and cannot truly provide the deeper answers we seek. In order to move beyond the stage of superficial *know-how* and into an era of true *knowledge* and *understanding*, we need to move beyond passable models and explanations that break down under the scrutiny of our laws of physics and our common sense. If a theory violates our laws of physics, then it is flawed as a true physical explanation, as this is precisely why we have such laws in the first place. Also, if a theory violates our common sense and is largely incomprehensible and mysterious, then it is a mere abstraction that has failed as a true explanation since an explanation, by definition, *clarifies* an issue – it does not *confuse* the issue. So, if we have only flawed models and failed explanations today, then what *is* at the root of our observations of falling objects and orbiting bodies? What exactly *is* gravity? To answer this question we begin by considering a thought experiment involving a hypothetical two-dimensional world.

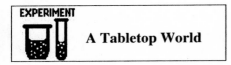

A Tabletop World

In 1884, Edwin A. Abbott (1838-1926) published a story called *Flatland*, about beings that lived, not in a three-dimensional world of length, width, and height such as ours, but in a two-dimensional world with only length and width. This would be as if these beings and their whole world were mere drawings on a flat tabletop. For such beings, only other two-dimensional objects drawn onto their tabletop world would be recognizable and comprehensible; the three-dimensional objects of our world beyond their tabletop would be as incomprehensible to them as the concept of a four-dimensional physical object is to us. The only way for a three-dimensional object to interact with this flat tabletop world is via a cross-sectional "slice" of itself as it intersects with this two-dimensional world.

Let's now extend this concept and explore what would happen if all objects were actually cone-like three-dimensional objects passing through the tabletop from above, allowing these flat beings to experience each slice of the growing cross-sections of these objects as they continuously advance through the tabletop (Fig. 2-1).

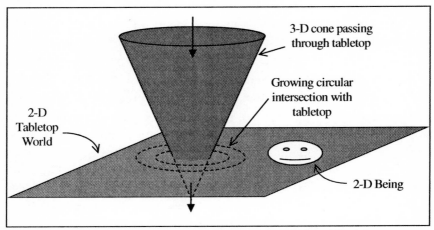

Fig. 2-1 Cone Passing Through Tabletop World

From the perspective of these beings, if numerous unseen three-dimensional objects were passing through their world in this manner from an incomprehensible third dimension, the cross-sections of these objects would appear as regular flat objects in their tabletop world. And if the objects had cone-like shapes in the third dimension, as shown in Figure 2-1, their cross sections would be continuously growing as they passed through the tabletop – for some unknown reason as far as the tabletop beings were concerned. Once one of these object cross-sections grew enough to reach one of the beings, the being would be pushed by the growing object and would experience a constant force as a result. This description of objects effectively approaching one another, as well as causing a constant force for beings in contact with them, may sound somewhat like our experience of gravity. In our world all objects are somehow attracted to one another (presumably due to Newton's gravitational force emanating from all objects), and the particular large object that we call our planet causes a constant force for those in contact with it.

However, the parallels would seem to end there. In this thought experiment the beings would see the objects around them growing in size continuously and eventually becoming gigantic relative to them, yet the objects in our world remain a constant size. Also, just as the third dimension is inconceivable to these two-dimensional beings, it is equally difficult for us to imagine that objects in our world may somehow be the result of bizarre objects from an inconceivable fourth physical dimension that project themselves into our three-dimensional world.

But what if the tabletop beings were *also* growing, and at the same rate as all objects around them? Then everything would remain the same *relative* size and, therefore, no growth would be directly noticed, yet everything still continues growing regardless, still taking up more and more of the available space between them. A scenario where everything appears to remain a constant size but the space between all objects continuously diminishes can be more simply described as a situation where ordinary objects are attracted to each other for some inexplicable underlying reason. The tabletop beings would simply take it for granted that objects in their universe somehow attract one another in everyday experience, and various hypotheses might arise from time to time in an attempt to explain this phenomenon – much as is the case in our experience and history with gravity. As mentioned earlier, Descartes, Newton, and Einstein each made attempts to explain this effect in our world. However, this tabletop analogy still leaves us with a sizable problem: the underlying explanation for this attraction is based in a mysterious extra dimension somehow *outside* normal experience, implying that familiar objects in our world would have a true form that is mysteriously beyond our normal three-dimensional comprehension. However, what if this other dimension were instead right here *within* our regular three-dimensional world? What if it existed *within every atom*?

A New Property of the Atom

Imagine, for a moment, that our current theories about internal atomic structure are wrong, and we are about to learn that below the outer surface of the atom is an inner dimension completely foreign to us, with a physics unlike anything in our atomic models today. Further, imagine that it is the nature of this foreign inner dimension to continuously expand outward into our dimension, literally *creating* what we call atoms – but atoms that are *continually expanding* from within this foreign inner dimension.

If this were the case, it would mean all of our atomic models over the past century have overlooked this phenomenon. As odd as this may seem, the reason this could have occurred is because we have always assumed that the inside of the atom is much like the world outside the atom. But what if we let go of this mindset? What if, just as the third dimension is entirely foreign and unknown to the two-dimensional tabletop beings mentioned earlier, so the inner dimension of the atom has an entirely foreign nature than we conceive of today? Then, just as it would be inappropriate for the tabletop beings to assume that their regular two-dimensional spatial experiences apply to the mysterious third dimension, it would be equally inappropriate to assume our regular three-dimensional notion of space applies to the inner dimension of the atom.

In actuality, our concept and experience of three-dimensional space is entirely a result of the world *outside* the atom, which exists only because the atom *first* exists to define it. Unless we undergo tremendous effort to split the atom, the inner dimension of the atom has nothing to do with our world, which is entirely defined by the external characteristics of atoms, and the space between atoms (not *within* them). Therefore, it does not necessarily follow that the three-dimensional physical world we know, which is based on the pre-condition that atoms *somehow* exist in the first place, can then be turned around and applied to the dimension *within* these pre-existing atoms to explain their existence. Atoms essentially create or define the three-dimensional space outside them, but do not necessarily contain that same spatial definition within.

In fact, we have *tried* to model the inner dimension of the atom after our familiar three-dimensional experiences, and have ended up with very unsatisfactory results. A very popular early model of the atom was the Rutherford-Bohr model, named after Ernest Rutherford (1871-1937) and Niels Bohr (1885-1962), in which electrons orbit the nucleus somewhat like planets orbiting the sun, only with electric charge holding it all together. Although this is still a useful introductory model of the atom, today it is considered far too simplistic to be the proper physical description of this inner dimension. Instead, we now have mysterious quantum-mechanical theories of subatomic and atomic behavior that even the creators and practitioners of these theories readily admit are bizarre and inexplicable. As we might expect, attempts to explain the dimension within the atom as if it were part of our familiar three-dimensional experience *outside* the atom may well result in theories that never quite make sense, since they might never properly describe a pre-existing dimension that may be truly unique.

Yet, this does not mean the foreign dimension within the atom cannot be understood; it just means this inner dimension cannot be understood from a perspective that does not acknowledge that it may have a *totally unique nature*. Even though the dimension within the atom may be beyond our normal experience – and even outside our existing science – it is not necessarily beyond comprehension. In fact, the nature of the inner dimension within expanding atoms is clearly explained in an exploration of the atom in Chapter 4, but for now, this new principle in physics is simply introduced as a *universal atomic expansion rate upon all atoms*. This concept will be referred to as *Expansion Theory* from this point on:

Expansion Theory

Every atom in the universe is expanding at an identical universal atomic expansion rate.

Can It Be?

A first response to *Expansion Theory* may well be that this is an interesting idea, but of course we certainly would have noticed by now if atoms were expanding. Countless experiments have been done with atoms and molecules, and we even have microscopes powerful enough to identify individual atoms, yet no atomic expansion has been seen or even hinted at. However, just as in the previous thought experiment, if everything were expanding at the same rate, then everything would remain the same relative size, and no growth could be seen. Since every material object is composed of atoms, if every atom were expanding at the same universal atomic expansion rate then so would every material object, as would *we* as well. Therefore, not only would we not *notice* other objects around us growing, but we would also be fundamentally *unable* to observe their growth directly – even if we deliberately looked for it. We would be "trapped" in the same universe of expanding matter as every other atomic object, and would be fundamentally unable to get *outside* that universe to observe its expanding matter. Beings that are made of expanding atoms within a universe entirely composed of such expanding atoms would never actually see other objects or individual atoms growing relative to them or relative to other objects. Even rulers and other measuring devices would grow in the same manner, and would therefore indicate constant sizes.

There would, however, be a noticeable effect from all this underlying expansion – objects would take up more and more of the space around themselves and, therefore, between one another as they grew, eventually touching and continuously pushing against one another due to their ongoing expansion. But does this actually happen in our world? Indeed it does. This is the very reason Newton introduced the notion of an attracting gravitational force – to explain the fact that all objects seem to pull toward one another, and experience an ongoing force holding them together once they touch. Newton explained this effect as a mysterious attracting force emanating from all objects, but *Expansion Theory* shows that no such force is needed to explain this effect once we consider atomic expansion.

Consider the case of a dropped object. The object is said to fall due to Newton's gravitational force emanating from the Earth and from all objects. However, imagine that the gravitational force does not exist or is turned off somehow, resulting in the dropped object simply floating above the ground. Now, what would happen if both the object and the Earth started *expanding*? Even if both expanded by a tiny percentage, the enormous size of the Earth means even this tiny percentage would be a very large *amount*, causing it to rapidly expand into the floating object, striking it with great force and continuously pushing against it as the planet continues expanding. From our perspective standing on such an expanding planet, we would have seen the dropped object effectively accelerate down toward the ground, where it appeared to remain "stuck" to the surface as if by some attracting force as the planet continued expanding against it. In fact, while standing on this expanding planet we would also experience this same ongoing expansion force against us from the planet below as if it were pulling us down as we struggled against it to remain standing – an effect that would result in what we call our weight. The whole while, we would maintain the same relative size compared with both the dropped object and the planet – since everything, including us, would be expanding at the same universal atomic expansion rate (i.e. the same percentage).

Despite this example, it might still seem that the concept of expanding atoms could be readily disproved by many other common gravitational experiences in everyday life. For example, what about the fact that we can literally *feel* the gravitational force pulling on objects as we hold them in our hands and feel their weight? As compelling as this is, it was just illustrated that we would feel our own body weight simply because we must struggle to stand up on a planet whose expansion continuously pushes against us from below. Likewise, if we picked up an object and held it in our hands, we would actually be pulling it along with us as we were pushed upward by the planet, rather than letting go and dropping it (i.e. leaving it behind to float in mid-air as the ground approaches it). The muscular strain we would feel by essentially dragging the object along with us – as we are accelerated upward – would cause the effect we call weight. In fact, a similar effect is commonly felt in elevators when they initially accelerate upwards from standstill, causing more strain on our legs to support our own weight, and more strain in our arm muscles as any objects we are holding become momentarily heavier. Likewise, in a universe of expanding atoms, our very weight – and the weight of any objects we are holding while simply standing on the ground – would be caused by a continuous "elevator effect" as our planet expands beneath us, causing a continuous upward acceleration.

NOTE

 In fact, it is just such an "elevator effect" – known as the *Principle of Equivalence* – that is widely known to have inspired Einstein's own theory of gravity, as will be discussed shortly, though Einstein did not consider a literal expansion of atoms and instead chose his far more abstract notion of *General Relativity Theory.*

Although the weight of objects in our hands and the accelerating free-fall of dropped objects may be explained by an expanding planet, there still remains the issue of our own *experience* of gravity when we fall. After all, it certainly doesn't *feel* as if we merely float comfortably while the ground approaches. Instead, we feel an alarming falling experience internally as we plummet, and we feel the wind rushing past as we fall.

How can this be explained without a gravitational force pulling us to the ground?

It is actually a well-established biological fact that the alarming feeling of falling is a survival response, triggered by a combination of visual cues and bodily sensors that register the altered physical situation of free-fall. The visual cues involve seeing buildings and cars and the ground below all rushing toward or past us as we fall, yet this is precisely what we would expect to see if the ground were accelerating toward us as we merely floated in space. A major bodily sensation is the rush of wind as we fall, yet this too would be expected if the planet and its atmosphere were speeding toward us – just as we feel a rush of wind as a subway train races into a station pushing a cushion of air ahead of it.

Another major bodily sensation is the alarming falling feeling generated when acceleration sensors in our body detect that we are no longer on solid ground, but are falling. Yet this too would be expected if these sensors register the constant expansion of the Earth against us as normal, and the removal of this expansion as abnormal and alarming. We have evolved as beings that are constantly in contact with the expanding Earth, and these sensors register that everything is fine as long as that situation continues. Stepping off a cliff removes this expansion force from beneath us as we are then floating in space, triggering the alarm mechanism, while the ground also rushes toward us. Astronauts in space feel this same falling sensation, despite the fact that they are floating in space and not falling at all. This psychological alarm mechanism exists so that we will rapidly reach for something solid to hang on to before it is too late. *No objectively verifiable pulling force can be shown to exist during the free-fall of either objects or people.*

Incidentally, according to today's gravitational theory, astronauts in orbit about the Earth – on the space shuttle for example – are considered to be in constant free-fall toward the planet while their orbital velocity compensates, rather than merely floating, as *Expansion Theory* states. However, if this were truly the case then it might be expected that astronauts on the Apollo missions to the moon would have immediately lost this falling feeling as they left the free-fall of orbit and truly floated in deep space on the way to the moon. Yet this is not the experience reported back from such missions. Some astronauts do report eventually adjusting to this falling feeling, and some never do, but this is the case whether they are in a "free-fall" orbit about the Earth or floating on the way to the moon. While, from the perspective of Standard Theory, this fact – that orbital free-fall is essentially the same experience as floating – might be a bit surprising, this is expected in *Expansion Theory* since there is no such thing as falling (i.e. being pulled downward by a "gravitational force") – there is only floating. Falling on Earth, orbiting the Earth, and floating in deep space all *feel* the same because they *are* the same – they are all *floating* experiences according to *Expansion Theory*.

An expanding planet also explains many other aspects of our planet and environment. For example, the Earth is known to have tremendous crushing pressure at its center, which is currently attributed to the Earth's total gravity pulling everything inward. Yet, once again, this would mean gravity has been powerfully squeezing the Earth with no known power source and no diminishing in strength for over four billion years, in violation of the laws of physics. However, this inner pressure would be expected if the planet were expanding outward in all directions – an expansion resulting from the very nature of the inner dimension of each atom composing the planet. All of this outward force in all directions would bear down on the center of the planet, creating the tremendous crushing pressure at its core.

Also, our atmospheric pressure would be easily explained if our planet were constantly expanding outward, compressing the thin layer of gas surrounding it (our atmosphere) to create a constant pressure within it. The atmosphere has nowhere to go to escape from the expanding planet beneath since it is a gas that entirely surrounds the Earth; therefore, it would be essentially trapped by our planet's outward expansion, with a resulting overall atmospheric pressure distributed throughout. Currently, atmospheric pressure is explained as the Earth's gravity pulling on each atom in the air to hold them all to the planet's surface.

The Earth's constant expansion also leads to a natural explanation for rainfall. Since clouds are simply areas of more concentrated humidity within the atmosphere, they also would be accelerated upward by the expanding Earth along with the atmosphere.

As more and more water gathers in the clouds, the inertia of the water (i.e. its natural resistance to being continuously accelerated) would make it increasingly difficult for the water to remain within the upwardly accelerating clouds. This is the same effect as the inertia of heavier people making it more difficult to push them on a swing. So eventually it would become too difficult to keep accelerating the water along with the wispy clouds, and the water would begin to be left behind in mid-air as the clouds continue upward, with the ground approaching this floating water from below. To beings on the ground this would appear as water falling down, as if a gravitational force somehow reached up and pulled it from the clouds. This effect is easily demonstrated by holding a sponge saturated with water until it stops dripping, then rapidly raising it upward. Water is left behind in mid-air, just as rain would be left behind by saturated clouds as they were accelerated upward.

Mysteries and Violations Resolved

Expansion Theory now provides a clear answer to how gravity might "pull" objects to the ground. While our science still has no clear answer to this question centuries after Newton's gravitational force was proposed, it can now be seen that there *is* no pulling force to explain. Objects are *not* pulled to the ground, but merely remain in mid-air while the expanding Earth approaches. With this new perspective, science can now end its search for mysterious "graviton particles," which many particle physicists hypothesize to exist and to be moving rapidly between objects to somehow cause the attracting force of gravity. This is a sizable unresolved issue in physics today, which is resolved and shown to be entirely unnecessary once expanding atoms are considered. The same is true of the search for "gravity waves," which many astronomers currently hypothesize to exist and to somehow ripple through the four-dimensional space-time continuum of Einstein's *General Relativity Theory*.

A related issue that was mentioned in the previous chapter is the ongoing mystery that no workable "gravity shield" that would allow objects to effortlessly levitate has ever been put forward. There have been many claims to this effect, but none have ever proven to be valid. *Expansion Theory* provides a clear understanding of the nature of gravity for the first time, showing why such ideas are fundamentally flawed from their inception. It is pointless to attempt to block the gravity field, since there *is* no gravity field. We can find ways to continuously push an object upward to keep it ahead of the expanding planet – via magnetic repulsion or the downward thrust of a rocket for example – but there is no "gravity shield" that will allow objects to effortlessly levitate.

The description of falling objects provided by *Expansion Theory* also solves a great mystery mentioned in the previous chapter – the fact that all objects, from a tiny golf ball to a massive ocean liner, are somehow accelerated by gravity at the same rate and fall at the same speed. Such immense differences in mass and inertia are apparently inconsequential to gravity, as it accelerates all objects with equal ease while also causing no apparent stresses upon the accelerated objects – feats that are otherwise unparalleled in physics. The mystery vanishes, however, if we consider that objects don't actually fall at all, but remain floating in space while the ground expands upward to meet them. In that case the ground would approach all floating objects equally of course, regardless of their mass (Fig. 2-2).

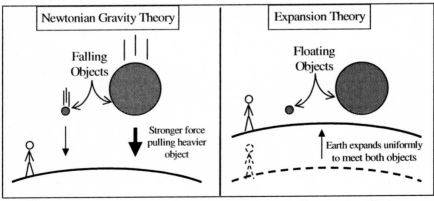

Fig. 2-2 Equally Falling Objects: Newton vs. Expansion Theory

This shows why, in the previous chapter, falling objects were said to undergo an *acceleration effect* instead of an *actual* acceleration due to a gravitational force. The masses of the objects in the right-hand frame of Figure 2-2 have nothing to do with how fast they "fall" in a universe of expanding atoms, since objects only *appear* to fall in such a universe – they actually float in space while the ground approaches them all equally. This provides a clear resolution for this mystery, with no unexplained forces pulling across space, while Newtonian gravitational theory states that even the most massive objects are somehow immediately and rapidly accelerated from standstill by an attracting force when they are dropped.

The *weight effect* can also now be seen simply as the inertia or resistance of an object to being moved as it is pushed from below by the expanding planet – an effect that increases with mass just as a more massive person is harder to push on a swing. The difficulty in pushing a more massive person on a swing is the same phenomenon behind the effect we call *weight* in the parallel situation of the expanding planet pushing massive objects upward. No appeal to a gravitational force is required to explain the mass-dependent *weight effect* felt here on Earth.

Also, since objects are not actually accelerated toward the ground by a force when they are dropped, there is no power source requirement to explain this effect. Indeed, our observation that objects "fall" downward is merely a *geometric effect* caused by the planet expanding up toward them, and not an actual *force-driven motion* of the objects at all. This does not mean falling objects are merely an illusion of course – dropped objects obviously *do* hit the ground and experience a contact force with the Earth – but their apparent accelerating approach downward toward the ground is actually due to the pure geometry of our planet expanding up toward them. This removes a serious violation of the *Law of Conservation Of Energy* pointed out in the previous chapter, since no power source is required if no force is accelerating dropped objects. Likewise, the fact that objects appear to be forcefully held to the ground by gravity can now be explained as the Earth expanding against them. Once again this removes the law violation of a force endlessly expending energy to hold objects to the ground with no known power source, also removing the need to use the *Work Function* inappropriately as an excuse or diversion.

It is important to point out that atomic expansion does not simply replace the mystery of a gravitational power source with a new mystery regarding the power source for expanding atoms. There is a crucial difference between these two scenarios, and an important step forward in knowledge and understanding provided by the concept of expanding atoms. Currently, we have no explanation for the existence and behavior of Newton's gravitational force, so we invent the term "gravitational energy." Here, the term "energy" is simply a generic term for something we cannot explain and know nothing about, since the actual cause – expanding atoms – is unknown to science today. We do not *understand* "gravitational energy" at all; this term is merely a euphemism for an observed attraction between objects that we cannot truly explain even three-hundred years after Newton's proposal of a gravitational force, and a century after Einstein's alternate proposal. As the preceding discussion shows, this euphemism can now be replaced by the proper physical understanding – that all atoms are expanding at a universal atomic expansion rate.

This further implies that there is no such thing as non-expanding matter in the whole of the universe. If an object exists, it also expands – material existence and expansion are one in the same. Of course, it is still natural to further wonder why this might be the case – why do we live in a universe whose atoms all continuously expand? The answer cannot be "energy" since we just saw that "energy" was not an answer to our gravitational questions either, but merely a name given to our lack of understanding of expanding atoms. So then, it is no more meaningful to reintroduce "energy" as the answer for expanding atoms either. Instead, a completely new concept of the atom and its inner structure and nature according to *Expansion Theory* is proposed in Chapter 4 to answer this question.

The issue that Newtonian gravity violates the speed-of-light limit now also has a clear and simple resolution. As was just shown, an object dropped from a given height on Earth is not a case of forceful downward acceleration toward the ground, but actually one of *pure geometry* between two mutually expanding objects (the planet and the dropped object) at a distance from each other. And since this would be the case between *all* objects at a distance, the dynamics between objects in the heavens and throughout our universe would also be purely geometric effects between mutually expanding objects. This leaves only the question of how quickly geometric effects travel, which is readily answered.

If an object is removed from an environment, so is its geometry. If its geometry is an expanding one, then its expansion – and expansion effect – is also removed instantly along with it. In other words, just as an object floating in space at *any* distance from Earth would be considered to be falling toward our expanding planet, if the Earth suddenly vanished, the object would immediately regain its identity as simply an object floating in space. And if there were another expanding planet remaining in the universe, the object would then immediately be considered to be falling toward this planet instead, no matter how far away it was. There is no time lag involved since the geometry of expanding matter essentially "travels" across any distance in the universe instantaneously.

This is considered a mysterious and impossible feature of Newtonian gravity, for which a solution was invented by Einstein in his *General Relativity* theory of gravity, as mentioned in the previous chapter. However, we can now see that Newton's model of a faster-than-light force is fine, as long as it remains a mere *model* for the actual instantaneous geometric effects of atomic expansion. This speed-of-transmission issue raises some obvious questions about *General Relativity Theory*, such as whether its stated speed limit on gravity can be checked by experiment, and what it might mean for this theory if such a speed limit is either disproved or deemed unnecessary in view of *Expansion Theory*. As mentioned earlier in Chapter 1, no definitive, undisputed experiment has ever determined the speed of gravity.

As will be shown shortly, the concept of expanding atoms also allows for many more long-standing questions in physics to be clearly and definitively answered, where neither our current theories of gravity nor our best experimental attempts have provided any clear answers. New possibilities that could not be considered today are also shown to arise when Newtonian gravity is replaced by atomic expansion. However, first we return to the "elevator effect" of our expanding planet, mentioned earlier as the cause for weight. We will see that this concept already exists in our science in the form of the *Principle of Equivalence*, though its implications for expanding atoms have been overlooked, leading instead to Einstein's far more mysterious and abstract *General Relativity Theory*.

LAW

 The Principle of Equivalence

It is impossible to distinguish between the force of one's weight due to gravity and the force that would result from being pushed upward through space with an equivalent acceleration.

Though not technically a law of nature, the *Principle of Equivalence* is an experimentally established fact that Einstein pondered in an attempt to uncover the true physical nature of gravity. He did not believe Newton's gravitational force was the final word on gravity, and had an intuition that there was something important to be learned from the equivalence principle. He wondered why the weight and acceleration effects on Earth, supposedly caused by a phenomenon as mysterious and unexplained as the gravitational force, could be easily duplicated simply by accelerating an object through space. His now famous thought experiment that he used to follow this idea, and to eventually create his *General Relativity Theory*, is as follows:

EXPERIMENT

Einstein's Elevator

Einstein imagined a person floating in space inside a box far from Earth. He then imagined that this box was an elevator that was somehow pulled upward through space. As this happened, the floor of the elevator would come up and hit the person floating inside, carrying them along with it as it was continuously pulled upward. Einstein noted that, according to the *Principle of Equivalence*, if the elevator were pulled upward with a constant acceleration of 9.8 m/s^2, the effect would be identical to the acceleration effect of 9.8 m/s^2 experienced on Earth (mentioned in the previous chapter), currently attributed to Newton's gravitational force. That is, the person in the elevator would effectively fall to the floor (as the floor actually rose upward) and would also experience precisely the same body weight effect as on Earth thereafter (as the elevator continued its upward acceleration). *There would be no difference whatsoever between being in an elevator accelerating upward through space and simply standing on the ground on Earth.*

Einstein felt that there was something very important to be learned from this fact, which would lead to a true physical understanding of gravity. In fact, as shown in *Expansion Theory*, our expanding planet is *precisely* the elevator in Einstein's thought experiment. As the universal expansion rate of all atoms causes our overall planet to constantly expand outward, it essentially pushes us upward through space with a constant acceleration.

NOTE

 However, our expanding planet is not a mere *equivalent* to gravity, as in Einstein's elevator thought experiment; it *is* gravity.

This discussion is graphically illustrated in Figure 2-3 below, where the *Principle of Equivalence* between Newtonian gravity and Einstein's elevator is shown in the first two frames, and our expanding planet is shown in the third.

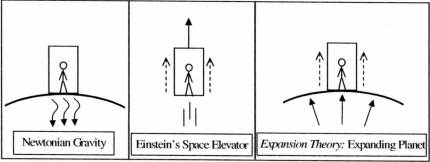

| Newtonian Gravity | Einstein's Space Elevator | *Expansion Theory:* Expanding Planet |

Fig. 2-3 Progression of Ideas Leading to Expansion Theory

The first frame shows a person standing in a box on Earth, with Newton's proposed gravitational force pulling downward to cause the weight effect we are so familiar with in everyday experience. The second frame shows the same person in the same box, but this time the box is in deep space and is essentially a space elevator that is being pulled upward at a constant rate that exactly mimics the gravitational experience on Earth. The person in the box would have no way of knowing the situation had changed from the first frame on Earth. This is exactly the reasoning Einstein went through as he pondered the true nature of gravity a century ago. Finally, the third frame shows how both concepts in frames one and two can be combined to achieve the clear, physically viable explanation for gravity found in *Expansion Theory*. The apparent downward force of gravity would be literally *created* by the upward acceleration of our very own space elevator in the form of an expanding planet. That is, this would not merely *mimic* gravity; it would actually explain the *true physical origin and nature of gravity itself*. The third frame of Figure 2-3 quite literally shows the answer to this age-old question pondered by generations of scientists and thinkers – in a clear, simple diagram.

However, Einstein did not follow his thought experiment to the conclusion of an expanding planet, which would, in turn, have implied the existence of expanding atoms. Instead, he took a very different turn. To Einstein, the equivalence shown in the first two frames meant that the mass of our planet must somehow warp the surrounding space, except that it is not merely three-dimensional space that would be warped, but a four-dimensional realm known as "space-time". That is, much as flat, tabletop beings would only experience objects in two dimensions (length and width) despite their unseen three-dimensional quality, Einstein proposed that our experience of three dimensions actually arises from an unseen four-dimensional quality of physical reality, once *time* is included as the fourth dimension. In this four-dimensional abstraction of our three-dimensional reality and experience, i.e. length, width, and height – plus *time*, Einstein envisioned that the mere presence of matter somehow warps this four-dimensional "fabric" of the universe, called space-time, to cause falling objects and orbiting bodies. As profoundly different as this view is from the concept of an expanding planet suggested above as the logical conclusion to this thought experiment, this warped space-time conclusion is the direction taken by Einstein, resulting in his *General Relativity* theory of gravity, published in 1915.

At this point it is suggested that Einstein's path down the road of *General Relativity Theory* based on this thought experiment was a wrong turn that led him far off the mark – that is, if his intention was to arrive at the one true physical, scientifically viable explanation for gravity. As yet another *model*, Einstein's theory offers a useful perspective on certain elements of observed gravitational behavior, as will be discussed later, but it does not provide the much needed breakthrough in understanding that would be necessary to resolve the many gravitational mysteries and violations mentioned thus far.

There are many reasons why *General Relativity Theory* can only be yet another model. The main reason is that it is, in essence, a pure mathematical abstraction that offers no clear physical understanding or resolution to the problems of Newtonian gravitational theory. Is it the logical conclusion that would follow clearly and rationally from the *Principle of Equivalence* and the elevator thought experiment just presented? Can we truly accept that merely because time passes as events unfold in our three-dimensional universe, that this literally transforms it into a four-dimensional physical universe that is mysteriously warped by the mere presence of matter in some unexplainable manner? Is it reasonable and satisfying to explain falling objects on Earth or their weight in our hands in such terms? Do we truly achieve a better understanding of orbits by considering objects to somehow follow the curvature of a warped four-dimensional space-time fabric surrounding the Earth? How do we even begin to evaluate the issue of a power source for falling objects and orbiting bodies from such an abstract perspective? Where do our laws of physics even apply here? The questions are endless and without substance, and the answers are equally abstract and arbitrary. Again, it is suggested that this is an interim theory during the ongoing search for a true physical explanation for gravity, but cannot be considered as the true physical explanation itself. If it were, it would resolve the mysteries and problems with Newtonian gravitational theory, and would hail a new era of understanding and clarity. Instead, it does neither of these things, and in fact, introduces far more abstraction and mystery to the issue.

The preceding discussions have introduced the general concept of expanding atoms in *Expansion Theory*, and Chapter 4 delves into the inner dimension of the atom to explain this phenomenon in physical detail. However, between general conceptual introduction and final physical details lies an important area where atomic expansion requires a more concrete form and more precise definition. If all atoms are expanding at a universal expansion rate, what is this rate? Can it be determined, and if so, how? If atomic expansion is the true physical explanation behind Newton's equation of a gravitational force in his *Law of Universal Gravitation*, then shouldn't a replacement equation arise in *Expansion Theory*? These questions are answered below, beginning with a determination of the actual universal atomic expansion rate of all atoms.

EXPERIMENT

 Determining the Universal Atomic Expansion Rate

Expansion Theory proposes that all atoms in the universe are expanding – a principle in nature that underlies many of our observations and experiences, and which has been overlooked or misunderstood for millennia. Given this, it may seem that the universal atomic expansion rate might be a deeply held secret of nature; yet it turns out to be quite straightforward to calculate. Since all atoms, objects, planets, and stars maintain a constant relative size, they must all be expanding at the same atomic expansion rate (the same percentage each second). If this weren't the case, some objects would soon grow enormously relative to others, or shrink into oblivion. Therefore, every atom and object must be exactly matched to a single universal atomic expansion rate. So, if we determine this rate for any single atom or object, we will have arrived at the expansion rate of all atoms and objects throughout the universe.

A convenient object to use for this purpose is our own planet. If indeed a falling object is actually a free-floating object that is approached by our expanding planet, then the equal distance that all objects appear to fall in one second on Earth is actually the amount of expansion of our planet in one second toward these objects. Therefore, we simply need to determine the height from which all objects take one second to fall to the ground. This height would actually reflect the amount of expansion of our planet in one second, and since we know the size of our planet, we can figure out what percentage or expansion rate this represents. Once we perform this straightforward calculation, we will have arrived at the universal expansion rate of all atoms and objects throughout the universe.

It is well known that all objects strike the ground in one second if they are dropped from a height of 4.9 meters, regardless of their mass (assuming there is negligible wind resistance). Also, all expanding objects would expand outwardly in all directions from their centers – a radial expansion. Since we also know that the radius of the Earth is 6,371 kilometers (6,371,000 meters), we simply need to calculate what fraction 4.9 meters is out of 6,371,000 to arrive at the fractional expansion of the Earth in any given second (Fig. 2-4). This fractional expansion rate works out to slightly less than one-millionth each second. Therefore, every atom, object, planet, and star in the universe expands at the common rate of roughly one-millionth its size every second.

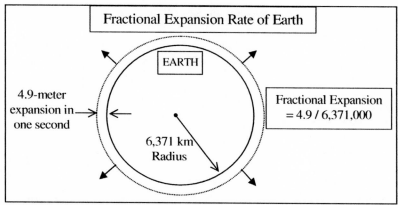

Fig. 2-4 Planet's Fractional Expansion Gives Atomic Expansion Rate

The precise calculation of the universal atomic expansion rate is shown below. Since precise calculation and use of the atomic expansion rate requires that it be given a symbol to refer to, it will be given the symbol X_A. The 'X' is used to symbolize expansion, and the subscripted letter 'A' refers to *atomic* matter (i.e. atoms and objects composed of atoms) since there are other forms of matter such as subatomic particles, which are not part of the discussion at this point. Further discussions of atomic vs. subatomic matter are dealt with in Chapter 4.

OPTIONAL MATH

(x, y) Calculation of the Universal Atomic Expansion Rate

Once the standard acceleration due to gravity on Earth of 9.8 m/s^2 is seen to result from the radial expansion of the planet, the *amount* of expansion of the planet in one second can be calculated. The equation for distance traveled due to constant acceleration was introduced in the previous chapter as:

$$d = \tfrac{1}{2} a t^2,$$ where a is the constant acceleration

t is the length of time the acceleration is applied

This gives the radial expansion *amount* of the Earth after 1 second to be:
$$d = \tfrac{1}{2}(9.8)(1)^2 = 4.9 \text{ meters.}$$

Dividing this one-second expansion amount of the Earth by its radius gives the universal atomic expansion rate:

$$X_A = \frac{4.9}{R_E}$$ where R_E is the radius of the Earth (6,371,000 m)

X_A is the universal atomic expansion rate to be calculated

$\therefore$ $X_A = 0.00000077$ (or 7.7×10^{-7} in scientific notation)

NOTE

 The value of the universal atomic expansion rate, X_A, for all atoms and objects is 0.00000077 per second, each second, giving the units /s^2 or per-second-squared.

This numeric result for the universal atomic expansion rate, X_A, means that we can now calculate the amount of overall expansion each second of any atom, object, or planet simply by multiplying their radii by X_A. Multiplying the radius of the Earth by X_A gives 4.9 meters, which is the distance an object "falls due to gravity" on Earth in its first second. So, in essence, we can now calculate the effective gravity of every atom, object, and planet simply by knowing their *size*. This is radically different from Newton's idea of gravity, which states that a gravitational force emanates from all objects in proportion to their *mass*. This point bears special note:

NOTE

 In *Expansion Theory* the gravity of an object or planet is dependent on its *size*. This is a significant departure from Newton's theory, in which gravity is dependent on *mass*.

This "size vs. mass" difference between *Expansion Theory* and Newton's theory of gravity explains why it was stated in the previous chapter that the masses of the moons, planets, and sun that we accept today are only approximations. They are based on the assumption that our observations in the heavens are a direct result of the *mass* of bodies in space as a gravitational *force* emanates from them, while *Expansion Theory* shows that these effects are actually a result of the *size* of these bodies as they *expand*. Therefore, according to *Expansion Theory*, a planet the same size as Earth would have the same gravity due to its identical amount of expansion each second, even if it were made of a material that caused it to weigh half as much as the Earth (half the mass). That is, orbiting objects would behave identically about either of these two equal-sized planets, and standard Newtonian calculations using today's *Newtonian Orbit Equation* would incorrectly determine the mass of the lighter planet to be the same as that of the Earth. So, if the materials composing the other moons and planets in our solar system are different in density to that of the Earth, then our mass calculations for these bodies will be off by a corresponding amount. This "size vs.

mass" issue will be discussed in greater detail in the next chapter, showing some important and surprising conclusions regarding our current assumptions about the moons and planets of our solar system.

This effort to provide a more concrete definition for the atomic expansion rate leads not only to a precise *value* for the expansion rate of atoms, X_A, but also to a precise *equation* for the effect this constant expansion has on the distance between objects. That is, while Newton overlooked atomic expansion, instead modeling it as if there were an attracting force between objects that pulls them toward each other, *Expansion Theory* shows that atomic expansion exists and is the true cause of objects tending to approach one another. This dynamic can be captured in an equation that calculates how the distance between any two expanding objects would decrease over time − i.e. the *Atomic Expansion Equation.*

The New Equation of Gravity

Before proceeding with the development of the *Atomic Expansion Equation*, a word should be said about the difference between this equation and the equation of Newton's *Law of Universal Gravitation*. Since both *Expansion Theory* and Newton's gravitational theory deal with the same effective attraction between objects, it may be tempting to expect the *Atomic Expansion Equation* to largely parallel the form of Newton's *Law of Universal Gravitation* equation. However, a close look at Newton's equation shows that there is no particular reason for such an expectation. First let's restate Newton's gravitational force equation from Chapter 1:

$$F = \frac{G \cdot (m_1 m_2)}{R^2}$$ where m_1 and m_2 are the masses of the two objects.

R is the distance (radius) between their centers.

G is a constant, called the gravitational constant.

While this equation stands as the symbol of Newton's theory of gravity, the fact is that it truly is nothing more than that − a mere *symbol*. There are several reasons for this. First, recall from the previous chapter that this equation was derived from a rock-and-string model, which was shown to be a flawed, simplistic attempt to *physically* equate this model with orbiting objects in the heavens. In fact, this model is not even *conceptually* correct, even if such a physical equivalence were justified, since the previous chapter showed that other models such as a rock-and-*spring* are arguably even more functionally similar to orbits. Secondly, a good look at Newton's equation shows that it actually serves no useful function by itself. That is, it does not provide a direct calculation of either the speed of an orbiting object or the decreasing distance between objects over time due to Newton's claimed gravitational attraction between them. In actuality, it merely stands as an arbitrary, static *claim* that a force exists between any two objects – a claim that has repeatedly been shown to be scientifically impossible throughout the discussions so far.

Also, although it is currently thought that our *Newtonian Orbit Equation* is an extremely useful and important derivation from Newton's gravitational force equation, this was shown to merely be the **Geometric Orbit Equation** in disguise, which essentially existed prior to Newton. In short, Newton's equation serves as little more than a mere symbol of his claim that this force exists, describing no true physical reality at all, but merely providing a comforting, intuitive explanation for observations in the absence of an understanding of expanding atoms. After all, without Newton's explanation there would have been a discomforting realization that science had no feasible explanation for some of our most common daily observations – a state of affairs that may well have remained for centuries to follow.

It is this purely symbolic nature of Newton's equation that explains why the flawed rock-and-string assumption leading to this equation has not led to clear problems in its use; after all, what problems could arise from an equation that has no direct functional utility? It has never been used to calculate the strength of a *verifiable* gravitational force at a distance across empty space – one that has actually been definitively shown to exist. It is impossible to actually feel or measure this force at a distance in a manner that cannot be refuted – as shown repeatedly by *Expansion Theory* – because *Newton's force does not exist.* These can be difficult facts to digest, given that the existence of Newton's gravitational force has been a forgone conclusion and such a powerful illusion for centuries, but this is the rational conclusion that must be drawn from the discussions so far.

In some ways it can even be more difficult to see this point today than in Newton's day. We have all seen the tremendous effort required to propel a rocket off its launch pad, and the diminishing effort as it coasts into orbit far above. At first glance this may seem to be displaying Newton's diminishing gravitational force with distance. But, of course, it is not a diminishing gravitational force, but the fact that the rocket soon turns horizontally, *speeding in an orbital trajectory around the planet,* that makes it *appear* to float. As well, although orbiting spacecraft *are* far above from our perspective, they have barely left the ground from an outside perspective of the overall planet, meaning that Newton's gravitational force, if it existed, would hardly have diminished at all at typical orbital altitudes. Such events from our space programs can leave us with the impression that we have even *seen* Newton's diminishing force in action, even though we have not.

In contrast to Newton's gravitational force equation, the following *Atomic Expansion Equation* is not a symbolic claim that a mysterious force exists, but a functional equation based on a solid underlying physical concept that violates no physical laws. As such, there is no particular reason to expect any direct similarities with Newton's gravitational equation.

The derivation of the *Atomic Expansion Equation* now follows in two parts. The first part is a conceptual derivation or thought experiment, which shows the logical and physical origin of this new equation using plain English and clear diagrams – no math is involved. This section gives an important feel for what atomic expansion is and how it affects us in our daily lives. The second part rewords this conceptual derivation in the language of mathematics in order to provide the actual *Atomic Expansion Equation* that describes the observations around us. This mathematical derivation that follows the thought experiment below can be skipped without missing any key concepts since it is largely a restatement of the conceptual derivation in the following thought experiment.

EXPERIMENT

 Conceptual Derivation of the *Atomic Expansion Equation*

The concept of atomic expansion involves the changing geometry of objects relative to each other as they *expand* – essentially the change in *distance* between their *surfaces* as they are effectively drawn toward each other purely by the dynamics of expansion. This is in contrast to the *attracting force* found in Newton's equation to explain why objects are drawn to each other, which is claimed to vary with the distance between the *centers* of objects. This surface-to-surface distance between objects in *Expansion Theory* (rather than center-to-center) speaks to the fact that the distance between objects typically means the distance between their surfaces, and objects interact via surface contact; the centers of objects have little to do with everyday events and experiences. Therefore, from the perspective of *Expansion Theory*, the following two effects occur:

- *Absolute Distance Decrease* – The distance between the surfaces of separate objects continuously decreases as a result of their ongoing expansion into the space around them and between each other.

- *Relative Distance Decrease* – This distance between objects continuously decreases even further since all distances effectively become shorter relative to objects and beings that continuously grow larger moment by moment.

Both of these points address the fact that all objects in our universe are effectively drawn to each other by some underlying physical phenomenon. As mentioned earlier, a number of explanations have been put forth by different scientists over the centuries to explain this effect, with the most widely accepted one today being Newton's idea of an attracting gravitational force. Regardless of the explanation, they all attempt to explain the fact that all objects seem to be attracted to one another, with free-floating objects tending to migrate closer together over time until they finally collide, after which they remain forcefully in contact. This is why dropped objects effectively fall to the Earth, and why our moon is essentially trapped in orbit about the planet, rather than flying away into deep space as it attempts to speed past the Earth. These observations tell us that *all* objects must be attracted to one another, and Newton's theory claims that this attraction is due to a pulling force – one whose strength is somehow determined by the amount of mass each object contains. However, the two bulleted points above show that *Expansion Theory* explains this effective attraction of all objects for one another in a very different manner.

The first point is quite straightforward. The space between two growing objects will become less and less until the two objects eventually touch each other. This could be considered an *absolute* space decrease since the space between objects literally becomes occupied by their physical expansion. The amount of physical expansion of an object into the space around it must necessarily mean that the space, or distance, between that object and any others is then reduced by that same amount as a result.

The second point is a little less obvious perhaps, but equally real and important. It refers to the fact that the distances between objects effectively become smaller still, *relative* to objects and beings that continuously grow in size, since the space between expanding objects does not grow along with them. One way to envision this is to imagine walking toward a destination a kilometer away. If every being and object suddenly doubled (while the distances between them did not), the destination would then be effectively only half a kilometer away. Each stride taken toward the destination would be twice as long and would cover twice the distance if we were twice the size, making the distance effectively half as much. Of course, just for illustration purposes, this assumes that the ground did not also expand along with everything else composed of atoms. In reality, the ground would also have expanded, of course, leaving the distance unchanged; however, this example is meant to illustrate the effect that would occur between two objects floating in space, since empty space is not a material object and so would not expand along with everything else. For objects floating freely in space, this *relative* component means that the distances between them would effectively decrease continuously as all objects expand; this is in addition to the *absolute* distance decrease caused simply by their physical growth into the space around them.

Although it may seem obvious that the empty space between free-floating objects is not a *"thing"* that can be stretched or expanded, it is important to state this unambiguously since many followers of theories such as Einstein's *General Relativity Theory* believe space (or "space-time") can be warped or stretched. There is no such abstraction as a malleable "space-time fabric" permeating the universe in *Expansion Theory*, where space is simply a measure of the distance between objects, just as common sense suggests. If objects grow continuously, then the distances between them in free space effectively decrease continuously due to both the absolute and relative effects mentioned above. This is the distance decrease between objects that Newton attempted to explain by suggesting that an attracting gravitational force must somehow be emanating from them, pulling them toward each other.

Continuing with the conceptual derivation, the *Atomic Expansion Equation* must embody both the absolute and relative distance decrease effects. It must calculate the *absolute* decrease in distance between the surfaces of two objects as their expanding matter fills the gap between them, then further reduce this result *relative* to the now larger objects. This dual effect is illustrated in Figure 2-5. Here, two objects that double in size (i.e. each grows outward from its center by one full radius) should reduce the distance between them by the absolute amount of these two additional radii in total – from the initial separation distance of 6 radii down to 4 in this example (top frame). However, the remaining 4 radii separation distance is actually only effectively 2 radii from the perspective of objects that are now twice the size (bottom frame).

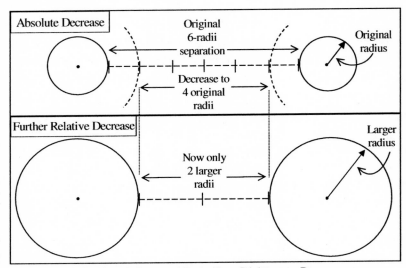

Fig. 2-5 Absolute and Relative Distance Decreases

The top frame of Figure 2-5 shows two objects originally separated by 6 lengths of empty space (the dashed line between them is only an imaginary distance reference, not an actual physical ruler). Each marked length along this imaginary reference line is the size of the radii of the objects to help simplify the diagram and discussion, but could have been the size of any other physical reference. If the two objects then double in size, they each consume an additional full radius of space around themselves, reducing the distance between them by 2 of the original radii in total – from 6 lengths down to 4. This is the *absolute* reduction of space between them as they grow into it.

However, the distance between objects is only meaningful in reference to such objects. In this example, the distance between the objects is evaluated in terms of the number of segment lengths separating them, each length being the size of the objects' radii. Therefore, since the radii of the objects have now doubled in size – as have all other possible physical references in the universe – the 4-length distance remaining would actually be evaluated as 2 lengths (two of the larger radii). This is the further *relative* reduction of space between objects, as it is judged to continuously shrink compared to continually growing objects. Therefore, as expanding beings ourselves, we would not actually see the larger objects growing into the fixed distance between them, as shown in Figure 2-5, but rather, we would see objects that appear constant in size but which seem to have drawn closer together (Fig. 2-6).

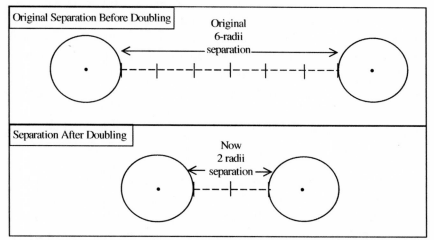

Fig. 2-6 Same Scenario as in Fig. 2-5 as it actually appears

Figure 2-6 accentuates the fact that atomic expansion is completely hidden in a universe where all beings and physical objects are also expanding. If we could somehow step outside of our existence as beings of expanding matter we would see unchanging distances being slowly filled by ever-growing objects, as in the top frame of Figure 2-5, where the distance decreases from 6 lengths to 4 in the given time period. But since we cannot step outside of our expanding existence, we instead see constant-sized objects with rapidly decreasing distances between them, as in Figure 2-6, where the distance decreases from 6 lengths to 2 in the same time period. Which scenario is the true reality? Both actually occur, though we are only capable of experiencing the latter case of Figure 2-6. They are two sides of the same coin, of which we will only ever be able to see one side. Without an understanding of this double-sided expansionary reality underlying our experiences we are left with mysterious explanations for the apparent attraction between objects, such as Newton's law-violating gravitational force.

To further illustrate that both the absolute and relative components are very real and important, we return to Figure 2-5. Consider the two objects in the figure to be asteroids that we must fly between in a spaceship that is also the same size as these asteroids. If we only considered the absolute component of object growth shown in the top frame of the figure, we would note that the 6-length distance between the asteroids was only reduced by 2 lengths due to their growth into the space between them. The remaining distance of 4 lengths between the asteroids would still fit 2 asteroids, and therefore 2 of our spaceships, leaving plenty of room to fly between them with a length to spare on either side.

However, this logic is incomplete since it does not take into account the effect of the double-sized asteroids and spaceship. Even though the original distance was reduced from 6 lengths to 4, those remaining 4 lengths now only measure 2 lengths relative to the now double-sized asteroids (and spaceship). And since both the asteroids and the spaceship will always be 2 radial lengths across no matter how much they expand, there is actually only room for our spaceship to just barely fit between the asteroids without a millimeter to spare. This can be clearly seen in the bottom frame of Figure 2-5, where only one of the larger asteroids, or an equal-sized spaceship, would fit between them. This is quite a different scenario from the ample space it seemed there would be before considering the relative distance decrease effect. As this example illustrates, this continuous relative scaling down of all distances as the objects in our universe continuously grow is quite real to the expanding beings and objects in our universe, and can have significant real-world implications.

The preceding conceptual discussion of two expanding objects in space can also be expressed mathematically, resulting in an equation that captures both absolute and relative components of the decreasing distance between expanding objects over time, as shown in Figure 2-5. This equation is the *Atomic Expansion Equation*, and will be derived in the following discussion:

OPTIONAL MATH

(x, y) **Mathematical Derivation of the** *Atomic Expansion Equation*

To derive the *Atomic Expansion Equation*, we first address the *absolute* distance decrease between objects due to atomic expansion. As shown in Figure 2-4 earlier, the physical expansion of the Earth into the space around it in one second is 4.9 meters, which can be seen by dropping an object from this height and noting that it strikes the ground after one second. Although Figure 2-5 shows that this distance also includes the expansion of the dropped object toward the Earth as well as a relative decrease in distance as all objects expand, we can ignore these two additional effects here, since they are negligible in this scenario. That is, since the atomic expansion rate, X_A, is the tiny fraction 0.00000077 in one second, such a tiny fractional expansion of a relatively small dropped object and equally tiny relative reduction in distance are negligible compared to the resulting 4.9-meter expansion of our enormous planet. Therefore, for all practical purposes, the 4.9 meter drop distance of all objects in one second is simply the absolute expansion amount of the Earth in one second, calculated by multiplying the atomic expansion rate, X_A, by the radius of the Earth, R_E. This is stated as:

$$X_A R_E = 4.9\text{m} \quad \text{-- Absolute Expansion of Earth in One Second}$$

Note that although the above expression technically gives the units of m/s^2, there is actually a hidden multiplication of $(1\text{s})^2$ which cancels the s^2 units; this will be clearly seen later in the derivation. Now, there are two ways that a planet might expand by 4.9m in one second – either at a constant speed of 4.9m/s, or by accelerating from 0 to 9.8m/s to give an average speed of 4.9m/s and thus a resulting growth amount of 4.9m in that second. We know the Earth's expansion takes the latter option and accelerates since we feel a constant force while standing upon it. If the Earth were merely expanding at a constant speed, the effect would be the same as taking one's foot off the accelerator of a car and coasting – there would be no pushing force at all. So, the Earth's growth does not merely coast at the constant speed and grow by the constant amount of 4.9m per second, but instead, increases in speed by 9.8m/s every second, usually represented as an acceleration of 9.8m/s^2. This is shown in Table 2-1, along with the Earth's resulting expansion amount (the distance dropped objects are seen to fall) each second:

Time(s)	Speed Increase	Avg. Increased Speed	Add'l Distance
1	0m/s – 9.8m/s	4.9m/s	4.9m = **1(4.9m)**
2	9.8m/s – 19.6m/s	14.7m/s	14.7m = **3(4.9m)**
3	19.6m/s – 29.4m/s	24.5m/s	24.5m = **5(4.9m)**

:

Table 2-1 Speeds and Distances Attained by Earth's Expansion

As shown in bold print in the last column of Table 2-1, the additional distance each second can be expressed as integral multiples of the original one-second drop distance. Since this original 4.9m distance is the expansion amount of the Earth, $X_A R_E$, we can represent the total drop distance, or amount of the Earth's expansion after a given number of seconds (based on the last column of Table 2-1) as:

Time(s)	Add'l Dist. Each Second	Total Drop (all previous distances added)
1	1(4.9m) = $1(X_A R_E)$	$1(X_A R_E)$
2	3(4.9m) = $3(X_A R_E)$	$3(X_A R_E) + 1(X_A R_E) = \mathbf{4(X_A R_E)}$
3	5(4.9m) = $5(X_A R_E)$	$5(X_A R_E) + 3(X_A R_E) + 1(X_A R_E) = \mathbf{9(X_A R_E)}$

:

Table 2-2 Additional and Total Distances in Terms of $X_A R_E$

A convenient pattern now emerges for the total distance of the Earth's expansion after any given number of seconds, as shown in bold print in the last column of Table 2-2. The total drop distance simplifies to the original expansion amount of the Earth in one second, $X_A R_E$, multiplied by the square of the overall drop time (a fact that also underlies Galileo's constant-acceleration equation mentioned in the previous chapter). This total distance, d, can be written as the integral variable, n, squared and multiplied by the Earth's one-second expansion amount:

$$d = n^2 X_A R_E$$ -- Absolute Amount of Earth's Growth after n
                            Time Intervals

Now, consider two spheres similar to those in Figure 2-5 earlier and with radii R_1 and R_2 . The original distance between their surfaces, D, would be reduced to the lesser value D' by their combined *absolute* expansion amounts of $n^2 X_A R_1$ and $n^2 X_A R_2$ after n seconds of expansion. This simplifies to the equation:

$$D' = D - n^2 X_A \cdot (R_1 + R_2)$$

-- Decreasing Distance between Two Objects over Time due to their Absolute Expansion

Finally, we must address the *relative* distance decrease as our measuring devices grow. From the perspective of objects that have increased in size, the remaining distance, D', calculated above would be scaled down further by a factor equal to the growth amount of atomic objects over time. After all objects in the universe double in size, for example, even unchanged distances between them would still effectively be halved in size. The new growth amount of the Earth over time was just shown to be $d = n^2 X_A R_E$, which means the total overall size of the Earth including its original size would be $R_E + n^2 X_A R_E$, which simplifies to $(1 + n^2 X_A) R_E$. This shows that the Earth and all other objects increase in size by the factor $(1 + n^2 X_A)$. Therefore, the distances between objects effectively scale down by this same $(1 + n^2 X_A)$ factor over time as all objects continuously expand. This is similar to the effect of economic inflation on currency, where an inflation rate of x percent deflates the value of the currency, calculated by dividing the currency by $(1 + x)$. Similarly, distances are scaled down or divided by the same $(1 + n^2 X_A)$ factor that describes the expanding objects. Therefore, the full equation for the decreasing distance between expanding objects is:

NEW IDEA

The Atomic Expansion Equation

$$D' = \frac{D - n^2 X_A \cdot (R_1 + R_2)}{1 + n^2 X_A}$$

where $X_A = 0.00000077 \ /s^2$ (or $7.7 \times 10^{-7} \ /s^2$)

The *Atomic Expansion Equation* above calculates the changing distance, D', between two expanding objects of radius R_1 and R_2 over time. The top portion of the equation is the *absolute* decrease in the original distance, D, between the two expanding objects as they take up more space, and the bottom portion is the further *relative* decrease or scaling down of this distance over time in comparison to ever-larger objects. The variable, n, is the number of seconds that have passed since the original distance was measured between the two objects, and the value shown for X_A is the same universal atomic expansion rate calculated earlier – which never changes. This equation would be used for all falling objects and all objects floating in space as they effectively approach each other due to their mutual atomic expansion (an effect currently thought to be caused by Newton's attracting gravitational force).

We can now see that there are sizable differences between the equation of Newton's *Law of Universal Gravitation* and the *Atomic Expansion Equation*. In *Expansion Theory* the "attraction" between objects actually results from the objects growing due to atomic expansion, so the resulting equation is based only the size of the expanding objects – there is no mass and no attracting force as in Newton's equation. Another obvious difference is that Newton's equation states that the gravitational force diminishes with the square of the distance between objects, yet no such distance-squared term appears in the bottom portion of the *Atomic Expansion Equation*. These stark differences accentuate the fact that Newton's equation does not describe an actual force in nature, and so nothing of its form need appear in *Expansion Theory*. That is, Newton simply invented an arbitrary abstract model of an imaginary force in an attempt to explain observations that are actually caused by atomic expansion. This imaginary force has never been felt or measured at a distance since it does not exist, so there is no particular reason why the equation describing the *real* cause of gravity – i.e. the *Atomic Expansion Equation* – should have any resemblance to Newton's arbitrary invention.

One final observation about the *Atomic Expansion Equation* is that, for objects dropped short distances, it is nearly identical to the equation for dropped objects and projectiles in use today – i.e. Galileo's constant-acceleration equation introduced in the previous chapter as:

$$d = \tfrac{1}{2}\,at^2$$

Although not immediately apparent, it will be shown later in this chapter that for short drop distances the *Atomic Expansion Equation* approximates an equation merely for the absolute expansion of the planet alone – which is completely equivalent to Galileo's equation. Therefore, the constant-acceleration equation for falling objects in wide use today is actually an approximation for the proper and as-yet-unknown *Atomic Expansion Equation* – an approximation that ignores the absolute expansion of the falling object and the relative decrease in distance over time.

It should be noted that the absolute and relative distance decreases captured in *Atomic Expansion Equation* would only occur as stated in the equation if the two objects were floating freely in space. If they were sitting on the ground – which also continuously expands at the same rate and thus keeps the growing objects a constant distance apart – then no distance decrease would occur between the objects. This is much like two circles drawn on an expanding balloon – the balloon, the circles, and the distance between them would all expand equally, resulting in no relative size changes and no distance decreases as everything expands equally. In this case, as with actual objects sitting on the ground, the distance between objects is a material "thing" that expands along with the objects, unlike the empty space between free-floating objects. Also, if an object were in a stable orbit about a planet, which means the object's speed and direction counteracts the expansion of the planet to keep a constant distance between them, no distance decrease will occur here either (orbits are dealt with in detail in the next chapter).

Therefore, although all objects in the universe are continuously drawn to each other – by a gravitational force according to Newton and by atomic expansion according to *Expansion Theory* – we do not normally see the attraction in the manner that either theory predicts. In fact, even the manner in which an object drops to the ground does not clearly distinguish between Newtonian gravity and *Expansion Theory*. This is because we actually use *Galileo's* equation for falling objects today – not Newton's gravitational theory – and the differences in drop times predicted by *Expansion Theory* require a long drop distance to be clearly seen, during which wind resistance interferes. This means we do not normally observe events that would cause us to either question Newtonian theory or suspect an alternate explanation such as *Expansion Theory*. The main indicators that Newtonian theory cannot be correct are its scientific law violations and the lack of any physical explanation for Newton's proposed gravitational force and its behavior. *Expansion Theory* provides a functionally equivalent explanation for observations, the proper and accurate mathematical description, and the clear physical explanation that truly underlies our experiences and resolves the mysteries and law violations that we are left with today.

Before moving on, let's apply the *Atomic Expansion Equation* to get a feel for how it might be used. As we can clearly see in the expansion scenario of Figure 2-5, an original separation distance of 6 radial lengths, or $6R$, would be effectively reduced to $2R$ as the objects double in size – due to both absolute and relative effects. This expansion of the objects by one full radius would take 19 minutes, as we can see by taking the absolute expansion of a single object in isolation, shown earlier in the derivation of the *Atomic Expansion Equation*:

$$d = n^2 X_A R \qquad \text{--- Expansion Amount of an Object of Radius } R$$
$$\text{after } n \text{ Seconds}$$

Since d is the distance decrease between objects caused by the expansion of an object of radius R into the space around it, this decrease would be a full radius (R) when the object doubles as in Figure 2-5. Substituting the full radial expansion R into this equation gives:

$$R = n^2 X_A R$$

which simplifies to:

$$n^2 X_A = 1$$

Recalling that $X_A = 0.00000077 \text{ /s}^2$ gives the final result:

$$n^2 = 1/(0.00000077)$$

This solves to a value for n of 1,140 seconds, or 19 minutes. Therefore, although we can't see it directly, all objects in our universe double in size every 19 minutes due to atomic expansion.

Now we have all the necessary information to substitute into the full *Atomic Expansion Equation* to mathematically calculate the decrease from 6 radial lengths ($6R$) down to 2 lengths ($2R$) as shown in Figure 2-5. Note that the two object radii, R_1 and R_2, in the *Atomic Expansion Equation* are both replaced by R below since both objects are identical in size in this example. Substituting all of our known values into the *Atomic Expansion Equation* gives:

$$D' = \frac{6R - (1140)^2 \cdot (0.00000077) \cdot (R + R)}{1 + (1140)^2 \cdot (0.00000077)} = \frac{6R - 2R}{2} = 2R$$

So, the *Atomic Expansion Equation* calculates that the original separation distance of 6 radial lengths reduces to only 2 radial lengths after the doubling expansion of the two objects, just as shown in the bottom frame of Figure 2-5 and in Figure 2-6. Note that although the object radius, R, used in the original separation distance of $6R$ is a smaller radius than the one in the $2R$ result since the objects have doubled in size by the end of the calculation, this same symbol, R, is still used in both cases. This is correct since no size difference can be noticed when all objects expand equally. As we expand over time we cannot tell that we have grown, and our before-and-after sizes are identical for all intents and purposes.

New Revelations and Possibilities

Until now, we have been shackled by arbitrary, abstract models of gravity that have appeared to work well enough, but have provided no clear, scientifically viable understanding. As a result, we not only have many unexplained mysteries *within* our own theories, but these same problematic theories leave us fundamentally unable to answer *additional* questions about our universe with any certainty. For example, do we know whether anti-gravity is a valid scientific concept? Is it possible to achieve? And if so, how would it be done? What principles would cause an anti-gravity force to push objects away, and for that matter, what principles cause objects to pull toward each other now (according to Newton's theory)? Also, do all objects truly fall at identical rates regardless of their size or mass? This has been a long-standing question in science ever since Aristotle (384BC – 322BC). Today we believe that all objects do fall identically, but this is difficult to verify definitively by experiment and is not satisfactorily answered by our current theories of gravity. One of the reasons we seek a deeper understanding – and indeed the true Theory Of Everything – is to finally arrive at a theory that clearly embodies the true physical reality so we can answer any and all such questions definitively. And indeed, as shown below, *Expansion Theory* provides such definitive answers to many of our long-standing questions, while also showing new possibilities that have not been conceived of as yet.

 NEW IDEA

The Two Components of Gravity

Currently, gravity is thought to be the force that holds us to the planet, as well as the force that holds the moon in orbit about us. This is one of the defining contributions of Newton's theory of gravity. Until Newton, there was no agreement about a force holding us to the planet, and even less certainty about what keeps moons and planets in orbit. The concept of a singular gravitational force causing both effects is a Newtonian creation. However, *Expansion Theory* shows these two scenarios to result from two very different phenomena. The effect that holds us to the ground, causing our weight, is literally a continuous, forceful push from our expanding planet below, while falling and orbiting objects experience no force at all; falling and orbiting are due entirely to the *geometry* of expansion until they are in contact with the planet. This is a very important distinction between *Expansion Theory* and Newtonian theory, which can be shown by a simple thought experiment:

EXPERIMENT

The Styrofoam Planet

Consider a planet the same size as the Earth, but composed of a very lightweight material, such as Styrofoam. Both planets, regardless of their mass or chemical composition, would expand at the same universal atomic expansion rate as all matter, and therefore would approach any falling or orbiting objects at the same rate due to their equal size. So, an object dropped from the same height on either planet would "fall" at the same rate and hit the ground at the same time on both planets. From a purely observational standpoint, both planets would have the same effective gravity from a distance.

However, once the objects are on the ground, their weight is caused by the force of the expanding planet pushing against them. And the outcome of this process is not the same for both planets, but is partially dependent on each planet's mass. This is much like two people facing each other on roller-skates. If the first person pushes the second person away while leaning against a brick wall for support, the second person will roll away rapidly with the full force of the push. However, if the support wall is removed, both of them will roll away from each other, sharing the force of the push and reducing the speed and distance that the second person rolls compared to the first scenario.

Likewise, the enormous inertia of a massive planet is much like a brick wall in space to push against as the planet expands and pushes objects that are in contact with its surface. The objects essentially feel the full force of the Earth's expansion supported by the enormous "brick wall" of inertia beneath the ground. But if this "brick wall" in space were made of Styrofoam, and thus had less inertia, it would move backward to a greater degree in response to the push, essentially taking more of the pushing force away from the object on its surface. Therefore, the less massive the planet, the more the expansion force is split between accelerating the object and accelerating the planet. So, an object on our Styrofoam planet would not experience the full expansion force of the planet, even though the planet expands by the same amount as the Earth each second. The Styrofoam planet would be pushed backward by its own expansion to a greater degree than the Earth would when in contact with another object, reducing its ability to accelerate the object, and giving the object a somewhat lesser weight.

This thought experiment shows a very important dual nature to gravity that is currently overlooked. Newton's claim that the same gravitational force lies behind weight as well as falling or orbiting objects has forced an artificial numeric equivalence between the observed rate of fall and the measured weight of objects on Earth. There is no harm in making this numeric equivalence between falling and resting objects on Earth if we wish since it is a completely arbitrary equivalence of convenience; there is no particular reason why we couldn't make any arbitrary mathematical relationship between these separate components that we may desire. However, problems arise when we apply such an artificial assumption to another planet or moon that may have a different density than that of the Earth. As shown in the above thought experiment, the gravity experienced at a distance does not necessarily indicate the gravity at the surface, as Newton's arbitrary equality claims. As shown in the following chapter, this issue is an important factor in understanding the gravity of our moon and other planets.

 All Objects Do Not Fall Equally

As mentioned earlier, this question has been pondered for thousands of years. Aristotle believed that the rate of fall of objects was directly proportional to their weight – so heavier objects fell faster in proportion to their greater weight. This was widely accepted for roughly two thousand years until Galileo challenged it experimentally. Galileo claimed that it was only wind resistance that made some objects fall faster than others, showing that most objects only differed moderately in drop times even if they differed enormously in weight. With some clever experimentation he concluded that all objects fall at the same rate regardless of their weight (assuming the absence of wind resistance). This has stood for centuries as a reasonable experimental conclusion and a fair assumption theoretically, as captured by Galileo's constant acceleration equation, $d = \frac{1}{2}at^2$, still in wide use today. However, intuitive assumption and limited experimentation are not the same as solid understanding. Galileo never proposed to understand *why* objects were apparently accelerated equally toward the ground, but only made this assumption or hypothesis and found reasonable experimental evidence to fit his belief. In contrast, *Expansion Theory* clearly shows every element involved in falling objects, allowing us to see the flaws and approximations present in our legacy equations and characterizations of falling objects today.

Expansion Theory provides an equation that clearly captures the precise physical mechanism behind gravity. This equation shows that Galileo's equation, widely used today for describing falling objects, is only an approximation. In actuality, objects neither fall at rates determined by their weight, as Aristotle thought, nor do they fall equally, as Galileo thought and as we believe today. Objects fall *nearly* equally, since their "fall" is actually the ground expanding upward to meet them as they float in space – *plus* their tiny amount of expansion toward the planet at the same time. This additional tiny expansion contribution from the "falling" objects themselves is purely based on their *size*. So, all objects fall *nearly* equally, with larger objects falling slightly faster than smaller ones. *Expansion Theory* answers this age-old question effortlessly and definitively because it is not merely a blind hypothesis that seems to have reasonable experimental support, but is the *actual physical description of the underlying physics* – for the first time in human history.

A closer look shows that Galileo's equation is actually the *absolute* distance decrease found in the numerator, or top portion, of the *Atomic Expansion Equation* – except that it only takes into account the expansion of the Earth and overlooks the expansion of the falling object. That is, Galileo's equation for falling objects, $d = \frac{1}{2}at^2$, is effectively the same as that for the Earth's expansion alone, shown earlier as $d = n^2 X_A R_E$ in the derivation of the *Atomic Expansion Equation*. Both equations have a "time-squared" term, and the remaining terms – the $(\frac{1}{2}a)$ term of Galileo's equation and the $(X_A R_E)$ term of the expansion equation – both give the same 4.9 m/s^2 value as well. Both of these equations give precisely the same numeric result since Galileo's equation is simply an abstract way of representing the Earth's expansion. However, since Galileo didn't realize that it is the expansion of our planet causing the 4.9 meter "drop" of all objects in the first second, he couldn't represent it as the universal atomic expansion rate multiplied by the radius of the Earth ($X_A R_E$). So instead, Galileo considered the *object* to be speeding toward the ground for some as-yet-unknown reason, giving him an equation of accelerating objects rather than an expanding planet.

This was Galileo's best educated guess as to how falling objects behave, based on his intuition and his observations – a conclusion that remains even today. However, *Expansion Theory* not only shows that Galileo's explanation is not the literal truth behind "falling" objects, but also shows that Galileo's equation overlooks the fact that the drop time would vary slightly for objects of differing sizes as each expands differently toward the ground during their fall. *Expansion Theory* further shows that Galileo's equation also completely overlooks the additional *relative* decrease found in the denominator of the *Atomic Expansion Equation*. However, since both of these overlooked components are negligible for small objects falling short distances – and are also soon overwhelmed by wind resistance for any sizable falls – this inaccuracy has gone unnoticed for centuries.

In fact, even today it would be extremely difficult to show Galileo's equation to be inaccurate since we would need to drop objects in a perfect vacuum for an extended distance before the differences in drop times could be reasonably and definitively measured. Here we can see the power of *Expansion Theory*; since it directly captures the underlying physical cause in its equation, there is perfect clarity and understanding. We know that our current constant-acceleration equation contributed by Galileo overlooks two very specific components, and we can calculate precisely how large an oversight this is. We no longer need to wonder whether there are subtle drop-time differences between objects, and if so, why this might be, since the *Atomic Expansion Equation* now captures every element of the actual underlying physical reality. There is no particular need to conduct elaborate experiments to attempt to measure any such subtle discrepancies from current theory, and to try to figure out what such discrepancies might mean. There is no longer any room for uncertainty and doubt since *Expansion Theory* clearly shows the mechanism behind falling objects, as opposed to today's mysterious and largely unexplained "gravitational force" beliefs.

In fact, without *Expansion Theory* we would not even know to look for these elements of absolute and relative distance decrease in any experiments we may perform to test for any subtle discrepancies from our current theory of falling objects. We believe so strongly in Newton's theory of gravity today that any experiments would very likely be looking for different fall times between objects of differing *mass*. It is unlikely that we would even consider differing *size* to be the key factor – much less the *only* factor. Without this understanding we may even incidentally differ the sizes of the dropped objects in our experiments when we select different masses, not realizing that it is *size* that causes the tiny drop-time variations we might measure – not mass, further confusing the issue and exasperating our efforts.

Indeed, there is a classic lab experiment known as the Cavendish Experiment that claims to show the effect of a mass-based gravitational force between objects via an apparatus invented by Lord Cavendish, which he used to measure the gravitational constant of Newton's gravity equation in 1798. An analysis of this experiment is provided from the perspective of *Expansion Theory* in the next chapter, showing how our current lack of awareness of the true underlying expansion principle is the likely explanation for the widely differing results obtained by various groups performing this experiment around the world. This is also suggested to be the reason why statements have been made by the experimenters to the effect that further study is required to explain certain aspects of the experiment.

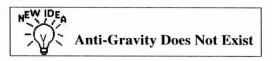

Anti-Gravity Does Not Exist

Currently, the concept of some type of anti-gravity force or device is considered a serious possibility in our science, and appears regularly in science fiction in one form or another. Given our complete lack of scientific explanation for the apparent gravitational force all around us it is understandable that we might also entertain the possibility of discovering or inventing a material that exhibits an *anti*-gravity repelling force as well. After all, gravity appears to simply emanate endlessly from matter with no known power source, so why not tinker with the atom until an equally mysterious *anti*-gravity force somehow emerges? Such theories of anti-gravity are no different in principle than our current theories of gravity, and no less scientifically feasible – both are equally mysterious and unexplained.

However, *Expansion Theory* shows that Newton's gravitational force is a mere fabrication. There is no such attracting force endlessly emanating from matter, and it is equally fanciful to expect to encounter an endless repelling force mysteriously emanating from matter as well. Today's gravitational beliefs are the result of a misunderstanding of our observations of expanding atoms; therefore, the notion of anti-gravity is based on an equivalent misunderstanding. While it may be somewhat disappointing to realize that we will never levitate with the aid of an anti-gravity device, if that is the reality of the universe we inhabit it is of great benefit to know this so that we don't naively focus on the impossible. Likewise, with our solid understanding we can begin to focus on *new* possibilities that would not have been considered without *Expansion Theory*, such as the following idea:

New "Artificial Gravity" Device

We have had a simple, viable artificial gravity device available to us ever since an idea of space visionary Wernher von Braun's was popularized in Arthur C. Clarke's novel, *The Sentinel*, later appearing in the movie adaptation, *2001: A Space Odyssey*. The concept is to spin a very large, circular or cylindrical spaceship on its central axis so objects inside the ship will be flung to the side, just like clothes in a spinning dryer. If the sides of this huge spinning spaceship were considered to be the floor, the furniture and astronauts would be held to this "floor" by the outward spinning force (centripetal force). The size of the ship and the speed of its rotation could be designed such that the amount of force holding everything to the sides (now the floor) is equal to the strength of gravity felt here on Earth. This is considered to be *artificial* gravity today, since it is created by a mechanical method that *mimics* Newton's mysterious gravitational force.

Expansion Theory shows that *all* gravity is artificial since even the Earth's gravity is created by a simple mechanical expansion of the planet; there is no such thing as "true gravity" – in the sense of a mysterious attracting force – anywhere in our universe. We can also now contemplate additional types of "artificial gravity" devices based on the principle of expanding atoms. One simple idea is a platform mounted at the end of a very long tower in space. If the tower were 6000 km in length (roughly the radius of the Earth) then its amount of expansion each second would be the same as that of the Earth – i.e. just under 5 meters each second. So a space station sitting on a platform at the top of the tower would experience the same acceleration force beneath it as if it were sitting on the Earth, creating an "artificial gravity" identical to that of our planet.

This is a concept that would be inconceivable with our current understanding since it is currently believed that a gravitational force emanates from objects according to their *mass*. In that case, the tower would have to be as massive as the Earth to create the same gravity, and there would be no reason to fashion this mass into a huge tower when a round lump of material would do the job. Of course, this would essentially mean creating a second Earth – not a very practical alternative. However, *Expansion Theory* shows that it is not *mass* that matters; the 6000-km tower in space can be very lightweight as long as it is made of a strong enough material to withstand the internal compression forces as it expands (just as there are compression forces within the expanding Earth).

One additional element would be required for the tower concept to work properly. It would need a counterweight on the opposite end from where the platform is mounted, equal in mass to that of the overall tower. Otherwise, since expanding objects must expand from their inertial center, or center-of-mass, the tower would expand from its center, essentially creating two 3000-km sections expanding in opposite directions from where they meet in the center. This would reduce the effective gravity at the platform to half the intended strength. However, with a massive enough counterweight on one end, the center-of-mass would be shifted over to that end, and the whole 6000-km tower would expand outward from the counterweight, creating the full gravity effect at the platform (Fig. 2-7).

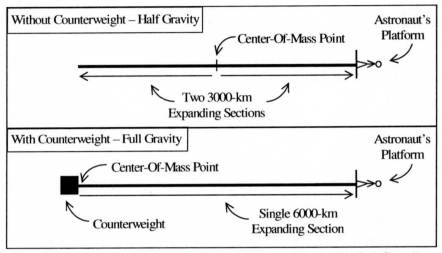

Fig. 2-7 Expanding Tower in Space Causing Half and Full-Gravity

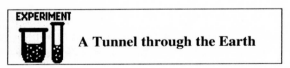

EXPERIMENT

A Tunnel through the Earth

There is a classic thought experiment presented in many physics texts and classrooms that explores what would happen if an object were dropped into a tunnel dug straight through the center of the Earth from one side to the other. Newton's gravitational theory claims that the object will be accelerated by the gravitational pull of the Earth below – strongly at first, then more weakly as it falls since the amount of mass below the falling object would become less as it approached the center of the planet. Since wind resistance is neglected in this thought experiment there is nothing to slow the falling object all the way to the center of the planet, so by the time it reaches the center it will be traveling at a tremendous velocity by Newtonian reasoning. It would experience no further gravitational pull at that point due to the equal amount of surrounding matter in all directions from the center, but its tremendous speed would supposedly carry it past the center and up the other half of the tunnel to the opposite side of the Earth. Again, according to Newton's theory of gravity, the object would be slowed as it headed up the tunnel to the opposite side since there would be more and more of the Earth's mass beneath it as it traveled, giving an increasing gravitational force pulling back toward the center. So, the object would then slow to a complete stop once it reached the tunnel entrance on the opposite side of the planet, and would begin to fall back toward the center again.

ERROR

    An Impossible Perpetual Motion Machine

This Newtonian analysis once again shows the clear flaws in Newton's proposed gravitational force. Since Newton's gravitational theory does not incorporate any energy drain from a power source for his force, the object continues to oscillate back and forth endlessly from one side of the planet to the other – a clearly impossible perpetual motion machine. Energy is required to accelerate the object to tremendous velocities then decelerate it again over great distances repeatedly, millennium after millennium, with no consideration for a depletion of energy from the power source that must be driving such a process. This prediction would also result from Einstein's theory of gravity. Despite this unmistakable violation of the laws of physics, this solution is taught generation after generation as the correct application of current gravitational theory, and as the literal outcome that would be expected if such an engineering feat as a tunnel through the planet were ever undertaken.

Expansion Theory shows that quite a different result should be expected, and one that does not violate the laws of physics. In this view, the dropped object does not actually get pulled down the tunnel by a force, but rather, both the planet and the object merely expand in place as they float next to each other in space. The tremendous size of the planet means that it expands by a far greater amount each second than the object (but the same universal percentage of course), essentially swallowing the object into the tunnel as it expands (the *absolute* component of expansion). Since all other objects are also expanding, including all measuring devices, the distance to the center of the planet would be essentially shrinking in comparison as well (the *relative* component of expansion). The result is shown in Figure 2-8.

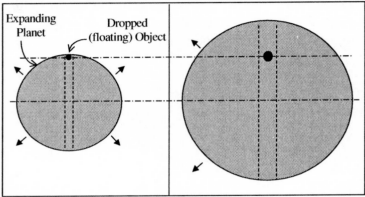

Fig. 2-8 Planet and Object Locations Unchanged during Free-Fall

From the perspective of the dropped object, it enters the tunnel as the mouth of the tunnel accelerates past it and off into the distance above it, while the center of the planet also draws closer as everything grows, scaling down the distance to the center even further. Since there is no atmosphere in this thought experiment no wind resistance is felt, and since there is no actual gravitational force no pull is felt either. The object merely floats while it effectively "drops" down the tunnel toward the planet's center.

As this continues, both the planet and the object eventually grow so much larger than their original sizes that the original distance to the center of the planet eventually shrinks to zero (the object eventually falls to the center of the planet). This can be seen in Figure 2-8, where the original distance between the object and the center of the planet is technically unchanged as the object "falls," yet is effectively continuously reduced as expansion continues throughout the universe.

Once the object and planet grow enough to effectively reduce the distance to the center to zero, the object remains floating at the center of the planet for all time. This is because two expanding objects, one centered inside the other, remain geometrically unchanged relative to each other no matter how much they expand from that point on.

From the perspective of *Expansion Theory*, the dropped object never actually experienced a pull toward the center of the planet, never actually gained speed or momentum in any absolute inertial sense, and never experienced a decelerating force as it eventually slowed to a standstill at the center of the planet. All of this relative motion did occur, but it occurred purely through the relative geometry of mutually expanding objects. There is no unexplained attracting force, and no perpetual motion back and forth in violation of the laws of physics. It is also a simple matter to use the *Atomic Expansion Equation* to arrive at this same conclusion mathematically:

OPTIONAL MATH

(x, y) **Expansion Calculation of the Tunnel Through the Earth**

Inserting the radius of the Earth, R_E, the radius of the object, R_O, and the fact that the original distance between their surfaces is zero (i.e. the object is initially held at the opening of the tunnel at the Earth's surface) into the *Atomic Expansion Equation* gives:

$$D' = \frac{0 - n^2 \cdot X_A \cdot (R_E + R_O)}{1 + n^2 \cdot X_A}$$

Now, for the end result we allow an infinite amount of time to pass ($n = \infty$):

$$D' = \frac{0 - (\infty)^2 \cdot X_A \cdot (R_E + R_O)}{1 + (\infty)^2 \cdot X_A}$$

which simplifies to:

$$D' = -(1) \cdot (R_E + R_O)$$

This calculation shows that the original zero-distance between the surface of the object and the surface of the Earth as it was held at the tunnel entrance changed to a negative distance (i.e. inside the planet) of $R_E + R_O$. This places the center of the object at the center of the Earth, floating there for all time (Fig. 2-9).

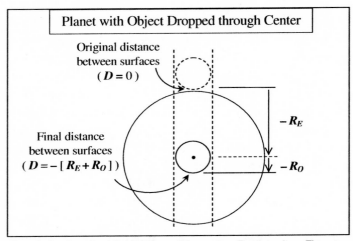

Fig. 2-9 Central Location Given by Expansion Theory

So far we have seen that, despite the apparent success of Newton's theory of gravity, it remains a theory that violates one of the most fundamental laws of physics and relies on the existence of a completely unexplained power source and mechanism of transmission and attraction across space. Einstein's attempt to provide an alternate theory of gravity resulted in his even more mysterious and far more abstract *General Relativity Theory*. Yet, the simple concept of expanding atoms removes all of these mysteries, abstractions, and violations, and explains all known gravitational observations. Many of our long-standing, classic questions about everything from the rate of fall for various objects to the concept of anti-gravity are also clearly and definitively answered for the first time by *Expansion Theory*. It was also shown that the deep physical reality of expanding atoms can actually be readily seen in something as common as a dropped object since this event is actually the result of the expanding planet beneath us as it approaches the object. And, this universal atomic expansion rate is also easily calculated from this same simple event, working out to a tiny fractional expansion rate of just under one millionth each second. The equation embodying this deep physical truth in our universe is also easily derived, serving as a definitive guide to understanding many of the questions that have been posed in the past – and that might be posed in the future.

Yet, despite this compelling introduction to the very real possibility of a crucial scientific discovery that has eluded us for centuries, there are many more issues to deal with in a truly thorough exploration of atomic expansion. After all, despite the serious problems with Newton's theory of gravity, it has been widely accepted for centuries, and it has appeared to serve us well over that time and even in large part today. Therefore, *Expansion Theory* must not only alleviate the law violations of Newtonian theory and explain the many everyday experiences just discussed, but must also explain orbits, ocean tides, and many other celestial observations currently attributed to a gravitational force. If orbits do not result from passing bodies being forcefully held in circular paths by some type of attracting gravitational force, and there is no gravitational tug from the moon to give rise to our ocean tides, then these events must have a solid explanation in *Expansion Theory*. Indeed they do, along with many other celestial observations, as will be shown in the next chapter. There are also additional mysteries in the heavens that currently have no explanation in our science and stand as well-known, published "gravitational anomalies," which are resolved in the next chapter as well. So, we now move on to explore the heavens from the perspective of atomic expansion.

- 3 -

Rethinking

Our

Heavenly Observations

The previous chapter proposed a new concept in physics called *Expansion Theory*, showing that if our universe is one where all atoms continuously expand by one-millionth their size every second, we would have the first clear physical explanation for gravity ever produced. It would eradicate numerous conceptual and scientific flaws in our current gravitational theories, while also providing the first explanation to remove unexplainable mysteries and abstractions from our science – rather than adding to them as our current alternate theories do today. There would no longer be any need for an unexplained attracting force somehow emanating from all matter, nor a mysteriously warped four-dimensional space-time continuum permeating our universe. This new theory of expanding matter – *Expansion Theory* – was shown to explain every familiar effect in the world around us that we currently attribute to Newton's gravitational force. As the discussions to follow show, this same new principle in nature also explains observations related to the heavenly bodies – planetary orbits, ocean tides, interplanetary space travel, etc. – without appealing to a gravitational force that violates our laws of physics. This chapter completes our exploration of gravity, showing that not only are our observations of the heavens explainable entirely in terms of expanding atoms, but that these explanations resolve significant questions that still remain largely unanswered today.

The most elementary of all dynamics in the heavens is that of orbits. In fact, it could be said that orbits form the foundation of celestial structure and order – moons orbit planets to form self-contained planetary systems, planets orbit stars to create stable solar systems, and countless stars orbit together in immense swirls that form the galaxies composing our universe. This extremely important celestial dynamic known as the *orbit* will now be explored from the perspective of *Expansion Theory*.

Orbits

Newton's Flawed Description of Orbits

We have been brought up with the belief that the moon is forcefully constrained in its orbit by Newton's gravitational force; if it weren't for this force, we are told, the moon would no longer be constrained in orbit and would speed past the Earth and off into space. Yet, as the previous chapters have shown, the forceful constraint described by Newton is actually an arbitrary invention whose foundation in the rock-and-string analogy is both conceptually flawed and scientifically impossible. Instead of this "gravitational rope" forcefully holding the moon in orbit, it was shown that, in principle, the *Geometric Orbit Equation* pre-dates Newton's proposal, and describes a purely geometric orbit effect that involves no masses or forces. And, now that the principle of expanding atoms has been introduced, this new principle can be shown to be the actual physical mechanism underlying this purely geometric orbit effect in the heavens. However, before exploring this new concept of orbits, it is important to take a good look at our current beliefs, beginning with a thought experiment proposed by Newton to explain orbits using his "gravitational force" idea. This thought experiment can be shown to have sizable flaws and oversights that have gone unnoticed for centuries – a state that remains even today.

WATCH FOR...

- Newton's "cannonball" thought experiment for orbits is not the general description of orbits that it is commonly thought to be, but only applies to a very limited special-case scenario.
- Newton's thought experiment also contains a fatal flaw that makes even his special-case orbital scenario impossible.
- The corrections to these problems provided by *Expansion Theory* require that one of Newton's most fundamental laws of motion – *Newton's First Law* – be rethought.

EXPERIMENT

Newton's Orbiting Cannonball

Newton described orbits as a logical extension of the path taken by objects thrown horizontally on Earth; they all fall toward the ground in a parabolic arc, and the faster they are thrown, the farther they go before

striking the ground. Newton reasoned that if an object had enough speed
– perhaps if shot from a tremendously powerful cannon on a hilltop – it
would travel far enough along a theoretical straight-line path as gravity
pulled it down that it would actually fall in a circular path around the
planet (Fig. 3-1). This continuous circular path about the planet, taken by
a speeding object under the pull of gravity, would presumably explain
the orbits of objects such as our moon. In fact, this explanation of
Newton's is the reason orbits are considered to be a state of constant
free-fall today – orbiting objects are thought to be constantly falling
while effectively regaining any lost height by speeding past the curved
planet, maintaining a constant orbital altitude.

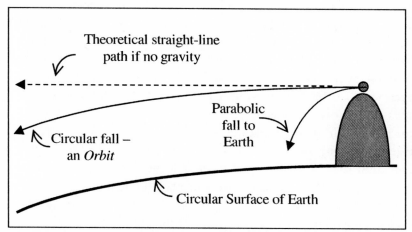

Fig. 3-1 Newton's Orbit Explanation: Speeding past Curved Surface

---

ERROR

✗   Newton's Description Does Not Explain Orbits in General

This is Newton's explanation of orbits, which remains as our
explanation even today. Yet, there are sizable problems with this concept
on closer examination (in addition to the fact that the Newtonian
gravitational force in this thought experiment violates our laws of
physics). One of the additional problems can be seen once we consider
faster and faster orbiting objects. Objects in near-Earth orbit, such as the

space shuttle, complete an orbit in roughly 90 minutes. As we know, both from experience and from the orbit equations, as objects are put into ever-higher orbits they travel more and more slowly around the planet. In fact, they can be positioned at high enough altitudes that they take a full 24-hour day to complete each orbit – the same rate as the daily rotation of the Earth. This orbital altitude is known as a *geostationary* or *geo-synchronous* orbit because the orbiting object effectively remains positioned over the same spot on the planet continuously – *synchronized* with the planet's 24-hour rotation rate below. This is a very useful orbit for certain types of communication satellites, for example, which must remain fixed overhead at all times to provide continuous coverage over a given area on the planet.

However, if an object can orbit over the same location on the surface of the planet, then we can no longer characterize it as speeding past the curvature of the Earth; it does not *pass* the surface at all, but effectively hovers over the same spot continuously. Likewise, we can no longer say that the orbiting object is in *free-fall* either since an object that is falling over a fixed location must obviously fall to the ground below. Yet both of these features – speeding past a curved surface and constantly falling all the while – are the two pillars of Newton's explanation for orbits. It is not acceptable that Newton's fundamental explanation for orbits applies to *most* orbits but completely breaks down for others. It must either apply to *all* orbits, or it is a fundamentally flawed explanation. Let's examine these points in further detail.

Geo-synchronous orbits show us that the curvature of the planet can be completely irrelevant to an orbit, and therefore, *must* be equally irrelevant to *all* orbits since all orbits must be based on the same fundamental physics principles. There cannot be one set of physics principles underlying orbits below geo-synchronous altitude, and an entirely different set of principles once an object reaches this "special height." In fact, geo-synchronous orbits show us that the planet could be *any* arbitrary shape whatsoever – even square – without having any impact on the orbit. That is, while an object attempting to circle a square planet would strike one of its corners, this would not occur if the planet rotated at the same speed as the orbiting object. In that case, the orbiting object would never get any closer to the corner up ahead and would maintain a geo-synchronous orbit indefinitely despite the non-curved surface of the planet (Fig. 3-2). Therefore, since hovering over the same spot on a rotating square planet is a perfectly valid orbital scenario – yet is quite different from Newton's description of speeding past a curved surface – Newton's description of orbits cannot be the true explanation, as currently thought.

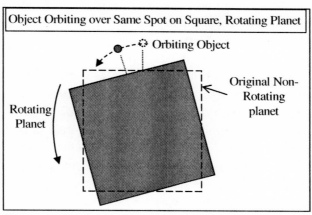

Fig. 3-2 Geo-synchronous Orbits do not require a Curved Planet

This issue shows us that while Newton's explanation of orbits in Figure 3-1 offers a simple, intuitive explanation for orbits based on his idea of a "gravitational force," it is actually merely a *model* of a *special case*, which breaks down as a true physical explanation for orbits in general. The special case model referred to here is that of a *spherical* and essentially *non-rotating* planet. The *spherical* assumption arises from the fact that moons, planets, and stars all naturally form into spheres, and the *non-rotating* assumption follows from Newton's starting point of an object shot horizontally from a fixed location on the Earth, such as a hilltop. Considering the hill to be a fixed feature from which the cannonball is fired essentially ignores the fact that the hill is located on a rotating planet. This spherical, non-rotating special case scenario provided a convenient backdrop for Newton's gravity-based explanation, leading him to the erroneous conclusion that the true physical explanation for orbits is that of traveling in a free-falling arc past the (spherical) surface of (non-rotating) planets.

Newton was able to ignore the important element of planetary rotation in his attempt at a general theory of orbits because planets are generally spherical in shape. From typical orbital heights of hundreds of kilometers any jutting surface features are inconsequential, leaving an essentially smooth, unchanging curved surface whether or not the planet is actually rotating below. Therefore, we could still use Newton's special case scenario of a free-falling circular arc about a fixed, curved surface even though the planet may actually be rotating at an arbitrary rate that leaves the object hovering over one spot. The special-case shape of a smooth sphere is what allowed Newton to ignore the rotation of the planet in his explanation of orbits. However, as shown in the case of the square planet in Figure 3-2, it is not valid to assume that planetary rotation can be universally ignored in developing a general theory of orbits. To do so is to overlook an important ingredient in understanding orbits about any arbitrarily shaped body. Geo-synchronous orbits expose this oversight in Newton's explanation by accentuating the issue of planetary rotation, leaving the object hovering over one location of an arbitrarily-shaped planet, which violates Newton's two key elements – speeding *past* a *curved* surface. Thus, in the common (though still special-case) scenario of a spherical planet, Newton's thought experiment may *appear* to offer a viable explanation for orbits, but this clearly cannot be since it does not explain orbits in general.

ERROR

  Gravity-based Circular Orbits are Impossible

The second problem with Newton's thought experiment is that the trajectory of the speeding cannonball that finally achieves orbit in Figure 3-1 is represented as a *circular* trajectory, yet the slower cannonball that hits the ground follows the typical *parabolic* trajectory of *all horizontally-fired objects*. The parabolic trajectory of horizontally fired objects occurs because all objects undergo a *constant downward acceleration* as they fall; therefore, horizontally-fired objects drop at a *constantly increasing speed* while still coasting horizontally at their original speed. This dynamic always creates a *parabolic* path that curves ever more rapidly toward the ground as the object proceeds – not a *circular* path. This is taught in all elementary physics courses, and is embodied in Galileo's constant-acceleration equation mentioned in the previous chapters as $d = \frac{1}{2}at^2$. Yet, we can see in Figure 3-1 that both the Earth's surface and the path of the orbiting cannonball follow parallel *circular* curvatures away from the theoretical straight-line extending out from the hilltop. This circular curvature away from the horizontal is an *impossible* trajectory for projectiles, which does not emerge from our equations of projectile motion.

Therefore, it is impossible for the speeding cannonball in Newton's thought experiment to travel in a stable circular orbit – according to our widely accepted models of objects in free-fall. The longer the object is in free-fall near the planet (as said to be the case when orbiting), the faster it must fall toward the ground, according to gravitational theory, which cannot be balanced into a stable orbit simply by coasting past the circular curvature of the planet. The circular orbital path in Figure 3-1 is clearly just drawn arbitrarily because many such stable circular orbits are known to exist in our solar system, but *they could never actually occur according to Newtonian theory* – no matter how fast the cannonball was fired. In mathematical terminology, no matter how much a parabolic curve is stretched it will *always* be parabolic – never circular.

In fact, even if we assume Newton intended for his special-case non-rotating sphere example to be extended to include more general orbital scenarios such as a rotating square planet, it still breaks down due to the non-circular trajectory of all horizontally fired objects. This is because a planet that rotates about its central axis rotates in a perfect circle, which can only be matched by an equally circular orbital trajectory. But, as just discussed, an object in the free-fall of orbit experiences a constant gravitational pull, according to Newton, which speeds it ever-faster toward the ground in an ever-steeper parabolic arc, *not* in a smooth circle that matches the planet's rotation. This can be seen in Figure 3-3, where the actual observed geometry of geo-synchronous orbits is shown on the left and the Newtonian prediction for such orbits is shown on the right.

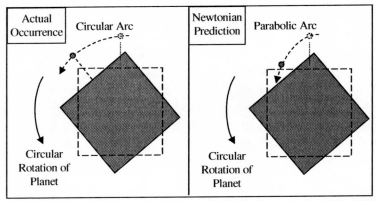

Fig. 3-3 Circular Geo-synchronous Orbit vs. Newtonian Prediction

The *circular* arc of the orbiting object in the left frame means that, as the planet also proceeds in its circular rotation at the same rate, the object effectively remains the same distance above the ground – in geo-synchronous orbit. This is what we actually observe, for example, with many of our communications satellites. However, the *parabolic* path of the object in the right frame, as predicted by Newtonian theory, does not match the circular rotation of the planet. This means that even if the planet rotates at the same speed as the orbiting object, the object still falls ever faster toward the ground all the while. Objects that are subject to plummeting in parabolic trajectories toward the ground cannot maintain a stable orbit either around a spherical planet or above a circularly rotating planet of any shape. Yet, Newton's theory of gravity requires that all objects fall at an accelerating rate, and therefore, follow this plummeting parabolic trajectory. Therefore, the above discussions show that not only is Newton's explanation of orbits merely a limited special-case scenario and not a truly general purpose orbital principle, but his own theories of gravity and motion show that the very premise upon which his circular orbital explanation rests is an impossible invention.

The second point above – regarding the impossibility of circular object trajectories – raises a very serious and far-reaching question. Whether we consider falling objects to undergo an *actual* acceleration caused by a gravitational force, or a geometric *acceleration effect* caused by our expanding planet, the same acceleration is still effectively present, and therefore, horizontally fired objects still effectively plummet in parabolic paths to the ground either way. So, if these parabolic paths are the same regardless of the cause, how could *Expansion Theory* – or *any* theory – possibly make any difference in attempting to explain orbits? How can we possibly explain the circular orbits of objects, as shown in the left frame of Figure 3-3 and as seen in our moon's orbit overhead, if objects are *fundamentally unable to follow circular paths as they fall*? As we will see shortly, the answer to this question does indeed lie in *Expansion Theory*; however, it also requires that we rethink another cornerstone of our scientific legacy – *Newton's First Law of Motion*.

Rethinking Newton's First Law of Motion

LAW

 <u>Newton's First Law of Motion</u>

An object in motion continues to travel in a straight line unless acted upon by an external force. Likewise, an object at rest remains at rest unless acted upon by such a force.

Newton's First Law marks the beginning of what is known as *Classical Mechanics* – the everyday dynamics of objects in motion – and is the first of Newton's three elementary laws of motion. Although this is one of our most fundamental, long-standing, and undisputed laws of nature, it must be remembered that it is a law that was developed without an understanding of the even more fundamental principle of expanding atoms. As such, it should not be surprising if the discovery of such a fundamental new principle as expanding atoms has implications for even such time-honored scientific ideas as Newton's laws of motion.

In fact, *Newton's First Law* does not truly stand alone, as it appears, but is intimately paired with his theory of gravity since Newton claimed that all objects in the universe pull on all other objects via his attracting gravitational force, which must then alter their otherwise straight-line paths. Although *Newton's First Law* presents a *theoretical* straight-line path for objects in motion, in actuality, this could never occur because two objects passing each other in space must be pulled toward one another by the external force of gravity from each other. So, even by Newtonian standards, *Newton's First Law* is merely an idealization that would presumably occur in a universe without gravity, but *never actually occurs in nature*. Even an object that rolls along the ground in a "straight line" is actually rolling *around the circular Earth*, constrained in this circular path by the gravity of the planet. The fact that two objects will roll along the ground side-by-side gives the appearance that they are proceeding in straight lines according to *Newton's First Law*, but in the bigger picture they are actually on separate *curved* paths around the planet, which happen to parallel each other. While this may seem like a trivial point from the Newtonian perspective, it is actually a highly significant point from the perspective of *Expansion Theory* – a point that will allow us to understand how orbits are possible.

Newton's idea of orbits is one where a passing object literally *possesses* a velocity and momentum as it speeds by, and a gravitational force also literally reaches out and forcefully pulls it toward the planet, constraining it in an orbit. Therefore, Newton had to invent the notion of theoretical, absolute straight-line paths "possessed" by objects, from which gravity would pull them, forming an orbit. Yet, a universe composed of expanding atoms requires neither the concept of a gravitational force nor an absolute straight-line trajectory somehow possessed by objects, since objects would always orbit one another naturally in such a universe. There would be no need to appeal to a "gravitational force" and no law of absolute straight-line paths for objects. There would simply be the *natural orbit effect of expanding objects*:

The Natural Orbit Effect

For the most effective explanation of the natural orbit effect of expanding objects, it is necessary to begin with a brief discussion of the fundamentals of object motion. Newton's three laws of motion compose our current understanding of object movement; however, Newton was intent on absolutes, and conceived of a universe with the following attributes:

- Composed entirely of passive, non-expanding lumps of matter
- A never-ending gravitational force emanates from all matter
- Objects possess absolute speed and momentum as they move through space

While the first two points have now been thoroughly addressed, showing that matter is not passive but is actively expanding, and that Newton's gravitational force does not exist, the last point regarding objects possessing absolute momentum requires further discussion. In actuality, contrary to Newtonian physics, *no such absolute momentum exists.*

The fact that motion is not absolute – that it is arbitrary whether one object is considered to be in motion past another or vice versa – is not new. Galileo worked on this *Principle of Relativity,* which then took a back seat to Newton's ideas that moving objects possess an inherent speed and a definite straight-line forward momentum, and are pulled off-course by an ever-present gravitational force from other objects. However, once we see that Newton overlooked atomic expansion, the fact that objects do not possess *absolute* speed and momentum, but have a *relative* nature of motion, becomes clear. To illustrate this point, first consider the speeding object shown in Figure 3-4.

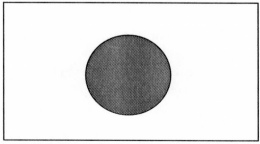

Fig. 3-4 Lone Object Speeding Through Deep Space

This object is the only object in the known universe and is speeding along through space – or is it? How could we possibly tell if there are no other objects with which to compare its motion? It is tempting to simply imagine the object moving across the frame of the diagram; however, the frame would not exist in a universe that contained only that object and nothing else – there would be no frame of reference at all. We may picture it speeding past *us* as we float nearby in space; however, *we* do not exist in this scenario to provide a reference point either. A lone object cannot possess an absolute velocity or momentum – nor have any specific velocity at all – without there being another object that serves as a stationary reference point. Now consider the case where this object is speeding past a reference object (Fig. 3-5).

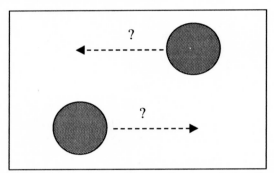

Fig. 3-5 One Object Speeding past Another

Which object in Figure 3-5 is moving, and which is stationary? If these are the only two objects in the universe, it is impossible to tell. If we consider one to be stationary then it is the other that is in motion, but this selection is completely arbitrary. In fact, it is not just that we cannot *tell* which is actually in motion, but that neither object actually *possesses* any *absolute* velocity or momentum; there is only *relative* motion between them. *Either* object could be considered to be moving while the other is stationary, or both could be moving toward each other at any of an infinity of speed combinations that add up to the overall speed of their approach toward each other. In fact, both objects could even be moving in the *same* direction – say, to the right – with the one behind simply moving faster and thus catching up to the other. Therefore,

Newton's First Law is already in trouble since the speed and momentum of each of the two approaching objects in Figure 3-5 is completely arbitrary. Neither object individually *possesses* either an absolute velocity or an absolute state of rest. Therefore, it is impossible to solidly apply *Newton's First Law* here. Which object will continue in its straight-line trajectory unless forcefully altered? Which will remain at rest unless pushed by an outside force?

Now consider that both objects of Figure 3-5 exist in a universe of expanding matter. Figure 3-6 shows their expansion on the left as we would see it if we were somehow outside of this universe of expanding matter ourselves, mapped over to the right side, which is what we actually see as we expand along with everything else, and so cannot see the size increase. On the left, we can see that the growing objects get closer and closer as they expand while passing each other, and on the right is what we actually see – constant-sized objects that approach each other as they pass. This gives a natural curvature to the paths of the two passing objects relative to one another – *a natural orbit effect.*

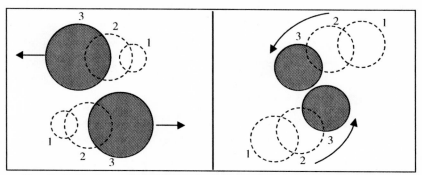

Fig. 3-6 Natural Orbit Effect of Expanding Matter in Motion

In the first time period (labeled 1 in both left and right frames) the objects are some distance apart. By the second time period they have grown closer due to their mutual expansion as they pass each other nearly side-by-side in both frames. By the third time period they pass each other and are quite close. The left frame shows the underlying reality of two passing objects expanding toward each other, however our observation and experience is that of two constant-sized objects somehow spiraling toward one another. If we lived outside our universe of expanding matter we could observe objects growing in size as they pass each other, as in the left frame, but we are equally expanding beings *within* this universe, and thus see the natural orbit effect of constant-sized objects in space instead.

Now, recall from Figure 3-5 that two objects in space have no absolute velocity definition – their velocity and momentum are entirely defined by the relative motion between them. So, once we understand that all objects are composed of expanding atoms, we can see that the orbit effect in the right-hand frame of Figure 3-6 is the *natural behavior of objects*. This is the proper "first law of motion" according to *Expansion Theory*, and replaces *Newton's First Law*. That is, in our universe of expanding atoms, objects are effectively drawn to one another by their inherent expansion, resulting in a natural orbit effect as they pass each other in space. This curved trajectory does not represent a *deviation* from some absolute straight-line momentum possessed by all moving objects – as Newton claimed – but rather, it is the *natural* trajectory to expect for all objects in motion past one another. No force is required to explain this orbit effect since it is the natural condition of all objects in motion past each other in a universe of expanding atoms.

NEW IDEA

New "First Law of Motion"

Objects neither travel in isolated straight-line trajectories nor sit still in space, but rather, objects always either move *toward each other* or travel in *curving or orbiting trajectories about one another* due to their mutual expansion.

Newton's First Law is currently so deeply ingrained in our thinking that it can be very difficult to truly free ourselves of it to properly consider the new First Law introduced above. For example, even though Fig. 3-6 does clearly show the natural curved trajectories that would exist in a universe of expanding matter, it only illustrates a partial orbit resulting from three successive motion snapshots. This is a reasonably straightforward dynamic to see as we glance back and forth between the two sides of the diagram, but what if we continued with further motion snapshots? Our natural tendency in the left-hand frame, based on our Newtonian straight-line thinking, may be to simply continue drawing a progression of expanding circles along the same straight-line paths shown for the first three snapshots. And what would happen if we did so?

It is safe to say that the top object would always remain in the top half of the diagram, and the bottom object in the bottom half. Their centers would never cross over an imaginary horizontal line between them no matter how much they expanded while continuing in straight horizontal lines past each other, so how could the top object ever circle around below the bottom object, and the bottom object circle up over the top one? Similar logic shows that it would be equally impossible for the two objects to effectively reverse direction and head back toward each other. Yet, both of these impossible dynamics are required for the objects to circle around each other, continually changing places top-to-bottom and left-to-right in an orbit. So, how can we seriously consider a theory as the literal description of our world if it apparently breaks down for something as fundamental as orbits?

The answer is that it isn't *Expansion Theory* that is breaking down here, but the attempt to apply it. The above train of thought does not actually replace Newton's concept of orbits with that of *Expansion Theory*, but partially mixes *both* concepts, describing *neither* worldview properly. As stated earlier, Newton's unflinching straight-line momentum concept is purely a creation of his personal worldview, and one that is actually at odds with nearly every observation of object motion; a ball tossed "straight ahead", for example, actually curves toward the ground.

In actuality, the correct application of *Expansion Theory* requires that we completely let go of Newton's absolute straight-line momentum concept and realize that the partial orbit shown in Fig. 3-6 is a *fundamental unit* of expanding object motion that occurs *each moment*. In the tiniest possible passage of time the expansion of the two objects fills the otherwise increasing gap that would occur between them as they coasted past each other. And, logically speaking, two objects that coast a distance past each other while moving no further away can only be said to have partially orbited about one another at a constant distance. And so, in the next moment the two objects will pass each other in the same manner as before, but they are now a little further along in a curved orbit about each other, which also means the angles of their tangential paths past each other have now changed accordingly. The geometry has effectively reset itself for the next partial orbit, much as if Fig. 3-6 were about to recur, but with the left-hand side of the diagram slightly rotated counter-clockwise, in this example. This progression would continue again and again in a full and continuous orbit.

There is no reason why there must be an absolute, unflinching straight-line path of travel for the passing objects in the left frame. After each moment things have changed; the objects have expanded and coasted slightly past each other, and their trajectories have naturally followed a partial curved path as a result. A full orbit would emerge over time in the right-hand frame, which shows the effective result for objects and beings that are immersed in this universe of expanding matter.

Another way to view this same scenario is from the perspective of someone attempting to write a program to simulate an orbit according to *Expansion Theory*. Their first inclination may be to program horizontal "x = x + 1" steps for the movement of the expanding objects, repeated over and over again. Yet, doing so is the equivalent of building Newton's straight-line invention directly into the simulation, corrupting the expansion dynamics that they intended to simulate. Again, this would not be a simulation of *Expansion Theory*, but an odd hybrid of old and new ideas that simulates neither theory properly, bringing odd and meaningless results. We have been so powerfully programmed with Newtonian thinking that it is not uncommon to mentally attribute straight-line trajectories even to situations where none actually exist, and if we're not careful such assumptions can follow us even as we try to consider completely new views of our world.

Newton's straight-line idealization is an understandable conclusion since it *appears* to be the case from observation of many common situations, but it leads to great misconceptions in situations such as orbits. Billiard balls essentially travel in straight lines as they roll across a pool table, but this situation is far removed from the pure example of two objects in space. The short distance that billiard balls roll on a pool table masks the fact that they are essentially rolling along the curved surface of the planet (made artificially flat for a few meters by the level pool table). If they rolled far enough they would have to circle the planet, and would not travel straight off past the horizon and into space. Also, although two billiard balls would travel along beside each other as they circled the Earth, these two parallel paths only occur because the ground between them is expanding, keeping them apart. These parallel paths would not occur if the billiard balls were traveling side-by-side in deep space, but would converge as the balls expanded and effectively approached one another.

There are certainly many situations where Newton's straight-line idealization is a reasonable and useful approximation or idealization, but it does not represent the proper physical truth underlying the motion of objects in our universe. True straight-line trajectories of objects passing each other in space *never* actually occur – curves and orbits are the norm throughout the universe. Nor do objects possess absolute motion or momentum, which a "gravitational force" pulls into a circular orbit, since they naturally orbit about each other purely due to the geometry of their mutual expansion and relative motion past each other. This new understanding now allows us to see how circular orbits are possible despite the plummeting parabolic paths of falling objects. With this understanding we now revisit the issue of a general explanation for orbits.

The New Description of Orbits

 Orbits – According To Expansion Theory

The previous chapter showed that we do not live in a universe of passive lumps of matter from which a never-ending, attracting "gravitational force" emanates. If this were the case, every atom and object in the universe would violate the *Law of Conservation Of Energy* by expending energy endlessly, and without even having an identifiable power source. A universal law of physics cannot be considered as such if it is violated by every element of matter in the universe. Therefore, Newton's model of a gravitational force ruling orbits can only be a *model* – not the literal reality. Instead, we live in a universe where the very nature of matter is that it continuously expands, and this alone defines the mechanics of bodies in space. Just as there is no need for a gravitational force pulling falling objects to the ground once planetary expansion is considered, orbits also follow naturally from the dynamics of expansion alone. Let's begin with Einstein's efforts to re-think Newton's gravity-driven orbit concept.

EXPERIMENT

Einstein's "Space Elevator" Explanation of Orbits

Recalling Einstein's "space elevator" thought experiment from Chapter 2, which showed that an upwardly-accelerating elevator in space created the identical conditions to gravity on Earth, Einstein further imagined a beam of light passing across the elevator. He realized that, from the perspective of the person in the elevator, it would appear as if the beam of light were being forcefully bent toward the floor by gravity as it passed across the elevator while the elevator accelerated upward. It is only from a perspective outside the elevator that we would be able to determine whether gravity is bending the light beam, or whether the beam is continuing on in a straight line while the floor of the elevator approaches it (Fig. 3-7).

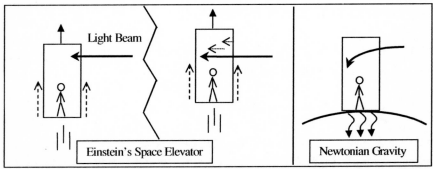

Fig. 3-7 Speeding past a Light Beam is same as Gravity Bending it

This thought experiment led Einstein to an alternate explanation of orbits, rather than Newton's purely gravitational notion of a force pulling objects into an orbit. However, just as in Chapter 2, although this thought experiment is clearly a direct parallel to the expanding Earth accelerating upward through space, Einstein did not follow the experiment to this conclusion. After all, since we *can* stand outside such an elevator here on Earth, we plainly see no acceleration through space, but an elevator simply sitting on solid ground. Einstein did not bring his thought experiment down the path of expanding matter, which would have shown that indeed the elevator *could* be accelerating through space as the planet expands, while paradoxically also appearing stationary on the ground as we are also accelerated upward by the planet.

In essence, there is no such thing as an outside observer in *Expansion Theory* since we are all immersed in this universe of expanding matter. Einstein, however, invented the far more abstract notion that a surrounding "space-time fabric" in the vicinity of the planet must somehow warp into a four-dimensional curvature around the planet, which all objects – including beams of light – have no choice but to follow. In this view, as the light beam follows the curvature of four-dimensional space-time, it appears to bend as if by a force in three-dimensional space. This is Einstein's concept of orbits according to his *General Relativity Theory*. The mysteries related to how and why such a "four-dimensional space-time curvature" would occur in response to the mere presence of matter still remain to be explained, although, as an abstract working model, this concept has provided some useful and thought-provoking results.

Expansion Theory takes Einstein's thought experiment to its straightforward, logical conclusion, while resolving related mysteries rather than introducing further ones. It shows that curving orbits *are* literally the result of an upward acceleration while an object (or light beam) speeds past – there is no need to invent further abstractions and complications when expanding matter is considered. Also, it explains the mystery of how Newton's "gravitational force" would be able to tug on a beam of pure light energy. Energy is presumably far more ethereal than matter, being composed neither of atoms nor even of subatomic particles. In fact, we really have no idea what energy truly is today (though it is clearly explained in the coming chapters); therefore, we have absolutely no explanation for how Newton's "gravitational force" would cause beams of light to bend as they pass a planet. *Expansion Theory* removes the need to explain such a mystery since beams of light would effectively curve past a planet for the same purely geometric reason that any object would – the *Natural Orbit Effect* that occurs when passing an expanding object. This effect is one of pure geometry involving no mysterious attracting forces or warped "space-time," and applies just as readily to a passing light beam as it does to a passing object.

The earlier discussion prior to the section on the *Natural Orbit Effect* posed the question of how circular orbits are possible. It was pointed out that all projectiles follow a plummeting *parabolic* path to the ground, and so could never orbit the planet in a *circular* trajectory. However, this conclusion was based on the Newtonian belief that objects following this parabolic trajectory actually *possess* a parabolic momentum as a "gravitational force" continuously accelerates them downward toward the ground. But once we realize that the parabolic trajectory of projectiles is *not* an absolute momentum possessed by these objects, but is merely the resulting *geometry* of objects that are coasting past an expanding planet, the whole situation changes.

Expansion Theory shows that the parabolic paths of these falling objects are, in a sense, illusory trajectories that result when an object travels too slowly past a planet to counteract the planet's tremendous expansion toward it. The object does *effectively* follow a parabolic arc toward the ground, but it does not literally undergo a momentum change from a straight-line path to a parabolic one due to the forceful tug of a "gravitational force." Nothing changes for the passing object at all – it merely continues coasting along while the Earth's expansion toward it creates an *effective* parabolic path on the way to a collision with the passing object (Fig. 3-8).

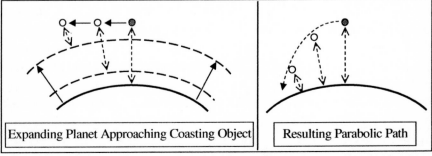

| Expanding Planet Approaching Coasting Object | Resulting Parabolic Path |

Fig. 3-8 Expanding Planet Creates Parabolic Plummets to Earth

The left frame of the diagram shows the underlying expansion of the planet as an object coasts past. As discussed in the previous chapter, the planet's growth *accelerates* as the planet gets continuously larger while expanding at the atomic expansion rate, X_A . This is somewhat like a bank account that grows faster over time due to compounding interest. This accelerating growth can be seen as a greater expansion amount over time in the left frame, while the object continues to coast along at a constant speed overhead. As indicated by the distance arrows between the surface of the planet and the object at each stage of expansion, the planet closes in ever more rapidly on the passing object. However, since we cannot actually see the expansion of the planet toward the object, we see the resulting parabolic plunge of the object toward the planet (right frame of diagram). This is a purely geometric effect caused by the expanding planet, which Newton mistook for a forceful momentum change imposed on the object by the pull of his "gravitational force."

This means that projectiles are not necessarily tied to this parabolic plummet to Earth under the "pull of gravity", as we currently believe today. Instead, they have an effective parabolic plummet toward a collision with the ground when traveling by too slowly to overcome the planet's expansion, and, as we observe in the heavens, an effective *circular* path known as an orbit when they *do* travel fast enough. That is, as long as an object speeds past an expanding planet rapidly enough that in any given moment it moves away from the planet by the same amount as the planet expands, it keeps a constant distance with an effectively circular orbital momentum (Fig. 3-9).

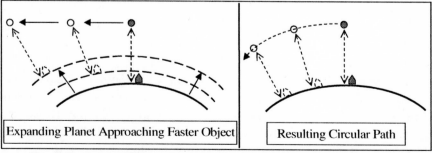

Expanding Planet Approaching Faster Object | Resulting Circular Path

Fig. 3-9 A Faster-Moving Object Produces a Circular Orbit

Figure 3-9 shows a faster-moving object, whose movement past the planet perfectly counteracts the planet's expansion toward the object in any given time period. This results in a constant distance between the object and the planet as the object continues coasting along – the very definition of a circular orbit. Note that the rapidly increasing acceleration of the planet's growth, shown earlier in Figure 3-8, is not seen in Figure 3-9. Instead, it appears as if the planet grows by the same amount each time period – not the compounding amount mentioned earlier. The reason for this is the fact that the object moves fast enough to counteract the planet's expansion moment-by-moment, as just mentioned. Therefore, after each second of time passes, the geometry is indistinguishable from that of the previous second; the object effectively continues coasting past at the same altitude all the time.

This is somewhat like continually withdrawing the interest from a bank account as soon as it is deposited. In that case, the bank account would not grow at an ever-increasing rate since the compounding effect is continually removed before it has a chance to accumulate. Likewise, an orbiting object effectively resets the geometry of the situation moment-by-moment, removing the compounding growth effect of the expanding planet (for orbiting objects *only* – not for those that are falling or sitting on the ground) and producing an effective circular path as shown in the right-hand frame. The object then behaves as if it has a continual circular momentum about the planet, though this is a purely geometric effect – not a true circular momentum possessed by the object and forcefully maintained, as Newton claimed. Both the parabolic and circular paths merely result from the pure geometry of expansion, and are free to change to whatever is dictated by the geometry of the moment. An object traveling more slowly past an expanding planet naturally exhibits an effective parabolic path toward the ground, and a faster speed exhibits a circular orbit about the planet.

Such a view of object trajectories would not be possible in Newtonian physics, since horizontally fired objects are said to fall due to a gravitational force, and thus possess a parabolic momentum. However, *Expansion Theory* shows that no such "gravitational force" and no such *absolute* parabolic momentum exists. Just as objects that fall straight down do not actually *possess* a falling momentum, as shown earlier, but merely follow an *effective* falling geometry while it is the planet that actually expands toward them, so horizontally fired objects do not *possess* a plummeting parabolic momentum either. Everything is merely relative geometry and can take whatever form may be dictated by the dynamics of the moment. Thus, stable circular orbital trajectories are impossible in Newtonian physics, where objects possess an absolute momentum, as stated in *Newton's First Law*; but such orbits are quite natural in *Expansion Theory*.

Finally, note that a small surface feature was provided on the ground in Figure 3-9, which moves along with the orbiting object and indicates that the planet is rotating in step with the object. In other words, Figure 3-9 shows a geo-synchronous orbit, where the object always remains over the same location as it orbits. There are several important points to draw from this situation. First, it is completely irrelevant whether the surface feature moves or not. It could just as well have been drawn in the same unchanging location throughout – indicating a non-rotating planet – without changing the geometry and dynamic of the orbit overhead in any way.

Therefore, while Newton's gravitational explanation of orbits required that the object *passes* the curved surface of a *spherical* planet, *Expansion Theory* shows that this is an unnecessary restriction or special case. Orbiting objects do not necessarily need to be *passing* surface features, but can also effectively hover over the same location if the planet happens to be rotating, since an orbit is not based on escaping a "gravitational pull" from below but on counteracting the geometry of an expanding planet. Secondly, although the classic *spherical* planet is shown in Figure 3-9, it could have been an expanding planet of *any* arbitrary shape – as long as the planet's rotation and expansion balance with the object's altitude and speed to produce a stable geo-synchronous orbit. This follows from the fact that if the object remains over the same surface feature as it orbits a rotating planet the remaining overall shape of the planet *cannot* matter. The object never passes over the remaining terrain of the planet, and never encounters the planet's overall shape; it simply orbits because the planet has a circular rotation that rotates away from the passing object at a rate that balances the planet's expansion toward the object (as shown earlier in Figure 3-2).

To reiterate and clarify these concepts, an orbit results once the object is traveling by fast enough to *counteract the planet's expansion*, which does not necessarily involve the special case of *passing the planet's curvature*, as in Newton's explanation. The *Natural Orbit Effect* of passing expanding objects dictates that an object speeding past a planet will naturally curve about the planet. It does not matter what shape the planet is, as long as the object continuously overcomes the planet's expansion. One way to achieve this is by speeding past the surface of a *spherical* planet, whose *shape* curves away from the passing object to counteract the planet's expansion. Another way is by passing a planet of *any* shape that rotates away from the object at the proper speed (a geo-synchronous orbit). That is, the typically spherical shape of planets, though very common in nature, is a special-case geometry whose roundness essentially simulates the circular rotation of an arbitrarily shaped planet. In this case, the planet does not need a circular rotation away from an orbiting object overhead if the planet's very shape already curves away from the object in a uniformly circular manner.

This geometry allows objects to orbit spherical planets at a wide variety of speeds and altitudes – not just specific geo-synchronous orbital speeds and altitudes that match the planet's rotation and expansion. This spherical special case, where a planet's rotation can be ignored, is the example used by Newton to explain orbits in general, with a "gravitational force" causing circular orbital trajectories past the curved surface while simultaneously causing *parabolic* plummets for other objects. Not only does this show an inherent contradiction in the gravitational explanation, but it also shows that this explanation does not cover more generic orbital scenarios such as geo-synchronous orbits, in which the object does not pass any surface at all.

Orbiting Irregularly-Shaped Bodies

The Illusion Of Remote Gravity Detection

Let's now look more closely at the more general orbital scenario of an arbitrarily shaped planet. If the object speeds past a planet of any shape at a great enough distance so that it is above even the tallest features on the planet, then it can orbit much as if the planet were perfectly spherical. This can be seen by simply drawing a circle around the planet so that it encompasses even the planet's tallest features. As far as the orbiting object is concerned, there is no difference between an arbitrarily shaped planet and a spherical planet that happens to have arbitrary surface features, such as hills and valleys. From a distance, all planets can be surrounded by an imaginary spherical shell and orbited as if they were truly spherical planets of the size of this imaginary shell (Fig. 3-10). Therefore, once again, Newton's spherical example in his explanation of orbits is a special case of this more general scenario as well, with the surface features reduced until the planet is literally a smooth sphere that perfectly matches the imaginary spherical shell around it.

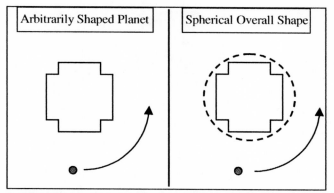

Fig. 3-10 All Planets have Spherical Orbits From a Distance

This orbital scenario, with the object at a great distance, allows the object to orbit a non-spherical planet of any shape without relying on the planet's rotation to match its orbit and create a geo-synchronous orbit. The object can orbit at a wide variety of altitudes and speeds from a sizable distance, passing variations in the planet's surface features as well as in its overall shape (square, oval, etc.) without regard for the planet's rotation. In this case, the orbiting object overcomes the expansion of the overall planet as if it were actually the larger imaginary sphere surrounding it, while its true shape and surface variations pass by within this imaginary sphere. As this occurs, it is possible to measure these variations from orbit. If the object were a satellite that bounced radio waves off the planet's surface as it orbited, the variations in the signal's return time as it bounced off of higher or lower surface features would indicate these height variations and allow a topological mapping of the hills and valleys below. Also, a much more gradual variation in return time measurements would occur as the gradual variations in the overall shape of the planet passed by below. This means that the satellite would essentially measure its altitude to be slowly varying higher and lower as the overall shape of the planet effectively receded and approached as it rotated by below (Fig. 3-11).

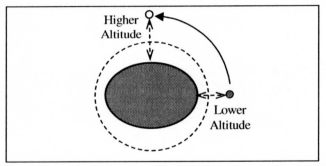

Fig. 3-11 Altitude Variations while Orbiting Non-Spherical Planet

These altitude variations are due to differences in the geometry of the orbit as the satellite passes over regions of greater or lesser planetary expansion due to the variations in the radius of an irregularly shaped planet. The longer axis of the oval-shaped planet in Figure 3-11 expands by a greater amount each second than does the shorter axis (though the same percentage of course), causing the satellite to effectively draw closer then recede further away as it orbits. However, since orbits are currently thought to be ruled by Newton's "gravitational force," such slow altitude variations are thought to be due, at least in part, to variations in the "gravitational field" of the planet. Such a mistaken belief would lead to misconceptions about the density variations of the mass within the planet, since variations in a planet's "gravitational field" would be directly associated with variations in the amount of mass within (from which this field is thought to emanate). That is, although

Expansion Theory shows that altitude variations tell us only about the overall *shape* of the planet, it is currently believed that altitude variations provide an indication of the *mass distribution* within the planet. This is a very important point which will be seen shortly to have sizable implications for our understanding of the gravity of our own moon – a lunar characteristic that *Expansion Theory* shows has so far been misunderstood and incorrectly measured.

The preceding discussions of orbits according to *Expansion Theory* have largely focussed on idealized *circular* orbits; however, circular orbits belong to the more general class of *elliptical* orbits. We will now briefly discuss elliptical orbits, showing that, once again, the gravitational explanation has a fatal flaw that *Expansion Theory* resolves.

Elliptical Orbits

Newtonian Theory Cannot Explain Elliptical Orbits

Although orbits are commonly represented as circular, in actuality perfectly circular orbits are very rare; most orbits are elliptical (i.e. oval-shaped). The degree to which an orbit is elliptical is known as its eccentricity, with an eccentricity of zero indicating a perfect circle. As can be seen in any standard astronomical table, none of the planets in our solar system have a zero eccentricity of orbit; nearly all orbits in nature are elliptical to some degree.

Yet, how do we explain elliptical orbits? Figure 3-12 shows a typical elliptical orbit, with the orbited body off to one side or focal point of the ellipse, as described by *Kepler's First Law* in Chapter 1.

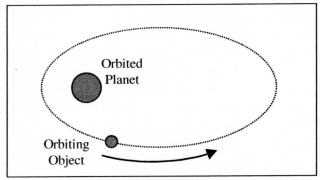

Fig. 3-12 Object Swings Around Planet and Away in Elliptical Orbit

From a gravitational perspective, this could only be explained as the orbiting object speeding past the planet and out to a distance before being slowed and pulled back by gravity, only to zip around the planet and speed off into the distance again in a continual cycle. At first glance this might even seem like a reasonable explanation, perhaps similar to how we might expect a rock to behave if swung around by a spring. The spring would stretch and slow the rock as it speeds off into the distance, then the kinetic energy lost by the rock and now stored in the stretched spring would pull the rock back, speeding it up as it swings around us, then off into the distance again.

ERROR

However, there is a fatal flaw in this logic on closer examination: *the gravitational force defined by Newton is nothing like a spring – in fact, quite the opposite.*

Newton simply invented an equation that calculated a theoretical strength for his proposed new force at varying distances. The distance-squared term in the bottom if his *Law of Universal Gravitation* equation shown in Chapter 1 means that if distance doubles, the strength of gravity drops to a quarter of what it was, and if distance triples, its strength drops to one-ninth, etc. Therefore, we can clearly see that Newton's gravitational force does not behave like a spring at all, and in fact, it behaves in a *dramatically opposite fashion*. A stretched spring pulls back with an *increasingly stronger force* that increases *directly* with the amount it is stretched – double the strength if it is stretched to double its original length, triple the strength for triple its length, etc. Yet, it was just mentioned that Newtonian gravity does not *increase* in strength like a spring, but pulls back with a *weaker* force as distance increases – and a *dramatically* weaker force at that. This is not like a spring at all, but more like a ball of chewing gum that becomes thinner and weaker and less able to pull back into its original form the more it is stretched. The chewing gum doesn't store the energy of being stretched in spring-like bonds that return all the energy later, but rather, in bonds that deform and largely remain so, eventually hardly pulling back at all as the gum is stretched ever further.

Similarly, Newton's gravitational force has no provision for somehow storing energy within its "stretched gravitational field." Neither does its power source become stronger or more "charged with potential energy" as the gravitational field is "stretched," since no power source has even been located for Newton's gravitational force. Such an appeal to a conversion of kinetic energy into "gravitational potential energy" as the object moves away – as commonly claimed today – is a mere abstract invention that has no known physical mechanism or scientific justification.

So, we have a situation where the orbiting object speeds off into the distance, losing kinetic energy all the while *but not storing this lost energy* for later spring-like return since the gravitational pull *weakens* – it does not strengthen like a spring that stores the lost energy. So then, where does the *additional* energy come from to pull the object back, and in fact, to pull it back at a *constantly accelerating rate*? It can only be said to come from the gravitational field – and this is precisely what gravitational theory *does* say. However, if a type of spring were invented that dramatically *weakened*, pulling with less and less force as it was stretched, then somehow managed to pull back with ever-greater force as it returned to its normal length, it would be a fascinating and unexplainable scientific curiosity. Unlike a normal spring, energy is not built up within this odd, hypothetical spring as it stretched, but is *lost* as the spring weakens instead, yet the lost energy somehow reappears within the spring as it recoils back with ever-greater strength. This is quite the opposite of the basic spring principle, and is physically impossible, yet this essentially characterizes the gravitational explanation for elliptical orbits.

Such an explanation provides an answer for an obvious occurrence in the heavens that would otherwise be a complete mystery, but it also *violates the laws of physics*. Energy cannot simply appear from nowhere, but must either be stored then later released or must drain some power source – according to the *Law of Conservation Of Energy*. Yet, there is no identifiable physical mechanism where the lost kinetic energy of the orbiting object is stored for later return as the object is accelerated back toward the planet, and no identifiable power source that is drawn upon to power this accelerating return. Nevertheless, this is the only explanation we currently have for orbits, and so we have invented the abstract justification that "gravitational potential energy" somehow builds as the object pulls away, then is returned as it accelerates the object back toward the planet. Precisely *where* this potential energy buildup exists, *how* it is stored, *how* it builds up simply as a result of the object moving away, and how the identical amount is later *returned* to the object to pull it back in a repeated cycle is never explained; no answers to such questions are to be found in our science.

Elliptical Orbits Explained

Expansion Theory provides a whole new perspective on this conundrum since it shows that orbits actually involve no forceful acceleration and deceleration of the orbiting object. There is no "gravitational spring" stretching and recoiling in physically and scientifically unexplainable ways. There is simply the geometry that results from expanding objects that coast past each other in space. When the object speeds away from the planet and off into the distance, it is merely coasting away at a constant speed while the planet continuously accelerates toward it as its growth compounds over time. Therefore, it is only a matter of time before the expanding planet begins to overtake the object, just as it does when an object is tossed into the air and eventually returns to the ground. This produces the *geometry* of an object that is slowing down if we do not recognize that it is actually the planet that is expanding, which creates the illusion that some type of attracting "gravitational force" is emanating from the planet to forcefully slow the object. This is the "pulling back" phase, as the object begins to return to the planet from a distance.

As the object then approaches the planet (or more accurately, as the planet approaches the object) the object does *effectively* gain speed, though again, just like a dropped object, this is merely an *acceleration effect* due to the geometry of the planet's compounding expansion. This effective speed increase of the object as it passes near the planet allows it to overcome the planet's expansion as it shoots past in a manner similar to the idealized circular orbits discussed earlier. Therefore, the object skims past the planet's surface at a relative speed that more than exceeds the planet's expansion, partially orbiting as it swings past and speeds off into the distance again, as in Figure 3-12 earlier. This final coasting trajectory *away* from the planet – like an object tossed into the air – rather than *past* it in orbital fashion allows the planet's expansion to catch up with it once again, bringing it back to complete another orbit. This perspective on orbits removes the many irresolvable mysteries that result from the gravitational perspective. Once we are able to see expanding matter at work, we can let go of the illusion of "gravitational forces" and objects with absolute momentum that lose and gain kinetic energy mysteriously in elliptical orbits. In fact, from the perspective of *Expansion Theory*, we can explain many more observations and mysteries in the heavens, as will be shown in the continuing discussions to follow. One such observation is a phenomenon called the *Gravitational Lens Effect*.

The "Gravitational Lens" Effect

As mentioned earlier in the discussion of orbits, not only do *objects* orbit about planets and stars, but even light is similarly affected as it zips past these bodies. Light does not actually *orbit* these bodies since it travels far too fast to be contained in such a tight circular path, but it is known to *bend* or *deflect* as it passes. We know this by observation, since distant stars appear to shift slightly from where they should appear when a large body, such as our sun, passes by their location in the sky. These stars don't *literally* jump around in the sky in this fashion of course, but *appear* to, as part of a temporary optical illusion due to their light bending as it passes near the large intervening body on the way to our telescopes on Earth. If starlight is bent before we detect it, then it appears to come from a slightly different location in the sky than it actually does, since we naturally assume that any light we see came directly from an object in a straight line extended back to the source. So, when the visual system in our heads naturally, though in this case incorrectly, assumes that the bent light rays have traveled in a perfectly straight line all the way from a distant star, we get an incorrect impression of the star's location (Fig. 3-13).

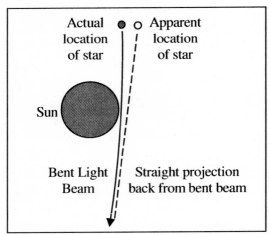

Fig. 3-13 Optical Illusion in Star Location Caused by Bent Starlight

This optical illusion is actually a common effect called parallax, which occurs any time light bends as it passes from one material into another – for example, from air into water, or vice versa. This is why a pencil appears to bend when it is half-inserted into a glass of water. The particular bending of starlight just described is known as a *gravitational lensing* effect since regular optical lenses – such as those used in magnifying glasses – have a similar bending effect on light. In essence, a large intervening object (the sun in Figure 3-13) affects light from distant objects very much like a large optical lens in space – though it is a lens that is considered to operate on gravity instead of regular optics. However, this description leaves us with an unexplained mystery:

 How Does Gravity Affect Energy?

Our current description of gravitational lenses provides an explanation for the optical illusions sometimes seen by astronomers; however, it suffers from a sizable problem – it is a *conceptual* idea without a solid scientific explanation. That is, we have good evidence from many observations of this effect that starlight sometimes behaves *as if* it were bent by the gravity of an intervening object, but we have no explanation for how this might literally occur. Newton would claim that gravity *somehow* bends the pure energy of light, and Einstein claimed that the four-dimensional space-time fabric surrounding the large object is *somehow* warped, causing the light to follow this curved path while passing the object. The problem with both descriptions in the previous sentence is the word *somehow*. Even today, we have no explanation for how or why Newton's proposed gravitational force would even pull on *material* objects, and we certainly do not know enough about the nature of *energy* to even begin to explain Newtonian gravitational effects upon light beams. Likewise, we have no clear explanation for how or why an object would cause a "four-dimensional warping" in its vicinity, as claimed by Einstein. Given the many problems discussed so far with both of these gravitational theories, we can only truly consider such explanations to be abstract models of an observation for which the actual physical explanation is yet to be discovered.

The explanation for this effect from the perspective of *Expansion Theory* has actually already been discussed. As shown earlier in Figure 3-7, the geometry of expansion of a planet or star toward a passing light beam would cause an effective deflection, just as it causes an effective orbit of a slower moving object or an effective fall of a dropped object. The geometry and underlying explanation are the same in all of these cases. Just as objects always move through space relative to other objects – all of which are expanding – so it is with light as well. And so, just as objects do not naturally travel in straight lines, as claimed by Newton, but undergo a natural curving or orbiting geometry with all other objects, so, again, it is with light. That is, contrary to popular belief, *light does not literally travel in straight lines*, but typically curves or deflects to some degree as it passes one expanding reference object or another. We can treat light as if it travels in straight lines as an *approximation*, especially over relatively short distances, but as we see in the heavens, this is far from the reality of the situation. We could only say that a beam of light truly traveled in a straight line if there were no matter of any kind nearby for the light beam to travel past on its journey from the light source to the detector. But, since there is nearly always some form of matter lying in the vicinity of any given light beam at least somewhere along its travels, a truly straight beam of light would be the exception – not the rule.

So, we can see that the gravitational lens effect seen in the heavens is not a particularly exceptional occurrence; nor is it the result of a mysterious interaction between pure light energy and Newton's "gravitational force," or an equally mysterious warping of Einstein's "space-time fabric." It is actually the natural and expected deflection of light that typically occurs as it travels across regular three-dimensional space, past the expanding matter that permeates our universe.

New Revelations About the Moon and Gravity

No discussion of orbits would be complete without addressing the most familiar example of an orbit known to man for millennia – that of our own moon. The moon has accompanied and fascinated us throughout the ages, spawning many myths and theories about its nature, behavior, and effect on us. Its far weaker gravity – one-sixth that of the Earth – has been experienced by humans who have actually walked on its surface. It always shows us the same face as it rotates on its axis at precisely the proper speed to match with its monthly orbit about the Earth. And it is believed to be the cause of our ocean tides, causing the oceans to rise then fall again as it passes by overhead. These are all aspects of our lunar companion that mankind has wondered about, and which we now believe we have answered. But have we really? The discussions in this and preceding chapters have shown that, on closer examination, Newton's proposed gravitational force has many flaws and quite likely does not exist at all. Yet, all of the lunar qualities just mentioned have explanations based on this force today. If the reality is that we live in a universe ruled by expanding matter rather than a gravitational force, how might this realization change our understanding of these lunar observations and experiences? As we will see shortly, *Expansion Theory* suggests that there are several new and surprising aspects of both our moon and our own planet that have been either overlooked or misunderstood.

We begin by taking a second look at the moon's gravity. Since we have visited the moon and walked on its surface, we know that its gravity is roughly one-sixth that of the Earth. However, if gravity is actually the result of atomic expansion, then we do not need to walk on the moon to know its surface gravity – we can easily calculate it simply based on its size. For example, if the moon were half the size of the Earth it would expand by half as much each second since the expansion amount each second is simply the atomic expansion rate, X_A, times its size. This would translate into half the gravity of the Earth. Therefore, we can easily calculate the expected gravity of the moon since we know that it is 1738 km in radius. Since the Earth is 6371 km in radius, this means the moon is just over one-quarter the size of the Earth. This also means that the gravity of the moon should also be just over one-quarter Earth's gravity. But this is much stronger than the one-sixth gravity that we have measured on its surface. So does this discrepancy mean that *Expansion Theory* is a fatally flawed theory of gravity?

In actuality, it is very common to find such initial discrepancies between theory and practice before examining an issue further. Even Newtonian gravitational theory initially predicts the same one-quarter surface gravity for the moon as *Expansion Theory* does. In fact, it can easily be shown that, in essence, Newton's theory actually predicts that the surface gravity of a body should vary directly with its *size* alone, and *not* with its mass or the distance from its center, just as *Expansion*

Theory states. This rather surprising fact can be demonstrated by the following example:

Consider a planet that is double the size of our Earth. Since volume increases with the cube of an object's radius, a planet twice the size of Earth would have eight times the volume (two cubed equals eight). This would further mean that this planet would have eight times the mass and, according to Newton, eight times the gravity since Newton's proposed gravitational force varies directly with mass. But a person standing on the surface of this double-sized planet would also be twice the distance from its center, and according to Newton the gravitational force weakens with distance squared. So, since two squared equals four, we must divide the eight-fold gravity increase that we just calculated by four to arrive at the proper predicted surface gravity according to Newton. This gives a final surface gravity of double that of the Earth for this double-sized planet. In fact, this example leads to the generalized result that even Newtonian theory predicts that the surface gravity of a body varies purely with its *size*, just as *Expansion Theory* states. So then, how is the measured one-sixth surface gravity of the moon rectified with the one-quarter value that would even be predicted by Newton for this one-quarter sized body?

The answer can be found in any standard table of astronomical data; it has simply been assumed that the moon must be composed of material that is significantly less dense than the material composing the Earth. That way, the mass of the moon would be even less than its volume would imply, allowing us to arrive at whatever lunar density is required to justify the discrepancy between the initial Newtonian prediction and actual measurement. This is how we have arrived at our current mass and density values for the moon today. Likewise, the same initial one-quarter surface gravity prediction by *Expansion Theory* can be explained by referring to a density variation within the moon, but in a very different manner that leads to some very important and surprising conclusions about our moon.

First, consider that in *Expansion Theory*, unlike Newtonian theory, halving the mass of our moon, for example, would not halve its surface gravity. Technically speaking, this *would* have the effect of reducing its surface gravity – as discussed in the 'Styrofoam Planet' thought experiment in the previous chapter – but the reduction would be negligible. That is, although it is true that a less massive moon would expand against an object on its surface with less force since a lighter moon is more easily pushed backward by this effort, the resulting reduction in the moon's expansion force would be negligible. This is because even halving our moon's mass still leaves it with billions of times more mass than any object on its surface, so it would still effectively be an immovable wall in space that hardly gives way at all as it expands full-force against objects of negligible relative mass.

So, today's solution to the Newtonian discrepancy – considering the moon to be of lesser density – does not resolve the same discrepancy arising from *Expansion Theory*. However, although a *reduction* in lunar mass is of little consequence, an internal *redistribution* of mass within the moon would readily explain this discrepancy. To see this, we need to consider two key points. First, the simple calculation of the expected one-quarter gravity for the moon makes an idealistic assumption – it assumes that the moon expands uniformly from its exact geometric center. Although this may seem like a reasonable assumption, we will see shortly that this is not at all a necessary condition of expanding objects. Secondly, we have not measured the surface gravity all around the moon, but merely on the near side facing the Earth. The manned Apollo missions all landed on the near side for practical reasons such as maintaining communications with Earth, which would have been impossible from the far side. These two key considerations suggest the possibility of a deeper story behind lunar gravity.

The first point – that objects do not necessarily expand from their geometric centers – was alluded to in the previous chapter during the discussion of the 6000-km tower in space. Recall that without a counterweight the tower would expand outward in both directions from its center, but with the counterweight it expanded entirely from one end. This is because the outward expansion of an object must have a central location where all of this expansion force comes to bear – essentially a location to push against as it expands. All sides of an object must essentially push against an internal location as they expand outward, which means there is one central location that all sides push against as they all expand outward from this center. However, this location is not necessarily the *geometric* center of the object, but rather, its *center-of-mass*. It is the *mass* of an object that provides the inertia or resistance to push against; therefore, the central pushing-off point within the object is the location where the amount of mass extending out to the edges of the object is equal in all directions.

In the case of the tower in space, without the counterweight it had the same amount of mass uniformly throughout; therefore, the center-of-mass was also at its geometric center. That is, from its geometric center there was the same amount of mass on either side. Therefore, it expanded from its geometric center, essentially creating two 3000-km towers expanding in opposite directions rather than one 6000-km tower expanding in a singular direction. However, with the counterweight that weighed as much as the whole tower, the center-of-mass was at the location where the counterweight joined with the tower. From that location the mass of the whole tower on one side balanced with the mass of the counterweight on the other side, allowing the whole tower to expand outward as one 6000-km tower from this pushing-off point. However, this center-of-mass expansion point was nowhere near the geometric center of the tower – a fact that is not immediately obvious from a distance.

This is not merely an academic exercise, but one of very practical importance since the effective gravity experienced at each end of the tower is very different in these two different scenarios. Without the counterweight, the geometrically centered expansion means that each 3000-km expanding section will present only about half the effective gravity of the Earth (which is roughly 6000 km in radius). However, with the counterweight the full 6000-km expansion of the tower would create roughly the same gravity at the far end of the tower as that of the Earth. And also noteworthy is the fact that the gravity for someone standing on the end of the *counterweight* would be far less.

If the counterweight were made of an extremely dense material that allowed it to have the same mass as the whole tower compressed into a size only 600 km across, for example, then it would have only one-tenth the gravity of the Earth since it is one-tenth the size. The difference between a geometrically centralized expansion and an expansion entirely from one end in this example is the difference between a tower that has one-half gravity on each end and a tower that has *full* gravity on one end and *one-tenth* gravity on the other. The overall appearance, size, and shape of the tower is nearly identical with or without the counterweight attached, yet the internal mass distribution makes a tenfold difference in the effective gravity experienced at a given end. In fact, if an external covering were placed over the whole tower so that one could not tell whether the counterweight were present or not, there would be absolutely no identifiable difference externally between these two scenarios until contact was made to determine the surface gravity at each end. The internal mass distribution is completely hidden, but has a dramatic effect on the gravity at either end of the tower. This has significant implications for our moon:

 Different Sides of the Moon Have Different Gravity

Similar to the tower-in-space discussion, *Expansion Theory* suggests that a non-uniform mass distribution exists within our moon. If the lunar material were more compacted and dense on the near side of the moon (the side facing the Earth) it would act much like the counterweight in the tower example, while the less dense far side of the moon would be much like the tower itself. This density difference would likely occur gradually, slowly changing from highly dense at the near side to least dense at the far side. This would be in keeping with one of the possible creation scenarios often suggested for the moon, as will be discussed shortly. It would also mean that the center-of-mass would be off-center toward the near side of the moon, which makes the near side expansion radius smaller than that of the far side. This further means the near side expansion amount each second would be less than the far side, and therefore, the effective *gravity* of the near side would be less than that of the far side (Fig. 3-14).

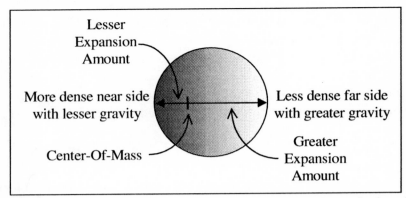

Fig. 3-14 The Moon's Non-Uniform Density causing Differing Gravity

Figure 3-14 shows how our moon could have a lesser gravity on the near side than would be predicted by a simple application of *Expansion Theory* that assumed a geometrically central expansion, leading to a correspondingly greater gravity than expected on the far side. Using the expected one-quarter gravity all around the moon predicted by *Expansion Theory* as a guide, this would mean that the one-sixth gravity of the near side would be accompanied by *one-third* the Earth's gravity on the far side, since one-sixth and one-third average to the expected one-quarter gravity.

NOTE

This means that, contrary to popular belief, the surface gravity on the far side of the moon should be fully *double* the one-sixth value that we expect today.

The possibility of such dramatically different gravity on the far side of the moon is not even considered today since we have measured the gravity variations all around the moon from space using satellites in orbit about the moon. While these satellites cannot literally measure surface gravity, the altitude variations they experience as they orbit are taken as indications of *gravity* variations that pull with differing strength on the satellite as it orbits. Based on this assumption, the fact that only slight altitude variations occur as satellites orbit the moon, combined with the measured one-sixth gravity on the near side, has led astronomers to the conclusion that the moon has a uniform one-sixth gravity all around. However, an earlier discussion exposes the flaw in this reasoning:

ERROR

 Lunar gravity measurements from orbit are based on a flawed assumption.

Recall that, as shown in Figure 3-11 earlier and in the discussion following it, altitude variations from orbit are merely due to variations in the height of surface features and in the overall shape of the orbited body, causing variations in the amount of expansion toward the passing satellite. It is these *size-and-shape-based* expansion variations that cause the altitude variations from orbit – not variations in Newton's still unexplained "gravitational force." Therefore, the nearly uniform "gravity" measurements assumed from the nearly uniform *altitude* variations from orbit simply indicate that our moon is nearly uniformly smooth and spherical in *shape*. These measurements do not actually indicate the *surface gravity* of the moon, but simply indicate its degree of *roundness and smoothness*.

A further important point regarding the non-uniform expansion of our moon is the effect it would have on objects at a distance, such as orbiting or falling objects. To explore this issue, let's return once again to the example of the tower in space. If we were floating in space next to the geometric center of the tower (without counterweight) we would not notice its underlying expansion – we would simply be floating next to the center of an apparently constant-sized tower (since we expand at the same rate). However, this would not be the case if the tower had the counterweight attached and was then expanding from one end instead of its geometric center. In that case we would see the tower expanding past us, much as if we had jumped from the tower on Earth and were falling beside it (Fig. 3-15). This is because the tower expands from an essentially fixed location at one end as it pushes against the massive counterweight, which means it must expand out from that end and past us as we float beside it. Eventually, the full length of the tower will have expanded past us, and we would remain floating at the base of the tower where it expands out from the counterweight (i.e. the center-of-mass point). Therefore, although any obvious physical differences between the counterweight and non-counterweight scenarios in terms of the tower's overall size and shape could be easily concealed, we would know which situation we were in by simply floating next to the tower in space.

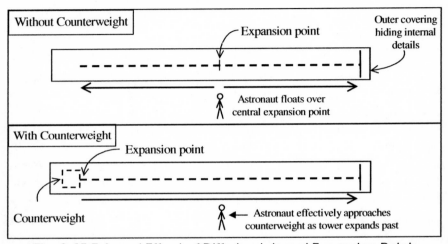

Fig. 3-15 External Effect of Differing Internal Expansion Points

Likewise, if our moon had the off-center expansion just mentioned, it might seem that this fact would be equally obvious from space. That is, if we positioned ourselves to be floating in space over the geometric mid-point of the moon, we would expect its off-center expansion from a center-of-mass point closer to the near side to cause the moon to expand past us. We would essentially find ourselves "falling" toward the near side as the less dense side of the moon expanded past us, and we would eventually end up floating over the center-of-mass point toward the near end (Fig. 3-16). Incidentally, we would also be effectively falling toward the moon's surface, as it would also expand outward toward us as we float beside it.

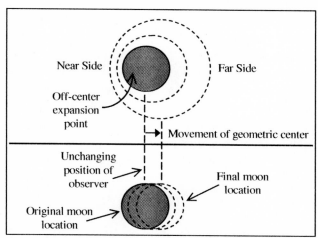

Fig. 3-16 Moon's Off-Center Expansion (Top) and Result (Bottom)

Figure 3-16 shows the moon's underlying off-center expansion in the top frame, and the effective result of this hidden expansion as we would experience it from the perspective of beings that are also composed of expanding matter. We would not see the actual growth of the moon, but would see a constant-sized moon that moves past us for some unknown reason (as we also effectively fall to its surface). We can see how the "unchanging" position of the observer does effectively migrate from floating over the geometric center of the moon to settling over the off-center expansion point closer to the near side of the moon in its final location. Since this effect would actually occur for an observer floating beside the moon, shouldn't it have some noticeable effect on our orbiting lunar satellites and spaceships? If we return to the earlier explanation of orbits according to *Expansion Theory*, we can see that an off-center expansion within the moon would have no effect on orbiting satellites.

Recall that there is no absolute momentum change in an object as it changes from a parabolic drop to a circular orbit. Both trajectories are independent geometric effects as the object travels at different speeds past an expanding planet. Likewise, the geometry and expansion dynamics of an observer floating at a distance from the moon are different and quite independent from those of a satellite speeding past the moon in an orbit. The floating scenario allows the effect of the underlying off-center expansion to unfold in plain view, while the orbiting scenario ignores this fact and simply sets its sites on orbiting a spherical expanding body in space. As long as the orbiting object has a speed and trajectory that overcomes the expansion of this sphere in space in any given moment, it is only this stable, sustainable geometry that matters. It does not matter that this sphere has an inner off-center expansion any more than it matters that objects typically follow parabolic paths to the ground and not circular orbits. The geometry of an orbit is an independent dynamic that – once properly set up – is only concerned with the shape and overall expansion rate of the body it is orbiting.

Therefore, a satellite orbiting the moon would be expected to experience the uniform one-quarter gravity all around, according to *Expansion Theory*, rather than one-sixth on one side and one-third on the other. This explains why we measure a one-sixth effective surface gravity on the near side and we also believe we experience a uniform gravity field from orbit. As a result, we do not even currently suspect that the moon might have double the surface gravity on the far side. It is only by referring to *Expansion Theory* that we would even suspect such a thing.

Interestingly, as the Apollo 11 astronauts first descended to the surface of the moon in 1969 they discovered that they were descending much faster than expected as they flew past the surface toward the intended landing spot. The additional descent speed was sizable enough that unplanned last-minute manual corrections were necessary, resulting in the lunar lander significantly overshooting the intended landing spot and causing mission control to consider aborting the landing altogether during those final crucial moments. Also, many unmanned Russian and American missions mysteriously crash-landed on the moon during the early days of our space programs.

According to *Expansion Theory*, the expected one-sixth "gravitational pull" on these lunar descent vehicles could actually have been up to *one-quarter*, depending on whether they followed more of a vertical descent straight down or a horizontal trajectory that skimmed past the surface in a somewhat orbit-like fashion as they descended. It would certainly be surprising to be executing descent maneuvers designed for one-sixth gravity in a scenario that involved up to one-quarter gravity instead – fully 50% greater than expected. A lack of understanding that the true nature of gravity is actually planetary expansion could spell disaster for any mission descending over a moon or planet, and may help explain other anomalies or even some complete mission failures in our space programs over the years.

Another observation that is currently considered a mystery by many astronomers is the fact that Io, one of Jupiter's moons, has a thin atmosphere while, our own moon does not. This is considered a mystery because Io is similar in size to our own moon and is thought to have a similar gravitational field – one that was too weak to hold an atmosphere on our own moon. So, if a one-sixth strength gravitational field is too weak to hold an atmosphere on our moon, why is it apparently strong enough to support an atmosphere on Io? As just discussed, *Expansion Theory* shows that a body the size of our moon should have an effective uniform gravity that is 50% greater than we currently believe exists on our moon – provided it has a central expansion point. If this is the case for Io, it might explain why a thin atmosphere still remains.

Conversely, since our moon has a much weaker effective gravity on one side, this would have the same effect as a slow leak in a balloon. Any early atmosphere that might have existed on our moon may have drifted off into space from the low-gravity side, while the higher atmospheric pressure on the high-gravity side would have continuously pushed the atmosphere toward the low-gravity "leak." This effect, combined with the natural tendency of any remaining atmosphere to diffuse uniformly across the moon and thus also continually move toward the leak, means the atmosphere would have continued to escape until none remained. Such an effect may explain this current mystery in astronomy.

Although *Expansion Theory* provides good reason to suspect a non-uniform density within the moon that varies from high density on the near side to low density on the far side, is there also good reason to expect such a density variation based on the likely creation scenarios of the moon? In fact, one well-known Earth-moon creation theory does support just such an inner density variation within our moon.

EXPERIMENT

Earth-moon Formation

One of the theories of our moon's origin is that our Earth-moon system arose in the same manner as it is believed our sun and planets did – condensing from a large rotating disk of gas and particles. Our whole solar system is believed to have once been such a disk before the sun and planets formed, possibly with smaller swirls here and there within the disk. Eventually, gravity would have caused the large rotating disk to compress into a large central spinning body, while the smaller swirls would become smaller orbiting bodies with their own rotation. This large central body is our sun and the smaller bodies are the planetary systems – i.e. the rotating planets and their orbiting moons. From the perspective of *Expansion Theory*, of course, it would not literally have been an inward-pulling "gravitational force" from the tiny gas molecules and particles that caused the compression of the disks and swirls, but rather, their outward *expansion* from within.

It is straightforward to see that, as such a disk becomes more compressed, it would have a dense center and would be progressively less dense toward its outer edges, much as soap in dishwater becomes more condensed toward the center of the sink as it drains. We can still see this effect in our solar system, with a large compressed sun at the center and progressively more empty space as we go further out toward the edge of the solar system. This condensing-disk explanation describes both the formation of our sun at the center of the solar system, as well as the formation of each planet at the center of its own smaller planetary system with orbiting moons.

This may well explain how our Earth and moon formed – from a smaller swirling disk within the larger solar-system disk, with the Earth forming as the central concentrated mass in this smaller disk and the moon condensing from the matter along the edge. Note that there are variations on this theme in some theories of the Earth-moon formation, such as the theory that the forming Earth was struck by a planet-sized object that flung matter into orbit around the Earth, creating a swirling disk from which the moon formed. This idea is suggested by the fact that the Earth is tilted relative to the moon's orbital plane, and rocks gathered on Apollo missions suggest the moon had a violent, molten past. Regardless, the general principles in the following discussion can be applied to either version.

Returning to the early Earth-moon disk, if the swirling material in the center separated from the outer surrounding material as it condensed into our rotating planet, this would leave a band of outer material to collect and condense into our moon. Such a process would give the Earth a uniformly symmetric internal density much like the center of the disk that it condensed from, further aided by the fact that it would have been spinning as it condensed from the swirling disk, helping to distribute the matter more evenly. However, this would not be the case for the moon.

The moon would have formed from the outer ring of particles, which would have had a density variation that was denser nearer the Earth and progressively less dense toward the outer edge of the disk. Also, there would be no central spin or rotation for the moon, just as the outer ring of particles from which it formed had no such dynamic. The outer ring would simply circle slowly about the forming Earth, and so would the moon that condensed from such a ring. This describes a process whereby our moon would be *expected* to have a denser near side and less dense far side, and would also be expected to have the same side constantly facing the Earth, since these are precisely the characteristics of the ring from which it formed. The fact that the daily spin of the Earth is now far faster than the monthly orbit of the moon is also expected since the Earth would naturally spin faster as it condensed, just as ice skaters spin faster when they pull in their arms.

This description shows that, contrary to popular belief, *the moon does not have a central spin* that happens to precisely match with its monthly orbital cycle so that the same side constantly faces us. This feature of the moon is not an odd coincidence, nor is it due to the Earth's "gravity" somehow forcefully maintaining this situation, as often claimed today. *Expansion Theory* shows that there are no forces whatsoever involved in orbits – they are perfectly natural geometric patterns as expanding objects pass each other in space. Therefore, our moon has no force-induced central spin any more than there was within the ring of particles from which it formed, leaving the density variation from the ring's inner edge to its outer edge to solidify within our moon, with no spin to distribute it more evenly (Fig. 3-17).

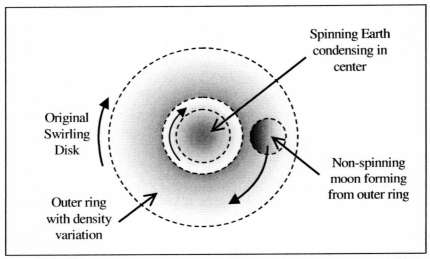

Fig. 3-17 Non-Spinning Moon with Density Variation of Outer Ring

To further illustrate how an orbiting object, such as our moon, could have no inherent spin yet could always manage to face the Earth as it circles, consider the stretchable orbiting rod in Figure 3-18. The rod is shown initially pointing perpendicular to the planet, and was given no initial spin at all. Now, we know both from observation and from the orbit equations of Chapter 1 that objects closer to the Earth orbit faster than those further away; therefore, the rod will stretch as it orbits, since the end closer to the Earth moves faster than its other end further away. As seen in the diagram, this also means it effectively begins to rotate and is no longer perpendicular to the planet.

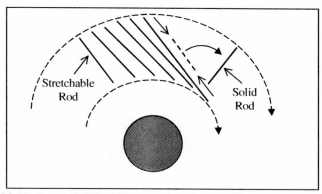

Fig. 3-18 Natural Orbital Geometry Keeps Objects Facing Earth

But what if the rod is made of a solid, non-stretchable material? In that case, it would be as if the stretchable rod was continuously pulled from within so that it did not continually stretch but always returned to its original length. But that means the near end would be continually pulled further away from the planet, and the far end would be continually pulled closer as both ends were pulled in from the center. And this means that the near end slows down as it is pulled to a higher orbit and the far end speeds up as it is pulled lower, causing the rod to effectively rotate back toward a perpendicular alignment with the planet as it orbits (also Fig. 3-18). Yet, since there was no external force to change the rod from its original non-rotating state, it maintains its original perpendicular alignment to the planet – equivalent to an object whose same side always faces the planet as it orbits – *without any inherent rotation at all.*

NOTE

Therefore, *non-rotating* objects naturally maintain a geometric alignment with the planet they are orbiting such that the same side always faces the planet.

As further evidence of this fact, consider our orbiting space shuttles and space stations. These spacecraft circle the planet extremely rapidly, typically completing an orbit in 90 minutes, and somehow always manage to face the planet in the same manner orbit after orbit for weeks, months, and even years. Can it really be that their orbits are so precisely orchestrated that they all have just the right amount of spin to perfectly match their orbits? The above discussion shows that they merely have to be flown straight into a *non-rotating* passing trajectory at the right speed for a stable orbit, and the geometry of expansion and the natural orbit effect takes care of the rest. Likewise, our moon continuously faces the Earth for the same reason. It has no internal spin and is not held facing the Earth by a "gravitational force" reaching out from the planet; in fact, only if it *did* have an internal spin would this natural geometry be upset, and we would see it rotate as it orbited.

NOTE

 Therefore, the non-uniform density within the moon that would predict a weaker gravity on the near side and stronger on the far side – according to *Expansion Theory* – is actually an *expected* outcome of this well-known Earth-moon creation scenario.

The Nature and Origin of Tidal Forces

One of the more compelling reasons for today's belief in Newton's gravitational force is the effect of *tidal forces* all around us. Tidal forces are obviously very real, causing the oceans to rise and fall here on Earth, and even believed to have pulled the comet *Shoemaker-Levy 9* apart as it passed Jupiter some years ago and eventually crashed into Jupiter in pieces in a landmark astronomical event. Although the gravitational force that we believe to be pulling between the Earth and moon has never been directly felt or measured, it is inferred from the fact that the oceans rise and surface gravity measurements on Earth reduce slightly as the moon passes overhead. It is understandably very tempting to conclude that this is due to Newton's gravitational force emanating from the moon and pulling on oceans and objects here on Earth as it passes overhead. However, if Newton's gravitational force does not exist, as *Expansion Theory* claims, then a solid alternate explanation would have to exist for each event that is currently attributed to gravitational tidal forces. As shown in the following discussions, such solid alternate explanations do indeed exist, beginning with an alternate explanation for our ocean tides.

Does the Moon *really* cause Ocean Tides?

Our oceans rise and fall roughly every half-day on Earth, giving us two high and low tidal cycles each day. This effect is currently attributed to the moon as it passes overhead, pulling on the oceans and causing a tidal bulge in the oceans that follows the moon as our planet makes a full rotation beneath it on a daily basis. A second related bulge also occurs in the oceans on the opposite side of the planet. Since these two bulges essentially circle the Earth each day as the planet rotates through them, we experience two occurrences of high and low tides daily.

Yet, as compelling as this explanation may be, it does not constitute conclusive evidence for Newton's gravitational force. There is undeniably a tidal force that causes oceans to rise and surface gravity measurements to drop whenever the moon passes overhead; however, the fact that this tidal force *coincides* with the moon's passing does not necessarily mean the moon *causes* it. In fact, if we return to the Earth-moon creation scenario just discussed, we can see good reason why the tidal force on Earth would exist *even if the moon suddenly vanished*. The tidal force coincides with the moon's passing for a very good reason, as we shall see, but this is still merely a *coincidence* – the tidal force is not actually *caused* by the moon at all.

 NEW IDEA

Ocean Tides Are Entirely Due To Earth's Inner Dynamics

Revisiting the formation of our Earth and moon in more detail, if the original disk was swirling about its geometric center, then its eventual division into the early Earth and moon means that the center-of-mass of the disk would become the center-of-mass of the Earth-moon system. This is because the center-of-mass of a system can only shift if an external force causes such a shift; otherwise, it remains at the center-of-mass of the overall system, whether that system remains a singular disk or separates into two spheres. This means that, since the Earth is far more massive than the moon – on the order of a hundred times more massive – the center-of-mass of the overall system would be well within the forming Earth, and would remain so as the two bodies separated. At this point, the Earth and moon would be two non-rotating bodies that have just separated and are slowly circling about this center-of-mass point across from each other at the speed of the original swirling disk (Fig. 3-19).

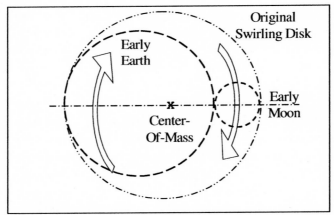

Fig. 3-19 Early Earth & Moon Circling Original Disk's Center-of-Mass

However, although this description of the original disk merely separating into two non-rotating, circling masses is correct, it changes somewhat when we consider each forming body separately. Considering the Earth in particular, although it is technically a non-rotating body that merely circles the center-of-mass point of the original disk, the location of this point *within* the planet means that the early Earth was *circling a point within itself.* An object that circles a point within itself is also effectively *rotating* about an inner point. Therefore, in the case of the early Earth, its circling about the disk's center-of-mass is entirely equivalent to a slow, off-center rotation or wobble within the Earth – much like the wobbling rotation of a slowly spinning egg. This is not the *central* daily rotation of the planet that we know today, but an entirely separate motion that would have existed before today's daily rotation even developed. The early moon, on the other hand, circles this center-of-mass point from a distance, and therefore, remains a non-rotating body that merely circles across from this point within the Earth, and at the same speed as the Earth's slow, wobbling rotation. Today this slow rotation is what we call the lunar month, as the moon waxes and wanes through its lunar cycle across from the Earth as the two bodies slowly rotate across from each other in front of the sun. Just as the original disk swirled at one singular rate, so its separation into two bodies results in the same circling rate about the original center-of-mass point for both the Earth and moon, leaving them also always facing each other.

However, although the moon does always face the Earth as it proceeds through its slow monthly orbit today, the Earth certainly no longer slowly rotates in unison with the moon – or does it? The rotation we are most familiar with today is our daily rotation – a geometrically centralized daily rotation that spins roughly 28 times for every orbit of the moon. This dynamic would have begun to develop soon after the early Earth and moon fully separated from each other. Both bodies would have effectively become smaller and more compacted in toward their centers due to gravity (actually the expansion of their component particles filling the space within), but for the Earth this would have brought a new inner dynamic. The condensing Earth would have begun to develop its own *geometrically centered* center-of-mass within the larger context of its slow, overall wobbling rotation, which would have tended to create an additional *centralized* spin within the Earth. This centralized spin would have increased in speed as the Earth became more dense and compacted, helping to further develop this new geometrically centered center-of-mass.

This leaves two rotational dynamics within our planet – a slow monthly wobbling rotation in unison with the moon, and a much faster centralized daily spin – while the non-rotating moon still slowly orbits across from the Earth in step with its wobble. The Earth's slow off-center wobble would cause objects and oceans to be flung outward over the axis of the wobble – a line that, if extended, would run through the Earth and intersect with the moon (also shown in Figure 3-19). This means that a tidal bulge (as well as reduced surface gravity measurements) would constantly exist in static alignment with the moon, while the central spin of the Earth rotates us past this line of tidal forces every day, giving us our daily high and low tides. So, tidal forces are very real and do *coincide* with the moon's passage overhead, but they are a simple consequence of the Earth's additional slow inner wobble, which has historical reasons for being synchronized with the moon's orbit.

One further point regarding ocean tides is that they are higher during full moons and new moons, which occur when the sun, Earth, and moon are all in a line. Full moons occur when the sun and moon are on opposite sides of the Earth, and new moons when they are on the same side. These alignments are currently thought to *cause* the higher tides due to the combined "gravitational pull" of the sun and moon in the same line, rather than these alignments simply *coinciding* with the higher tides for other reasons. Yet, the above discussion of tides can be readily extended to explain this effect as well. The discussion of the original swirling Earth-moon disk assumed that it had a perfectly circular shape; however, this assumption was made only to simplify the discussion. Instead, since the early Earth-moon disk would actually have been a swirl within the larger solar system disk, it would have experienced centripetal forces stretching it outward away from the forming sun as it was swung around while still physically part of the larger solar-system disk (Fig. 3-20). In fact, it is these same outward centripetal forces that would have caused the early solar system to flatten into a disk in the first place as it rotated.

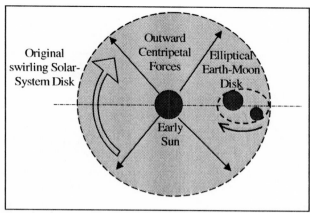

Fig. 3-20 Forces in Large Disk Cause Elliptical Earth-moon Disk

Therefore, our Earth-moon disk would likely have been an elliptical swirl with its elongated axis stretching away from the sun, while such an ellipse could also follow from the "violent impact theory" of the moon's origin, mentioned earlier, with the original impacting object likely speeding toward the sun. Such early dynamics could well have influenced the center-of-mass of the overall Earth-moon system to travel in an elliptical path. Therefore, the stationary center-of-mass point of the Earth-moon system shown earlier in Figure 3-19 would actually have had an elliptical wobble itself. Such a wobble-within-a-wobble can be seen when, for example, an egg is spun randomly instead of ensuring it is spun about its asymmetrical center-of-mass point nearer one end.

If these early dynamics caused such a wobble-within-a-wobble to develop within the Earth with the elongated axis of the wobble pointing away from the sun, this would explain the larger full-moon and new-moon tides. The peak of this additional elliptical wobble will always occur when the slow monthly wobble of the Earth passes through a perpendicular line extended to the sun (along the axis of the original elliptical Earth-moon disk – again in Figure 3-20). And, since the moon is always opposite the Earth's slow monthly wobble, it will also coincidentally be lined up along this perpendicular line when the peak tidal forces occur within the Earth (Fig. 3-21). Therefore, the sun, Earth, and moon will always be aligned when the peak tidal forces occur on Earth, but for historical reasons that are now mirrored in the internal wobbles and rotations within the Earth – not due to mysterious "gravitational forces" that are still unexplained by science.

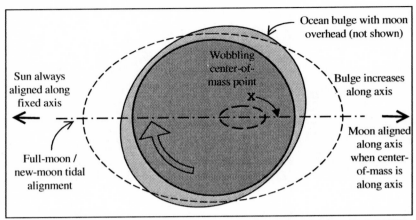

Fig. 3-21 Magnified Earth from Fig 3-20: Full/New-Moon Tidal Effect

The preceding discussion shows that tidal forces can be explained entirely by the internal dynamics of a rotating body. In particular, the type of internal dynamics that might be *expected* to develop within our planet based on the likely creation scenario just discussed would cause precisely the tidal effects that we see today, removing the need to defer to a "gravitational force" pulling from the moon. Newton's gravitational theory can provide a useful *model* to describe many celestial observations; however, once we literally consider it be a force that exists in nature, we run the risk of obscuring the true dynamics at work in the heavens. Even the complex rotational and orbital dynamics of our own Earth-moon system have either been overlooked or trivialized for centuries, since the belief in an *external* force of gravity acting upon the Earth has been the accepted model for the *internal* causes of the tides.

Do Planets *actually* tug on Passing Bodies?

Our current belief in the remote gravitational tug from distant planets even gives rise to periodic predictions of unusual increases in tidal effects on Earth. These predictions arise from an expected increase in the "gravitational pull" upon the Earth when a number of planets are due to align with the Earth in their orbits about the sun; yet, the predicted effects never seem to materialize. The reason nothing happens, of course, is because there *is* no gravitational force emanating from these planets to affect the Earth, and it is unlikely that the Earth has additional internal wobbles that would cause changes in our tidal forces to coincide with such arbitrary planetary alignments. Yet, there are still other observations that are commonly attributed to "gravitational tidal forces." What are we to make of these claims now that numerous flaws have been pointed out in gravitational theory, and such remote forces reaching across space do not exist in *Expansion Theory*? Let's now take a closer look at a well-known example of an apparent tidal-force effect in the heavens.

Comet Shoemaker-Levy 9

One of the most well known examples of apparent tidal forces involves the comet *Shoemaker-Levy 9*, which plummeted into Jupiter in 1994. The comet was actually composed of a number of separate pieces as it headed toward Jupiter, making a number of impacts when it struck the planet. It is believed that the original comet must have been torn apart by Jupiter's tremendous gravity on an earlier close approach to the planet. This is considered to be an example of a gravitational tidal force at work since Jupiter's gravity would theoretically pull stronger on the near side of the comet and weaker on the far side, thus pulling it apart. Yet, *Expansion Theory* states that there is no such thing as a gravitational force emanating from a planet to pull on distant orbiting objects. The comet would simply have coasted past Jupiter several years earlier at a rapid enough speed to overcome Jupiter's expansion, swinging past the planet due to the pure *geometry* of the situation but experiencing no "gravitational forces." And in fact, a closer look at this event shows that the gravitational explanation has a fatal flaw – again, in addition to the lack of scientific viability of such a force.

ERROR

Jupiter's Gravity Did Not Pull *Shoemaker-Levy 9* **Apart**

It is commonly believed that the gravitational field of Jupiter pulled the comet *Shoemaker-Levy 9* apart as it swung by on an earlier close approach; however, there is a clear flaw in this belief. To see this, consider the space shuttle, which circles the Earth roughly every 90 minutes. If the shuttle were truly constrained in orbit by a gravitational force, like a rock swung on a string, it might seem that there should be sizable stresses across the shuttle as it is so rapidly flung around the planet and continually forced into a circular orbit. Certainly an object swung rapidly on a string would experience such stresses, yet there is no sign of such a powerful force pulling on the shuttle. This is currently explained by the belief that gravity would permeate the shuttle, pulling on every atom so that the near and far sides of the shuttle would both experience nearly the same pull, with only a slightly weaker pull on the side further from the planet. Therefore, unlike a rock that undergoes great stress as it is pulled by an externally attached string, all of the atoms composing the shuttle are presumably immersed in the attracting gravitational field, resulting in only a slight differential strain across the shuttle.

If this explanation were true, then this small differential strain across the shuttle would be very tiny indeed. No signs of such a strain on the shuttle and its contents have ever been measured or noted – even after presumably acting for a week or more during a typical shuttle mission. Even free-floating objects show no sign of being even slightly disturbed by any such internal stresses pulling across the shuttle due to this slight differential pull of gravity. Therefore, it would be quite reasonable, if not generous, to say that if such a tiny differential force was actually pulling across the shuttle, it would be no greater than perhaps the force felt by the weight of a feather on Earth. Although the lack of evidence of any such force can be seen as a clear sign that the shuttle is actually on a natural force-free orbital trajectory as explained by *Expansion Theory*, let's see what happens when we apply this gravitational analysis to the scenario of the comet *Shoemaker-Levy 9*.

When the comet was first discovered in 1993, it was already fragmented. Attempts were made to determine how the comet broke apart by re-examining past observations. Although the evidence is sketchy, it is commonly reported that the comet was pulled apart by Jupiter's gravity during an earlier approach at a distance of roughly 1.3 planetary radii from Jupiter's center. That is, the distance of the comet above the surface of Jupiter as it flew past was roughly equivalent to one-third of the planet's radius. A standard calculation of the reduction in gravitational strength with distance – according to Newton's theory – shows that at that distance the comet would have experienced a gravitational pull that was 40% weaker than at Jupiter's surface. To put this in perspective, this represents a force on the comet that is only 50% stronger than the gravitational force that is *theoretically* constraining the space shuttle as it orbits the Earth (remember, no such force has ever actually been felt or measured).

Now, since we know that the net stresses across the shuttle in near-Earth orbit are imperceptible even when supposedly acting continuously for days, it is difficult to justify that a stress only 50% greater across the comet *Shoemaker-Levy 9* during a brief flyby would have torn it apart. The situation doesn't change even if we consider that there would have been a greater gravitational difference across the 2-km comet than if it were the size of the much smaller space shuttle. Each shuttle-sized segment of the comet's diameter would still have experienced a pulling force across it that is comparable to the weight of a feather, as mentioned earlier. Even with a hundred such segments across the comet, this total force of roughly the weight of a handful of feathers across a 2-km comet is many thousands if not millions of times too weak to tear it apart.

So, we are left with the mystery that Newton's gravitational force, even if it did exist, could not possibly have been responsible for the breakup of the comet *Shoemaker-Levy 9*, while there are no forces at all upon the comet according to *Expansion Theory*. However, this is not a complete mystery, as there are numerous additional explanations. Jupiter is known to have an immense magnetic field, which could have played a role in the comet's breakup. Alternatively, the comet could have collided with other space debris orbiting about Jupiter. Also, the comet would have undergone sizable alternate heating and cooling as it approached then receded from the sun during its travels, perhaps experiencing sizable blasts of plasma from sunspot activity as well. The comet could even have had a pre-existing fragmentation that was present for decades or even centuries, but was impossible to clearly resolve in earlier photos that may have contained the comet as a faint blur by chance in the course of other observations prior to its official discovery. Regardless, in the list of possible causes, it is clear that being torn apart by a "gravitational tidal force" could not be among them.

These discussions of tidal effects show that there is no clear evidence for the existence of "gravitational tidal forces" acting at a distance between orbiting bodies. In particular, the example of the comet *Shoemaker-Levy 9* shows how easily such verifiably impossible explanations of observations can become widely accepted in our science, eventually becoming unquestioned fact. Many of the ideas that we have inherited as a scientific legacy from centuries past have become so firmly ingrained in our thinking and belief system that they are often given unquestioned credit in situations where they clearly cannot possibly apply. Due to this process it is now readily accepted that an endless gravitational force reaches out into space, tearing comets apart and inducing ocean tides and volcanic activity on orbiting moons and planets. However, *Expansion Theory* allows us to take a second look at our inherited beliefs, and in the process, to see the clear physical causes at work that have been masked by such largely unquestioned beliefs as Newton's gravitational force.

The Slingshot Effect

One of the most compelling phenomena used in our space programs is that of the so-called *"gravity-assist"* maneuver, also often called the *slingshot effect*. This is a maneuver where a spacecraft catches up to an orbiting planet from behind, swings by the planet in a partial orbit, and then is flung away on a new trajectory at a faster speed. This is currently believed to be the result of the planet's gravity accelerating the spacecraft toward it, towing the spacecraft along briefly while swinging it around, then releasing it off into space again at an increased overall speed. This is a very real effect that many space missions rely upon to give fuel-free speed boosts to spacecraft that are sent across the solar system. Let's now take a closer look at this effect.

As with falling and orbiting objects, there is no question that the observed *effect* of the "gravity-assist" maneuver does occur; the question, though, is whether the current explanation in our science is at least logically sound – and further, whether it is scientifically viable and consistent with other heavenly observations. The discussions so far have repeatedly shown that the concept of a gravitational force at work behind many of our observations violates the laws of physics, while alternate, scientifically viable explanations for these observations have been presented according to *Expansion Theory*. This means that a gravitational explanation for "gravity-assist" maneuvers would now *stand alone* as a mystery based on a proposed gravitational force that is still *scientifically unexplained*. Therefore, even prior to deeper investigation, it can already be said that the current gravitational explanation for this effect is not scientifically viable, nor is it now even consistent with the other observations in the heavens (falling objects, orbits, tidal forces) – for which the gravitational explanation is highly questionable. The only remaining question is whether today's explanation for "gravity-assists" can at least be considered to be feasible *in principle*, regardless of the additional problems that arise when a "gravitational force" explanation is presented. The analysis to follow shows that, in fact, even the *logic* within the current explanation in our science does not stand up to scrutiny.

ERROR

Flaw in Gravity-Assist Logic

The basic idea of being pulled-in then flung off into space at a faster speed by gravity is a fundamentally flawed concept, since Newton's gravitational force is considered to be a purely *attracting* force. In order for the spacecraft to be flung off into space at an increased speed, the planet's gravity would have to "let go" of the spacecraft somehow, after pulling it in. Otherwise, the situation would be somewhat as if an elastic band were stretched between the planet and the spacecraft. The elastic band would pull the spacecraft in, accelerating it toward the planet, but then would *decelerate* the spacecraft again as it attempted to speed away. In somewhat similar fashion, the same gravitational force that supposedly accelerates a spacecraft throughout its approach to a planet would also continuously *decelerate* it as it traveled away, returning the spacecraft to its original approach speed as it leaves.

Yet, since spacecraft are clearly observed to depart with greater speed than on approach when this maneuver is performed in practice, logical justifications have been arrived at in an attempt to explain this effect from the only viewpoint available today – Newton's gravitational theory. The typical explanation in today's science does acknowledge the "gravitational elastic band" problem just mentioned, but states that there is an additional effect in practice when *moving* planets are involved – an effect where the spacecraft is said to "steal momentum" from the orbiting planet.

This concept begins with the idea that as a spacecraft catches up to and is pulled toward a planet that is orbiting the sun, the spacecraft would also pull the planet backward slightly. This would slow the planet in its orbit while the spacecraft gets a large speed boost forward due to its far smaller mass, essentially transferring momentum from the orbiting planet to the passing spaceship. Then, although it is acknowledged that the planet's gravity would pull back on the spacecraft as it leaves, slowing it back to the same relative speed it had with the planet before the maneuver, the spacecraft still leaves with a net increase in speed. This is said to occur because the planet is now traveling slightly slower in its orbit about the sun after being pulled backward, with this lost momentum now transferred to the spacecraft, speeding up the much lighter spacecraft by far more than the massive planet was slowed. Essentially, this explanation says that the spacecraft reaches ahead via gravity and pulls on the planet to speed ahead while slightly slowing the planet in exchange, thus permanently stealing momentum from the massive planet to give the tiny spacecraft a sizable lasting speed boost.

Although this explanation may seem feasible on first read, a closer examination shows that it suffers from the same fatal flaw mentioned earlier, where the gravity of a stationary planet would pull back on the departing spacecraft, canceling any speed increase that may have occurred on approach. The "momentum stealing" explanation simply creates the *illusion* that the situation is different when the planet is moving in its orbit. Let's now take a good look at this illusion.

First, taking the simpler scenario of a *stationary* planet approached by the spaceship, clearly a "gravitational elastic band" accelerating the spacecraft toward the planet would also equally *decelerate* it as it leaves, giving no net speed increase. This is what Newtonian gravitational theory would predict. The more complex scenario is that of a *moving* planet approached from behind by the spacecraft. Here, however, it is claimed that there is something fundamentally different simply because the planet is moving. It is claimed that the planet is pulled backward and permanently slowed in its orbit, giving a lasting "momentum transfer" and speed boost to the spacecraft that pulled itself ahead. This is where the illusion is created from flawed logic.

In actuality, there could be nothing fundamentally different with a moving planet – there would still be no net speed changes. To see this, we simply need to imagine ourselves coasting along with the moving planet, in which case the planet is no longer moving relative to us, and it is easier to see that the situation is essentially the same as with the stationary planet. The logical flaw in the current explanation is often overlooked because the additional issue of the planet being pulled backward in its orbit is typically only mentioned for the *moving* planet, making it appear as if a moving planet presents a fundamentally different situation than a stationary one. But in actuality, the stationary planet would also be pulled backward in the same manner by the "gravitational elastic band" as the spacecraft approached (Fig. 3-22); it is simply easier to overlook this fact with the stationary planet since the focus is on the motion of the spacecraft.

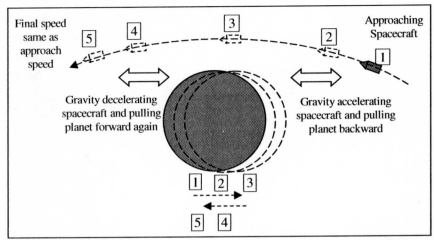

Fig. 3-22 Today's Gravity-Assist Explanation: No Net Acceleration

As today's gravity-based explanation in Figure 3-22 shows, the spacecraft would be accelerated forward by the gravity of the stationary planet, but would also pull the planet backward slightly in the process – just as commonly stated for the *moving* planet. Then, the situation would completely reverse itself after the spacecraft passed the planet. The planet would be pulled forward to its original position as it pulls on the departing spacecraft, slowing the spacecraft to its original approach speed as well. And, once again, there is no reason to expect this final situation to be any different with a *moving* planet – both the planet and the spacecraft would have no lasting speed change according to Newtonian gravitational theory.

A simple way to visualize this is to picture the whole diagram in Figure 3-22 moving to the left across the page. This would be entirely equivalent to the scenario of the planet moving in its orbit, with the spacecraft initially catching up to the planet from behind. It is clear to see that nothing fundamentally changes simply because the overall diagram moves across the page. Both the planet and the spacecraft still end up with no net speed changes. Likewise, nothing fundamentally changes in the "momentum stealing" explanation of "gravity-assist" maneuvers simply because the planet moves along in its orbit. Whether the planet is moving or not, there would be *no lasting slowing of the planet* according to Newtonian gravitational theory, and *no net speed increase imparted to the spacecraft* – in short, no "momentum stealing" by the spacecraft.

NOTE

According to Newtonian gravitational theory, gravity-assist maneuvers are impossible.

The belief that we understand the physics of this maneuver is a myth perpetuated by this flawed "momentum stealing" logic, which has gone uncorrected for decades. This has occurred because we have come to believe unquestioningly in Newton's gravitational force, and at this age of advanced science and technology it is almost inconceivable that a maneuver which is at the core of our space programs could be a completely unexplained – and *unexplainable* – mystery. Instead, we have simply learned to exploit a mysterious effect that obviously does occur, while attempting to invent logical justifications for it rather than allowing this mystery to stand in plain view, pointing to a deeper physical truth that is awaiting discovery.

NEW IDEA

The "Gravity-Assist" or Slingshot Effect is a Purely Geometric Effect.

Since it has just been shown that the "gravity-assist" maneuver cannot be explained using gravitational theory, the following explanation from the perspective of *Expansion Theory* will refer to this maneuver by another commonly used term – the *slingshot effect* – to make a clear distinction between the two explanations. This is also a more appropriate term to use in a discussion that shows this maneuver to actually be a purely geometric *effect* that does not involve any type of accelerating force upon the spacecraft as its speed *effectively* increases.

First, we must consider what a trip through the solar system means from the perspective of *Expansion Theory*. Just as every atom, object, and planet must all expand at the same universal atomic expansion rate to remain the same relative size, so must the *orbits* of the planets around the sun. If this weren't the case, the planets and their orbits would not maintain their current relative sizes, meaning their orbital distances from the sun would effectively either continuously increase or decrease, depending on whether the planets or their orbits were expanding at the greater rate. Therefore, the solar system could be thought of as a very large expanding "object" composed of equally expanding planetary rings centered on the sun, each maintaining a constant relative distance from each other as they expand. Therefore, the task of traveling *across* the solar system actually involves the geometry of *rising in orbit* about the expanding orbital rings of the planets. Let's see how this occurs.

The fundamentals of this principle can be seen even in the scenario of a spacecraft launched into orbit about the Earth. This is done by first rocketing vertically away from the ground, then slowly arcing toward a horizontal trajectory as the spacecraft is inserted into a coasting orbit around the planet. Now recall that, from the orbit equations in Chapter 1 (either the geometric version, $v^2R = K$, or our current gravitational version, $v^2R = GM$), each altitude corresponds only to one particular orbital speed. That is, for a given height or radial distance from the planet, R, there is only one velocity value, v, that will multiply to give the unchanging constant on the other side of the equal sign in the orbital equations, either K or GM . Therefore, if the speed of the spacecraft exceeds the orbital speed for that particular altitude when it turns to fly horizontally into an orbit, it will continue to coast upward in a rising orbit, eventually settling into a stable orbit further out. And, if the spacecraft is traveling fast enough, it will actually rise into a trajectory that escapes the planet entirely. In this case, it does not simply coast straight off into deep space, but moves into a rising orbit about Earth's enormous expanding orbital ring (Fig. 3-23).

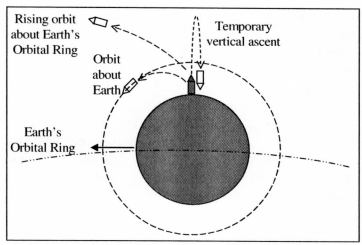

Fig. 3-23 Falling, Orbiting Earth, & Orbiting Earth's Orbital Ring

Although it would take tremendous speed to overcome the expansion of such an enormous "object" as the Earth's orbital ring, the spacecraft, of course, was already traveling fast enough to *equal* this enormous expansion even before launch. The Earth is essentially an object that "skims the surface" of its huge expanding orbital ring as it orbits the sun, speeding past at a rate that matches the outward expansion of this ring. Therefore, every object on the planet already has the required speed to *equal* the expansion of Earth's orbital ring, and the spaceship merely needs to fly fast enough to escape the planet in order to *exceed* the expansion of this ring, continuing away in a rising orbit about it. This is similar to one of the most common orbital maneuvers in our space programs today, known as the Hohmann Transfer Orbit, except that today's terminology assumes our spacecraft travel across the solar system by rising in orbit about the *sun* for *gravitational* reasons. *Expansion Theory*, on the other hand, shows this to actually be a rising orbit about the *nearest orbital ring* for purely *geometric* reasons.

The spacecraft's turn toward a speeding horizontal trajectory in Figure 3-23 altered its fate from one where it would have soon slowed in its vertical climb and fallen back to Earth (actually the Earth's expansion catching up to it), to one where it continues to coast in a rising orbit. From the perspective of *Expansion Theory*, this very same principle is involved in traveling across the solar system, with the spacecraft continuing in rising orbits about the *orbital rings* of successive planets.

Getting to Jupiter, for example, would first involve rocketing away from the expanding Earth, turning to rise rapidly in orbit about the planet, and eventually escaping the planet's expansion and moving on to a rising orbit about the Earth's enormous expanding orbital ring. Then, as the spacecraft coasted toward Mars, it would effectively lose speed as the Earth's orbital ring continued its accelerating outward expansion toward the spacecraft. However, just like the spacecraft that turns horizontally and enters a rising orbit about the Earth to avoid falling back to the ground, our interplanetary spacecraft encounters Mars, taking a similar turn as it effectively accelerates in a partial orbit around Mars, as shown earlier in Figure 3-22. But unlike Figure 3-22, the interplanetary spaceship does not have a decelerating trajectory relative to Mars as it departs; instead, before this can occur, the spaceship is effectively accelerated and launched into a rising orbit about Mars' orbital ring.

This occurs because the partial orbit about Mars defines a geometry where the spaceship is effectively accelerated as it heads toward the expanding planet and swings around it, though no forces are involved in this effective acceleration. This is not unlike the effective acceleration of a dropped object due to the planet actually expanding toward the object. However, this effective increase in speed causes the spacecraft to exceed Mars' expansion and escape into a definition as an object in a rising orbit about Mars' orbital ring, much like the initial escape from Earth.

Remember that neither *Newton's First Law of Motion* nor Newton's "gravitational force" actually exists. The dynamics in the heavens are entirely defined by the relative geometry of expanding objects. If this geometry defines an effective acceleration toward Mars, which immediately becomes a rapid escape from Mars into a rising orbit about its orbital ring, then this is what occurs. This is the natural way events unfold in the heavens, and there is no reason why this should not be the case. It is only our Newtonian thinking – with absolute momentum possessed by objects and unexplained gravitational forces – that turns this situation into a mysteriously unexplainable "gravity-assist" maneuver. Instead, *Expansion Theory* shows that it is simply a natural *slingshot effect*, just as there is a *natural orbit effect* as explained earlier. There is no "gravitational elastic band" and no "momentum stealing" that we must attempt to justify. The spacecraft simply continues on from this effective acceleration and launch into a new rising orbit, and coasts onward toward Jupiter (Fig. 3-24).

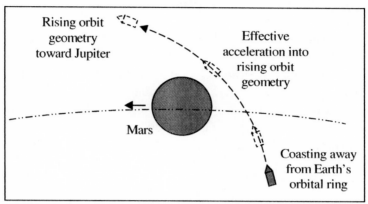

Fig. 3-24 Redefined Momentum: Expansion Theory Slingshot Effect

As further evidence of this, it has been verified that the Voyager 2 spacecraft experienced no forces, stresses, or strains due to the apparent "gravity-assists" it received as it traveled through the solar system, even when such maneuvers accelerated it to *over twice its original speed*. Although such stress-free acceleration is impossible in classical physics, it is a natural and expected result in *Expansion Theory*, since this is actually an *effective* acceleration due to the natural *slingshot effect*, which is a purely geometric effect that involves *no forces or absolute speed changes upon the spacecraft*.

Once again, *Newton's First Law of Motion* is not the literal truth – objects do not literally possess absolute momentum or speed, but only that which is defined by the expansionary geometry of the moment. This was seen earlier in the gravitationally unexplainable change from a *parabolic* plummet toward the ground, to a *circular* orbit about the planet simply because the geometry changed to one that continuously overcame the Earth's expansion once it passed a certain threshold in speed. Similarly, the slingshot effect changes the geometry from that of a slowing escape from the accelerating expansion of the Earth's orbital ring, to an accelerating partial orbit about Mars, which then immediately becomes a rapidly rising orbit about Mars' orbital ring on the way to Jupiter. Without the understanding that the dynamics in the heavens are purely due to changes in relative geometry as everything expands, we are left with today's unexplained boosts in the absolute speed of spacecraft by a scientifically impossible gravitational force as they pass planets.

The above discussion highlights the stark difference between Newton's universe of absolute speeds and forces, and that of *Expansion Theory*, which deals only with expanding relative geometry. The concept of our expanding solar system that was just introduced also helps to resolve an issue that has been an unanswered mystery for NASA scientists for well over a decade. This mystery has been widely published and discussed in journals and popular science magazines, becoming commonly known as the "Pioneer Anomaly."

The Pioneer Anomaly

 An Unexplained "Gravitational Anomaly"

The discussions throughout this chapter have shown that the current gravitational explanations of celestial events in our science today may serve as useful models, but cannot be the literal description of our observations. Therefore, since these models do not truly describe the underlying physics, it might be expected that difficulties and inconsistencies would arise that do not fit within such models. The inability of science to provide a viable explanation for the slingshot effect is one such example, though this has largely been hidden by flawed logical justifications; however, one example that does remain a clear mystery is the unexplained anomalies in the behavior of spacecraft that cross our solar system.

The complexities of traveling among moons and planets have tended to mask subtle deviations or anomalies that may exist in the behavior of our spacecraft as compared to standard Newtonian gravitational predictions; however, we have had a unique opportunity to see such effects much more clearly in recent years. This is because it has been over a decade since the Pioneer 10 and 11 spacecraft have sped past Pluto and continued out of the solar system. As such, there are no longer any moons or planets to encounter as they progress, so any anomalies that may exist in the motions of these spacecraft since leaving the solar system would stand out very clearly and consistently over time. Indeed, NASA scientists have noted and published observations of an apparent unexplained gravitational pull on both spacecraft back toward the sun, which exceeds the expected pull of gravity at that distance. This additional effective pull has been consistently recorded ever since the spacecraft left the solar system over a decade ago, having a constant unexplained decelerating effect on the spacecraft as they coast off into deep space. Attempts to explain this effect using all known or even proposed gravitational or physical theories have so far been unsuccessful.

However, when we look at this mystery from the perspective of *Expansion Theory*, the journeys of these spacecraft take on a very different quality. The situation now changes from that of two spacecraft being pulled back by an unexplained additional "gravitational force," to that of an *expanding solar system* and the effect it has on the *signals* sent back to us from the distant spacecraft. That is, the Pioneer spacecraft are not actually feeling the tug of the sun's "gravity" at all – and they never did all throughout their journeys across the solar system since such "gravitational forces" do not exist in *Expansion Theory*. Instead, the solar system was expanding outward all around them while they coasted through it, aided by the occasional slingshot effect as just discussed. And now the entire solar system is expanding outward toward them as they coast along billions of miles beyond its edge. Pioneers 10 and 11 are now somewhat like two stones that have been tossed up in the air on Earth and are still coasting upward as the Earth's expansion slowly gains on them, eventually bringing them falling back to the ground. Only, the "planet" is our enormous expanding solar system, and the "toss" that sent them flying away from it into deep space is the series of planetary slingshot effects they encountered throughout the solar system (since their initial rocket fuel alone would not have taken them much beyond Jupiter).

From this perspective we can begin to understand why the received signals here on Earth may hold some surprises. First, we must keep in mind how the speed and distance of the spacecraft are determined from the received signal. In essence, this is done by noting how long it takes for each signal blip to travel from the spacecraft to Earth. If we have just received a blip from the spacecraft and we know when it was transmitted, then we can easily calculate the distance to the spacecraft. Since we know the signal travels at the speed of light, simply calculating the time taken, multiplied by the speed of light, gives us the distance traveled by the signal, and thus the distance of the spacecraft when the blip was transmitted. Then, if we monitor a succession of such regular blips sent by the spacecraft, we can determine the spacecraft's speed as well. If the spacecraft is stationary, then the amount of time between received blips will be the same as the known time between the original transmitted blips at the spacecraft since nothing would have altered this timing as the signal traveled. But if the spacecraft is moving away, then each blip will be sent from a slightly further distance and will take slightly longer to reach us than the previous blip did. This means the time between received blips will be slightly greater than the known timing in the original transmission, telling us that the spacecraft is moving away at a certain speed. This simple analysis allows us to continually track the distance and speed of spacecraft in their travels.

Now, let's see what effect an expanding solar system might have on these timing measurements between signal blips. First, imagine a transmitter sitting on Pluto – the most distant planet in the solar system. Even though *Expansion Theory* states that both Pluto and Earth are moving rapidly outward on their expanding orbital rings, this underlying expansion does not alter the received signal blips on Earth. The blips would indicate that the transmitter is located at the expected fixed distance to Pluto. That is, the signal blips would take the expected amount of time to travel to Earth, and they would have the same timing between them as known to exist at the transmitter on Pluto. This is because *our system of measurements already takes into account our universe of expanding matter – whether we realize it or not.*

When we measure the fixed distance between two points on Earth, we do not consider the fact that the ground between these two points is expanding and pushing them apart as the entire planet expands. It was explained in Chapter 2 that this underlying expansion is hidden as everything in the universe expands equally, and so we simply call our measurement a constant distance. Also, when we measure the speed of light – for example, by timing its travel between two points on Earth – the value we arrive at is actually based on its travel between two points that are moving apart due to the planet's expansion. Therefore, the distance between Pluto and Earth is effectively every bit as "fixed" as it is between two points on Earth, and the travel time for the signal also follows from the same dynamics as for a light beam traveling between two points on Earth. As long as the dynamics of expansion create a scenario where two points are effectively fixed in space, they behave as fixed points regardless of the underlying expansion dynamics. Therefore, the natural orbit effect throughout our expanding solar system results in an arrangement of planets that are effectively fixed orbital distances apart in space, just as objects on our expanding planet are effectively fixed distances apart on the ground.

However, if the transmitter is on a spaceship heading through the solar system and off into deep space, the situation changes. This is no longer a scenario where the spaceship is naturally maintained at an effectively fixed distance from Earth by the underlying expansion dynamics between planetary orbital rings. Instead, the spaceship has broken this geometric balance, just as an object in a stable orbit about the Earth is in a very different situation than one tossed at the same speed straight up into the air. Both objects have the same initial speed, but one will coast indefinitely around the planet, while the other will soon crash into the ground. Similarly, the difference in the resulting geometry between a transmitter sitting on Pluto along its effectively fixed orbital ring and one in a moving spaceship heading out of the solar system means the transmitted signal experiences very different dynamics in each case. The signal from the spaceship must now contend with the underlying expansion of the solar system, just as an object that is tossed upward on Earth must contend with the expanding planet beneath it.

As the spaceship coasts away from the solar system each blip is no longer transmitting from an effectively fixed distance; the effective distance is now entirely dependent on how the underlying expansionary dynamics play out as it travels. So, when a blip is transmitted from beyond the solar system, Pluto's orbital ring expands outward toward the blip as the blip speeds toward Pluto. Therefore, the signal reaches Pluto sooner than would be expected due to the reduced distance (Fig. 3-25).

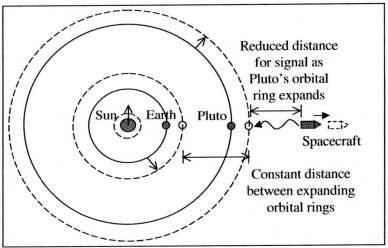

Fig. 3-25 Pluto's Expanding Orbital Ring lessens Signal Travel Time

Although the rest of the signal's journey continues normally between the effectively fixed orbital rings of the intervening planets on its way to Earth, the initial reduction in the expected travel time from the spaceship to Pluto means the overall travel time to Earth is slightly less than expected. Since we currently do not recognize the expanding orbital rings of our solar system, this reduced travel time from the spacecraft to Pluto would be interpreted as a transmission from *closer* than expected. Also, as successive blips continually arrive sooner than expected, the implication would be that the spaceship is traveling at a *continually slower speed than expected* – apparently due to an unexplained additional force pulling back on the spaceship. This underlying expansion-based effect is the likely answer to the currently unexplained "Pioneer Anomaly" noted by NASA scientists.

Orbital Perturbations

Planetary orbits are known to be slightly altered, or perturbed, by passing neighboring planets. In fact, it was an unexplained perturbation in the orbit of Uranus that led to the expectation and eventual discovery of our eighth planet, Neptune. While such disturbances in orbits caused by distant objects may appear to be the result of a gravitational force acting across space, such effects actually have a clear explanation in *Expansion Theory.*

 Dynamically-Centering Orbital Rings

Referring to the example of an object dropped into a tunnel through the Earth, from Chapter 2, it can be seen that an expanding object within a larger expanding object will effectively center itself within the larger object over time purely by the geometry of expansion. In this example, the dropped object eventually ended up sitting at the center of the Earth, effectively migrating and remaining there solely due to the expansion of the surrounding Earth, while maintaining a constant relative size due to its own expansion.

Further, we can now see that our expanding solar system is essentially a large expanding "object" with smaller expanding orbital-ring "objects" within it, all centered around the sun. Just as the dropped object in the tunnel through the Earth essentially sought the Earth's center and continuously maintained itself there via the dynamics of expansion, so it is with our planetary orbital rings centered within the solar system. However, these rings are not perfect idealized circles, but have significant irregularities, such as being somewhat elliptical in shape, with these shapes inevitably varying slightly over time and even sometimes slowly rotating as the years pass. These imperfections and variations would cause the orbital rings to deviate slightly from being perfectly centered within each other – a deviation that would constantly attempt to self-correct as the expansion dynamics continually seek the center. This process would occur in different ways and at different rates for each unique planetary ring, with each adjustment continually re-defining the geometry of each ring within its neighboring ring.

This description paints a picture of a solar system composed of irregular, sometimes even rotating rings that are constantly readjusting and re-centering themselves such that the whole geometry of the solar system is a dynamic feedback between these active rings. This "interaction" between orbital rings does not require a "gravitational force" acting across space any more than the object dropped through the Earth needed such a force to migrate toward the planet's center. The dynamics of our solar system occur purely by the geometry resulting from actively expanding matter. In fact, as mentioned earlier, this explains why fears of catastrophic effects on Earth due to increased gravitational forces caused by planetary alignments never materialize. There are no such "gravitational forces," nor any forces at all involved in the dynamics of our solar system.

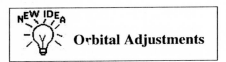

NEW IDEA

Orbital Adjustments

An interesting corollary to this discussion of expansion-based orbital balance is the adjusting or fine-tuning effect of our orbiting moon. Since it is not actually our expanding planet that is in orbital balance around the sun, but rather, our much larger overall expanding Earth-moon system, then altering this overall system could result in a change in our planet's orbit. For example, if we were to slow the moon slightly in its orbit about the Earth, the moon would move closer to the planet, resulting in a smaller Earth-moon system. This smaller system would have a lesser overall amount of expansion, and thus would be effectively less strongly "attracted" to the sun (actually to the expanding orbital ring of Venus) and would drift to an orbit further out. Such adjustments might even prove necessary eventually, since it is believed that the sun will eventually enlarge and become much hotter as it ages.

Although today's gravity-based beliefs would require that we significantly alter the mass of the Earth to perform such a task, *Expansion Theory* shows that it is within our grasp technologically to use the moon as a climactic fine-tuning device for our planet in this manner. Indeed, the Earth may have already benefited naturally from such adjustments since measurements indicate that the moon is receding slightly from the Earth year after year. This would continually enlarge our Earth-moon system, causing us to draw continually closer to the sun – a process that, over millions of years, may have contributed significantly to the temperate climate we enjoy today.

It should be noted that this discussion does not mean our planet's orbit can be altered by any arbitrary space rock that may come along and begin to orbit in the vicinity of Earth. If this were true, then a mere rock orbiting at twice the distance of the moon would double the size of our Earth-moon system and send us spiraling into the sun; of course, this does not and would not occur. Our Earth-moon system has evolved over billions of years to become a stable orbital system whose overall size (and expansion amount) also keeps it at a stable orbital distance from the sun. A space rock attempting to orbit the Earth further out than the moon would clearly lie *outside* this stable orbital arrangement. That is, the much larger orbital system that it *temporarily* represents would begin to drift toward the sun – beginning immediately with the *errant rock itself.* This simply means that the space rock would never even have a chance to become a stable part of our Earth-moon system, and would merely follow a chaotic, unstable trajectory as an independent object moving randomly in our vicinity, if not even one that soon spirals into the sun.

Galactic Implications

Where Does It All Go?

If expanding atoms result in expanding planets, expanding planets and their moons form expanding planetary systems, expanding planetary systems result in expanding planetary orbital rings, and finally, these expanding orbital rings produce expanding solar systems, what is the end result of all this expansion? Where does it all go? The answer lies at the next level of celestial structure – the galaxy.

We are part of an enormous galaxy composed of billions of stars, known as the Milky Way. Our galaxy, like many, is spiral in shape, with its billions of stars all swirling slowly inward toward its center. This is currently thought to be a gravitational attraction toward the galactic center, caused by the gravity of these billions of stars, and perhaps also by one or more Black Holes of immense gravity at or near the galactic center. However, there are many problems with this belief, one of which has already been pointed out in the scientific impossibility of a "gravitational force," and others which will be discussed in later chapters. Even today's astronomers readily acknowledge that this concept simply does not work.

According to today's gravitational theory, the gravitational attraction from all of the stars, and even the proposed Black Holes in our galaxy, is not enough to explain the structure and motion of the stars in the galaxy. In fact, it is often stated that our galaxy would require *ten times* more mass than is currently known to exist in the Milky Way in order to account for observations. This missing matter is currently an area of much investigation and conjecture, and is given the name Dark Matter because it is believed to be present but unseen. This has led to speculation that this "Dark Matter" might not only be massive unseen objects, but may even be some form of exotic matter that is a fundamentally different form of matter than is known today. However, *Expansion Theory* provides a very different perspective on this issue.

Expansion Theory shows that our galaxy is actually composed – not of billions of "gravity-generating" stars – but billions of *expanding* stars and solar systems. As all of these stars and solar systems expand outward, they take up more and more of the space between them and effectively draw closer to each other. But rather than simply growing together in-place until they form an enormous clump in space, they effectively spiral inward as they orbit the galactic center, with the distance between them continually decreasing. Just as water naturally spirals down a drain, taking up less and less volume as it drains, our galaxy naturally spirals toward its center as the expansion of its stars and solar systems consumes more and more of the space within – effectively pulling it inward as it spins. That is, an object whose internal space is being continually reduced is an object that is effectively *shrinking* (or in this case spiraling inward), especially in a universe where everything is expanding at the same rate, effectively hiding the expansion of the stars that cause this effect. Our galaxy effectively spirals inward for the same reason a dropped object on Earth effectively falls to the ground – in both cases these inward-falling scenarios are actually due to an outward expansion from within.

This galactic dynamic will be discussed further in the final chapter, where Black Holes, as well as the Big Bang creation event, are examined from the perspective of *Expansion Theory*; however, we can now see where all of this expansion goes. The expansion of the billions of stars and solar systems in our galaxy goes into the very spiral structure and motion of the galaxy itself, effectively causing it to shrink or spiral inward as its orbiting stars expand ever closer to one another. This also means that the nature of this inward spiral is not directly related to the total *mass* of stars in the galaxy, as currently thought, but rather, to the *sizes* of the expanding stars and solar systems. Although it remains to be seen if the visible stars and their likely solar systems suffice to explain the galactic dynamics when *Expansion Theory* is considered, it is quite possible that such a view will help greatly in solving the mystery of "Dark Matter."

Next, we take a look at a desktop experiment known as the Cavendish experiment. Although this does not technically fall into the category of a celestial observation, it is an attempt to observe the dynamics of objects in free space – brought down to the tabletop. Also, the analysis to follow requires a familiarity with many of the dynamics of expansion discussed so far. We begin with an overview of the experiment as it is currently interpreted – as a tabletop display of gravitational effects – then show the expansion-based explanation.

The Cavendish Experiment – What Does It Really Show?

EXPERIMENT

The Cavendish Experiment

Two centuries ago, Henry Cavendish performed an experiment in an attempt to show gravitational effects in the laboratory, which has been repeated in various forms ever since. The most classic form of these experiments suspends a small barbell between two larger fixed spheres in misalignment, then measures the slow rotation of the barbell as the tiny gravity of the spheres apparently pulls it into alignment (Fig. 3-26).

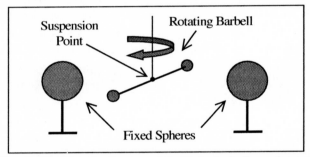

Suspension Point

Rotating Barbell

Fixed Spheres

Fig. 3-26 Classic Cavendish-Style Lab Experiment

While this experiment may appear to be a convincing display of gravitational fields in the lab, if this were actually the case we would have a completely unexplained mystery on our hands since it has now been clearly shown that Newton's proposed gravitational force has no physical or scientific explanation. Once again, it is not the *observation* that is in question, merely the current *explanation* of the event in our science today. Since we have seen that the concept of expanding matter has so far explained all events currently attributed to Newton's gravitational force, let's see what *Expansion Theory* says about this experiment.

The first step in understanding this experiment is to consider what would happen if the suspended barbell were not free to turn, but was rigidly connected to the fixed spheres. In that case, all elements of the experiment would remain fixed and unchanging. From the perspective of *Expansion Theory*, since atomic matter now rigidly connects all elements of the experiment, everything expands equally via the universal atomic expansion rate and the geometry does not change. This can be seen in Figure 3-27, where a top-view of one half of the apparatus is shown, but now with a connecting rod between the barbell and the fixed sphere, as everything expands.

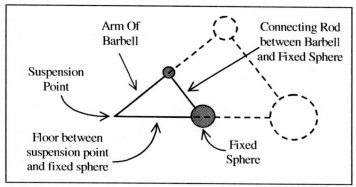

Fig. 3-27 Expansion Geometry of Rigidly-Connected Apparatus

As we would expect, there is no change in geometry as everything expands since everything is rigidly connected. The floor (and overall room structure) connects the barbell suspension point to the fixed sphere, and a rigid connecting rod has been placed between the suspended barbell and its neighboring fixed sphere as well. Now let's see what happens when the connecting rod is removed (Fig. 3-28).

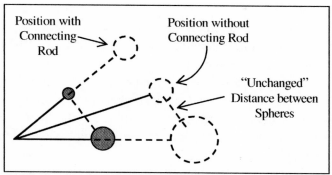

Fig. 3-28 Expansion Geometry with Freely Rotating Barbell

Here we see that the barbell has effectively rotated toward the fixed sphere. This has occurred because there was no connecting rod between the barbell and the fixed sphere, and therefore no expanding matter forcing them apart in step with all the other rigidly-connected, expanding elements in the experiment. So, while today's interpretation of this effect is to assume that some force (presumably Newton's gravitational force) pulled the barbell toward the sphere, *Expansion Theory* shows that the rotation occurred because of a *lack* of force. That is, there are inherent internal expansion forces present all around us within all atomic matter as it actively expands, and so, removing atomic matter from one axis of movement will free that axis for movement out of step with the rest of the world. A more obvious example of this is to pull the stand out from under the fixed sphere. With the rigid stand removed, the expanding planet below is now free to expand rapidly toward the sphere, and the sphere "drops to the ground." This is no different than removing the rod from between the freely rotating barbell and the fixed sphere, except that this frees the apparatus for expansion-based movement in the horizontal plane rather than the vertical.

But then, if the barbell is free to move out of step with the atomic expansion all around it, why doesn't it rotate in the same manner even if the fixed sphere is removed? The answer is the same reason why an object doesn't fall to the ground when it is floating in outer space. All object motion is purely defined by the geometry of expanding objects surrounding it. If there is no fixed sphere with a given distance between it and the barbell initially, then of course there is no final position with that same fixed distance between the expanded barbell and sphere as shown in Figure 3-28. There is no reason to expect the barbell to rotate to a position relative to an imaginary fixed sphere, especially since we could imagine a sphere of any size located anywhere in the room. The barbell has no idea what we may be imagining and only undergoes its expansionary dynamics relative to actual objects in its plane of rotation.

And what if we lowered the fixed sphere to the floor so that it is out of the horizontal rotation plane of the barbell? Newtonian theory would state that there is still a component of the gravitational force between the sphere and the barbell that causes the barbell to rotate horizontally until it is over the sphere, though the effect would be much weaker. Likewise, from the perspective of *Expansion Theory*, if we imagine a straight line between the sphere and the barbell, this distance represents empty space that is not entirely rigidly connected along that line. That is, there is not a rigid structure of atomic matter that prevents this distance from becoming slowly reduced by the expanding dynamics all around – not until the barbell is eventually positioned completely over the sphere below.

This discussion might help to explain why there are wide variations in the results of this experiment when performed by different teams intent on measuring Newton's gravitational constant, found in Newton's *Law of Universal Gravitation* from Chapter 1. Although one particular value for Newton's gravitational constant is generally accepted from a classic Cavendish experiment performed decades ago, attempts to refine this value by other teams in other labs have often yielded wide deviations from this accepted value. This is likely due to the fact that we currently believe that the masses and resulting "gravitational forces" of the spheres – both the reference spheres and those on the ends of the barbell – are responsible for the observed rotation. Therefore, calculations of the "gravitational constant" based on the masses and presumed gravitational forces may well give wildly different results from one lab setup to another since the observations are actually due to *size* and *expansion* – *not* mass and gravity.

This issue can become further confused since there is still reason to expect mass to play *some* role in the results. For example, if the masses of the small spheres on the ends of the barbell are altered, this change in the weight of the barbell would alter the tension and possibly the behavior of the suspending thread that twists as the barbell rotates. Yet, altering the mass of the fixed reference spheres on either side should have no effect since the geometry of the experiment remains exactly the same – that is, unless the *size* of the spheres was also incidentally altered in the process. So, the researchers may well note changes when *any* of the spheres in the experiment are altered, but the changes would be difficult to clearly characterize and explain due to a lack of understanding of expansion – and not a "gravitational force" – at work in the experiment. This is the likely reason why Newton's gravitational constant is considered to be the least accurately measured and most widely disputed of all the natural constants in nature. It is actually not a literal "natural constant" at all, but an arbitrary invention from a model that completely overlooks and misrepresents the true physical reality.

The Origin of a Natural Constant Revealed

The introduction in Chapter 1 stated that the new principle presented in these chapters – now identified as atomic expansion – is the basis for an entirely alternate "theory of everything," which is very likely the much sought-after *Theory Of Everything* that will truly explain our universe for the first time. One of the qualities that scientists expect of the *Theory Of Everything*, once it is found, is that it will allow us to truly understand the origins of many of our natural constants of nature, rather than simply accepting their measured values as we do today. As mentioned in the discussion of the Cavendish experiment, Newton's *gravitational constant* is considered to be one of these natural constants of nature; it naturally appears in our gravitational calculations and has a measured value, but cannot be explained beyond that point. The following discussion shows that *Expansion Theory* indeed *does* explain the origin of this "natural constant," showing that it is more of an arbitrary invention of Newton's than a true constant of nature.

 The Origin of Newton's "Gravitational Constant"

Since *Expansion Theory* suggests that Newton's proposed gravitational force is actually the continuous expansion of all atoms in our universe, the natural place to begin looking for the origin of Newton's gravitational constant is the most elemental atom in the periodic table – the Hydrogen atom. If Newton's gravitational model is actually an artificial overlay atop the true physical explanation – atomic expansion – then it should be valid to equate these two theories at the atomic level so that *Expansion Theory* can be used to *calculate Newton's constant from first principles*. That is, if *Expansion Theory* is the true physical theory for gravity, then it should be possible to dispense with an actual experimental setup such as the Cavendish apparatus, replace it with a purely theoretical calculation using *Expansion Theory,* and still arrive at the known value for Newton's gravitational constant. Therefore, we will equate the "acceleration due to gravity" that Newton would claim a single Hydrogen atom causes with the accelerating expansion of this atom according to *Expansion Theory.* This equality is illustrated in Figure 3-29.

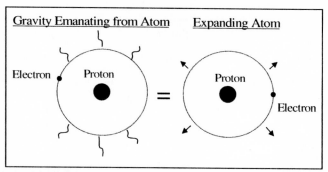

Fig. 3-29 Equating 'Gravity' of Atom with its Expansion

The Hydrogen atom is the simplest of all atoms, with only a single proton in its nucleus and a single orbiting electron. Using Newton's model of a gravitational force, we can calculate the "acceleration due to gravity" at the edge of the atom by plugging the atomic radius and the masses of the proton and the electron into Newton's equation from his *Law of Universal Gravitation.* The resulting gravitational force calculation would give the gravity at the edge of the atom. However, since we are now attempting to *calculate* the gravitational constant, *G,* found in Newton's equation, we must consider it to be an unknown parameter, and so we cannot complete the gravitational force calculation using the known measured value for *G* . Instead, we will equate the accelerating *expansion* of the atom according to *Expansion Theory* with Newton's equation for its gravity, leaving Newton's gravitational constant as the only unknown in the resulting expression. Solving this expression for the unknown parameter, *G,* will represent the first *purely theoretical* attempt to calculate Newton's constant from first principles – *based on an entirely new theory of gravity.* If the resulting value matches the known *measured* value from actual physical experiments, it will show that Newton's equation is not only a model of *observations* – such as falling objects and orbiting bodies – but that it is actually a model of the *atomic expansion* that underlies these observations. This point will be explored further following the optional mathematical exercise that performs the equality shown in Figure 3-29.

 OPTIONAL MATH

(x,y) **Calculation of Newton's Gravitational Constant**

The force required to accelerate a given mass, m_1, at a constant rate is given by:

$F = ma$ – *Newton's Second Law*

Rearranging this equation to calculate the acceleration experienced by the object due to this force gives

$a = F/m$

Therefore, to arrive at the *gravitational* acceleration of one mass upon a second we begin with the gravitational force,

$$F = \frac{G \cdot (m_1 m_2)}{r^2}$$

and divide by the mass of the second object, giving the acceleration expression:

$$a = \frac{F}{m_2} = \frac{G \cdot m_1}{r^2}$$

Now, if the first mass, m_1, is the Hydrogen atom, *Expansion Theory* provides an alternate calculation for this acceleration:

$a = X_A R$ where R is the atomic radius
           a is the effective acceleration of the atomic surface
           (the amount of expansion per second each second)

If the expanding atom concept is correct, equating these two acceleration expressions and solving for the only unknown, G, should result in the known value for this constant:

$$X_A R = \frac{G \cdot m_p}{R^2}$$ where $R = 5.29 \times 10^{-11}$ m   (atomic radius)

$m_p = 1.67 \times 10^{-27}$ kg (mass of proton in Hydrogen nucleus)

Recalling that $X_A = 0.00000077$, this gives: $G = 6.8 \times 10^{-11}$ m^3 / s^2 kg.
The measured Cavendish-experiment value is: $G = 6.6726 \times 10^{-11}$ m^3 / s^2 kg.

The essentially equivalent values for the calculated and measured values of G have several significant implications. One significant result is the validation of the expansion concept and the universality of the resulting atomic expansion rate, X_A, on all scales:

NOTE

 Recall from Chapter 2 that the value for X_A was arrived at by measuring a dropped object in relation to a hypothesized expansion of our *overall planet*, yet here this same value was borrowed and successfully applied to the proposed expansion of the *tiniest atom*.

Also, as mentioned earlier, this calculation shows that Newton's theory of gravity is actually an artificial overlay atop the true physical explanation of atomic expansion. This follows from the fact that experiments such as the Cavendish experiment exhibit the physical rotation of the barbell *regardless of the cause*, after which we *arbitrarily* apply Newton's proposed gravitational model to represent the results in terms of his equation and his gravitational constant. This does not necessarily mean that the resulting constant is a true constant of nature, but merely the value that must result from an equation that has been invented to contain the measurable masses and distances in the experiment, as well as an unknown *constant of proportionality*. Introducing a constant of proportionality into an equation is an extremely common procedure in mathematics, which has nothing in particular to do with natural constants. Newton's constant of proportionality, G, is considered to be a natural constant only because his equation is claimed to represent the true cause of the barbell's rotation. However, if it does not, then the resulting "gravitational constant" is merely a constant of proportionality in an arbitrary mathematical equation since it has been shown that Newton's gravitational theory was arbitrarily invented from a flawed rock-and-string analogy in an attempt to model otherwise unexplained observations. This theory did not follow from a solid, physically and scientifically viable foundation, but rather, it arbitrarily invents a scientifically unexplainable force that acts at a distance. Any number of such arbitrary equations with their own unique constants of proportionality modeling a proposed force at work in this experiment could be created if desired.

However, since *Expansion Theory* is not such an arbitrary proposal but instead presents a viable physical explanation for gravity with a clear physical mechanism behind it, arriving at Newton's constant by equating Newton's theory with *Expansion Theory* is very significant. Now, for the first time, we can arrive at Newton's constant *either* by equating his model with the actual physical experiment *or* by equating it with a mere calculation involving *Expansion Theory*. That is, the above calculation essentially replaces the *physical experiment* typically used to calculate Newton's constant (the Cavendish experiment), and shows that a *pure calculation* according to *Expansion Theory* does the same job. In other words:

NOTE

 The description of gravity provided by *Expansion Theory* is entirely equivalent to the *actual physical cause* of our gravitational observations – it *is* gravity.

If this were not the case, it would be impossible to replace an actual physical experiment, such as the Cavendish apparatus, with a pure calculation based on *Expansion Theory*, such as the atomic calculation just performed, and still arrive at the accepted value for Newton's constant either way.

In Summary

Although Newton's theory of gravity gave us a very useful and important model of a force that pulls passive chunks of matter around in the absence of an understanding of atomic expansion, it also created very powerful misconceptions about the fundamental nature of our universe. Newton's model works rather well as a predictive tool; however, from the perspective of the actual physics of our universe, Newton's universe does not actually exist. There *are* no lumps of passive matter to be found anywhere in the known universe, nor any absolute attracting forces working tirelessly between them. Although *Expansion Theory* may initially seem to be presenting a completely foreign universe from the current Newtonian model in our heads, in reality it is not foreign at all, but the universe that we have been living in all along without realizing it.

At this point, the apparent "gravitational" effects caused entirely by atomic expansion have been thoroughly explored. Now we turn to the structure and nature of expanding atoms themselves. The atom has been extensively studied over the past century, yet no mention of such expansion can be found either in today's atomic theory or in past theories of the atom. How can it be that atoms are continuously expanding, yet science has taken no note of this fact? What implications does atomic expansion have for the atom's component elements – protons, neutrons, and electrons? What type of structure and behavior might lie within the atom in order to cause and support such constant expansion? This entirely new principle in nature requires that we rethink our current beliefs and models of the atom, as well as the nature of the subatomic realm and even the behavior of protons, neutrons, and electrons. This is the subject of the next chapter, which investigates the expanding atom.

- 4 -

Rethinking the Atom
and its Forces

In Chapter 1, we saw that Newton's concept of a gravitational force can only be considered a useful mathematical model, but violates our most fundamental laws of physics when seen as a true force pulling on objects at a distance. And, in reality, this proposed force-at-a-distance has never been directly measured or felt – not by astronauts on the orbiting space shuttle, nor by falling objects (once the effects of wind resistance are discounted). The only solid gravitational force that can be felt and measured occurs exclusively when in contact with a large expanding body, such as a moon or planet. It was also shown that Newton's whole theory of gravity is essentially based on a flawed "rock-and-string" analogy for orbits, combined with a pre-existing, purely geometric orbit equation. The concept of expanding atoms was then introduced in Chapter 2, and named *Expansion Theory*. This new theory states that Newton mistook the behavior of expanding atoms for an attracting force somehow endlessly emanating from each atom, while Einstein interpreted it as a mysterious warping of a "four-dimensional space-time fabric" around atoms. Chapter 3 furthered *Expansion Theory*, dealing with the relative motion of large objects composed of expanding atoms, showing how the resulting *Natural Orbit Effect* lies behind the many celestial observations we currently attempt to explain using Newton's gravitational force.

These three previous chapters have explored many of the large-scale effects of expanding atoms, but the *how* and *why* of expanding atoms themselves have not yet been addressed; the *concept* of expanding atoms may be an interesting idea, but what of the physical atoms themselves? What does it mean to say atoms expand? What does this say about atomic structure as we know it, and the component protons, neutrons, and electrons? How can all atoms be continuously expanding outward in all directions from within, yet also sit passively alongside one another as part of solid objects? Before any further *external* effects of expanding atoms are explored, the discussion must turn to the *inner* nature of expanding atoms themselves to address these issues.

The Atom

Flaws in Current Atomic Theory

Although our concept of the atom has undergone repeated refinements over the past century, the atom is still generally considered to be a tiny spherical object composed of protons, neutrons, and electrons in a manner somewhat analogous to the structure of our solar system. The protons and neutrons are packed into a central nucleus with the electrons zipping about – an organization reminiscent of the way the planets orbit our sun. And also like planetary orbits held in place by gravity, the zipping electrons are considered to be held within the atom by the *electric charge force* attracting the negatively charged electrons to the positively charged protons in the nucleus. The simplest atomic model with these elements is the well-known Rutherford-Bohr atomic model shown in Figure 4-1.

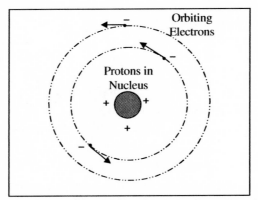

Fig. 4-1 The Rutherford-Bohr Atomic Model

Although this early model of the atom is considered overly simplistic today, its basic elements have remained essentially unchanged. Even in today's more complex quantum-mechanical models of the atom, the general concept of a nucleus containing positively charged protons and surrounded by negatively charged electrons still stands; it is largely only the details of how the electrons zip about that have been refined in today's atomic models. However, all of our atomic models overlook a number of serious problems with even their most elementary components.

VIOLATION

The Electric Charge Model Violates the Laws of Physics

Electric charge is considered to be a manifestation of the *electromagnetic force*, and as mentioned in the introduction of Chapter 1, this force is considered to be one of the four fundamental forces of nature. Yet, the concept of "positively charged" nuclear protons forcefully attracting "negatively charged" speeding electrons to constrain them within the atom is a physically unexplained abstraction – and one that also violates the *Law of Conservation Of Energy*. Despite our common use of the term "charge," we actually have no solid explanation for why electric charge would emanate from particles, why there are two types (positive and negative) that attract each other but repel their own, and what type of power source might be powering such behavior. In fact, the electric charge force between the protons and electrons in our atomic models holds atoms together without draining any known power source or weakening in strength – often for billions of years. Such a basis for atomic structure may provide a useful interim *model* of the atom, but cannot be considered the *literal* explanation without undermining our most basic tests of scientific credibility, as well as our most elementary laws of physics.

VIOLATION

The Strong Nuclear Force Violates the Laws of Physics

According to *Electric Charge Theory*, opposite charges attract one another, while like charges repel each other. Therefore, the close proximity of numerous positively charged protons in the nucleus of the atom should be impossible – the nucleus should fly violently apart due to the mutual repulsion of these positive charges in such close proximity.

This violation of *Electric Charge Theory* has been addressed by introducing yet another force into our science, known as the Strong Nuclear Force. This force is also considered to be one of the four fundamental forces of nature, and is said to be an attracting force between protons which only acts when protons are very close to each other, powerfully overcoming their large mutual repulsion within the nucleus. However, like electric charge, this proposed nuclear force comes with no clear physical explanation for its nature, has no known power source, and also acts endlessly for billions of years without diminishing in strength. Rather than resolving the problems with the electric charge model, the introduction of the Strong Nuclear Force only deepens the mystery, leaving us with *two* scientifically impossible claims supporting our current atomic models.

 The Mystery of Atomic Stability

Although various materials can differ in strength at the molecular level, the atoms that compose molecules are unimaginably strong and durable. Objects can be bent, melted, and crushed at the molecular level, but their component atoms withstand all but the most violent processes known to man – such as the explosion of an atomic bomb. Surely such amazing strength and durability could not arise from the "solar system" model of Figure 4-1, in which the electron's tendency to speed away is balanced by the pull of the electric charge from the nucleus – again, much like Newton's gravitational claim for orbiting moons and planets. If another solar system came speeding toward ours, we would not expect them to simply bounce off one another and maintain their separate stable structures. Instead, their delicate orbital balances would be completely thrown into chaos and disarray, which would also be the expectation for the "solar system" structure of the atom in Figure 4-1. This model cannot explain how atoms manage to bounce off each other or bond side by side within a solid object, often withstanding tremendous external forces without losing their delicate internal orbital balance.

Today, we usually replace this model with an updated quantum-mechanical model, which essentially states that electrons appear here and there about the nucleus with statistical probability rather than smoothly orbiting. This results in "electron probability clouds" or orbitals, which are regions where the electron might appear with a given probability (Fig. 4-2). The darker the region of the electron cloud or orbital in the diagram, the more likely the electron is to appear in that location. It is considered an inherent mystery how the electrons actually move within the atom, which does not help to solve the mystery of the atom's amazing strength and durability under crushing real-world conditions.

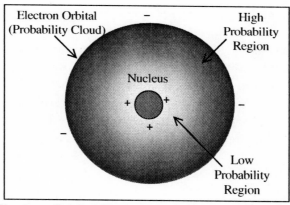

Fig. 4-2 Electron Orbital "Probability Cloud" in Quantum Theory

So then, if the concept of electric charge holding the atom together violates the laws of physics, a Strong Nuclear Force holding the nucleus together also violates the laws of physics, and our current models cannot explain the atom's strength, then what exactly *is* the physics of the atom?

A New Atomic Model

Expansion Theory states that all atoms are expanding by roughly a millionth of their size every second, but how does this translate into a scientifically viable atomic model? We know that neither electric charge nor a "strong nuclear force" are considerations as components of this new model, and that our current orbital and quantum-mechanical models are structurally unsound. Therefore, to arrive at a viable new atomic model, let's review the clues in the discussions so far:

- The concept of charged subatomic particles is not a scientifically viable explanation for atomic structure.
- The concept of a "strong nuclear force" holding the nucleus together is not a scientifically viable explanation.
- The orbital model of Figure 4-1 cannot explain atomic strength and stability, and quantum-mechanical modifications to this model introduce even more mysteries without resolving this issue.
- According to *Expansion Theory*, the overall atom is expanding by one-millionth its size every second.

Taken together, these clues suggest a new description of the atom as an extremely durable expanding structure composed of subatomic particles that have no electric charge, a nucleus with no Strong Nuclear Force, and a much more stable description of electron orbits. What type of atomic structure might fit this description? Let's begin with the all-important building blocks of the atom – its component subatomic particles.

Rethinking Charged Subatomic Particles

Since the atom is essentially nothing more than a stable organization of protons, neutrons, and electrons, it is critical that we understand the true nature of these subatomic particles if we are to arrive at the proper description of atomic structure. As mentioned earlier, we can no longer consider these particles to have "charge" since such a concept has no solid scientific explanation. Likewise, it is equally invalid to invent a "strong nuclear force" that somehow arises between nearby protons to hold the nucleus together. Therefore, we need some other explanation for electrons remaining in orbit about the nucleus, and for the nucleus itself holding together.

   A sizable clue can be found in the previous chapters, where orbiting objects and stable planets were explained by replacing the concept of a "gravitational force" with that of expanding atoms. This means that our overall planet is held together by the continual pressure of its countless atoms expanding against one another, while the overall expansion of the planet toward passing objects explains orbits. This same concept might also explain stable atomic nuclei and electron orbits without appealing to unexplained forces, provided that *subatomic* particles are similarly expanding. And, in fact, as we will see shortly, expanding atoms are merely a side effect of this deeper expansion principle – *the fundamental expansion of all subatomic particles.*

Subatomic Expansion

Subatomic particles are *charge-less* entities that expand at an identical *subatomic* expansion rate, X_S.

According to *Expansion Theory*, just as the atom expands at the universal *atomic* expansion rate of X_A, subatomic particles expand at the *subatomic* expansion rate denoted by X_S, while the concept of charged subatomic particles is simply a misunderstanding of this subatomic expansion principle. This would mean that protons, neutrons, and electrons, rather than being *charged* particles, are *expanding* particles. If this were the case, the mystery of nuclear stability would be solved since the protons would have no mutually repelling "positive charge," and their ongoing expansion against one another would cause a continual compressing force holding the nucleus together – just as *atomic* expansion holds our *planet* together. This would imply that today's belief in a "strong nuclear force" is merely a misunderstanding of the natural tendency for expanding protons and neutrons to cluster together in the nucleus.

Likewise, this same concept of subatomic expansion *could* be used to explain electron orbits – speeding electrons could orbit the expanding nucleus for the same reason speeding satellites orbit our expanding planet. However, there is one important difference between such orbiting electrons within the atom and orbiting satellites about a planet. We would not expect two colliding planetary systems to simply bounce off one another, with their orbiting moons and satellites remaining intact afterwards. Orbits are a delicate balance between spiraling downward and flying off into space, and are very easily knocked into chaos. Yet, atomic structure is extremely stable and robust, so even the expansion-based concept of orbiting electrons does not work. But this new principle of charge-less, expanding subatomic particles does allow another type of "orbit" that *would* be extremely robust, as discussed in the following section.

But first, we can now see that just as the concept of expanding atoms replaced the scientifically unexplainable "gravitational energy" in Chapter 2, the concept of expanding subatomic particles replaces the equally unexplainable energies related to "charged particles" and the "Strong Nuclear Force." These two mysterious forms of "energy" can now be seen simply as euphemisms for the previously unknown phenomenon of expanding subatomic particles. And, just as with atoms, subatomic matter *is* expanding matter – there is no such thing as non-expanding subatomic particles and no such thing as "energy" driving their expansion. These particles expand as part of their very existence – they quite literally *are* expanding particles and would cease to be protons, neutrons, and electrons if this were not the case. Material existence and expansion are one and the same for these particles, and the term "energy" (related to the concept of "electric charge" or "nuclear forces") is merely a label used in the absence of this understanding. We can now see that expanding subatomic particles answer the question of what drives expanding *atoms* (posed in Chapter 2), but it is equally natural to now wonder what drives expanding subatomic particles. For now, we can say it would be as meaningless to reintroduce the concept of "energy" driving *subatomic* expansion as it was for *atomic* expansion in Chapter 2. A deeper discussion of subatomic particles will be presented in Chapter 6.

Rethinking Electron Orbits

The new atomic model according to *Expansion Theory* suggests a new type of electron "orbit" based on expanding, rather than "charged," subatomic particles:

Electrons do not *orbit* the nucleus – *they bounce off of it.*

Let's consider a new type of atomic structure where the "orbiting" electrons don't actually orbit at all, but rather, bounce rapidly and repeatedly off the nucleus. This would be unthinkable in our current model of charged protons and electrons within the atom, but not in the new charge-free model. The mechanism behind this bouncing would be much the same as that for an actual bouncing ball on Earth. According to *Expansion Theory*, when a ball is dropped on Earth it actually floats in space while the expanding planet speeds toward it. When the planet strikes the ball it sends it flying off into space at a constant speed, only to catch up to it again shortly due to the planet's accelerating expansion, striking it again and sending it off into space once again. This repeated cycle is what we see as a bouncing ball as it *appears* to be repeatedly pulled back to Earth by a "gravitational force."

Likewise, if subatomic particles are expanding, the enormous nucleus (typically many thousands of times larger than an electron) would act just like our expanding planet, and the electron would repeatedly bounce off the expanding nucleus in the same manner as the bouncing ball just described. This would result in an extremely robust atomic structure. If such an atom is squeezed, the bouncing electrons would become confined to a smaller space, bouncing back and forth ever more vigorously the more the atom is squeezed. This would produce an increasing outward pressure to counteract the external force, resulting in a very stable, self-correcting mechanism that would keep the atomic structure intact under real world conditions. This simple and inherently robust atomic model is quite different from the delicate orbits of the Rutherford-Bohr model or the mysterious statistical nature of current quantum-mechanical models. Therefore, the *single concept* of subatomic expansion resolves the violations caused by the electric charge model and the Strong Nuclear Force, while also solving the mystery of atomic stability.

This concept of bouncing electrons also helps to explain our current quantum-mechanical model of electron "probability clouds," as shown earlier in Figure 4-2. We have arrived at this model as a result of probing the atom to find out where the electrons are and how they orbit at any given moment, finding that there seems to be no clear orbital path for these electrons. Instead, as mentioned earlier, it seems that they simply manage to appear in certain locations with greater probability than in others, but with no clear mechanism for how this occurs. However, once we consider the electrons to not be "charged" particles zipping *around* the nucleus, but *expanding* particles bouncing *against* it, we can see that our periodic probing of the atom is like "freeze-framing" a bouncing ball in the dark with a strobe light. In so doing, we would find the ball near the top of its bounce – where it is slowing down – far more often than at the bottom just before striking the ground. If we then plotted these results in terms of the probability of finding the ball at any given height without realizing that it is bouncing, we would have a mathematical "probability cloud" as in Figure 4-3, with a deep mystery as to how the ball moved from place to place. It is likely this same misunderstanding that underlies the apparent "quantum-mechanical" mystery of electron orbits within the atom, where our current theory of "oppositely charged" electrons and protons does not allow us to consider the possibility of the electron bouncing against the nucleus.

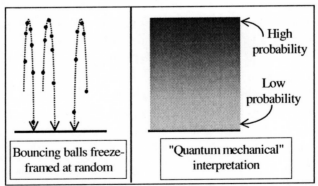

Fig. 4-3 Bouncing Electrons vs. "Quantum Probability Clouds"

Inner and Outer Dimensions

Since the proposed expansion rate of subatomic particles is unknown at this point, we are free to consider any expansion rate whatsoever – even one that vastly exceeds the universal *atomic* expansion rate, X_A. Later discussions will provide further evidence for this far greater *subatomic* expansion rate, which has the separate label of X_S. However, a far greater *subatomic* expansion rate would seem to introduce a sizable problem. It would mean that the nucleus would also be expanding at this tremendous rate, which further implies that the overall atom must expand at this rate as well in order for everything to remain in balance. However, we know that the atomic expansion rate, X_A, is a tiny fractional amount of roughly one-millionth each second. Such a tremendous disparity between inner and outer expansion would seem to be impossible – almost like the paradox of an inflating balloon that never grows in size. Logically, the insides of the atom cannot be bursting outward through space at a tremendous rate while the overall atom expands in space at a relative snail's pace. However, this is only a problem if the space within the atom is the same as the space *outside* atoms – *but what if it is not*?

Our current atomic models, such as in Figure 4-1 earlier, have a tiny nucleus surrounded by electrons at a distance, with a great deal of empty space within. So then, it may be tempting to picture expanding atoms to be growing ever larger in space, somewhat as if they were inflating balloons. However, the new atomic model presents a very different concept of the inner space of the atom. Studies of the atom over the years have indeed indicated that electrons are a great distance from the nucleus. Yet, modeling these large relative distances within the atom has led to the assumption that those distances are separated by empty space *as we know it*. That is, we tend to think of the space within the atom in the same manner as we think of the space that we move through in daily experience, but *Expansion Theory* suggests that this is not the case. To see this we must first take a step back and consider the nature of space itself.

EXPERIMENT

The Atom and Its Place in Space

A meter stick measures lengths and distances down to the finest divisions marked along its edge. Yet, we can measure even smaller distances with more sophisticated instruments and techniques, and in fact, there is no limit to how small a distance we can *conceive of* in our minds. This *conceptual* abstraction of dividing units into sub-units without end can be useful, but it also creates a somewhat false impression of space. Since we can imagine that space can be infinitely divided into ever-smaller segments, it seems that space-as-we-know-it must permeate the universe on all scales – even within the atom, as assumed in our current atomic models. However, as the following example shows, this is not necessarily the case.

For simplicity, let's assume we have determined that a typical atom is ten nanometers, or billionths of a meter, across (in actuality they are somewhat smaller still). In determining this ten-nanometer width we could not have physically measured ten nanometers across the atom as if we had 10 physical reference markers lined up across the atom, each being one nanometer in size, since the whole atom is already the smallest physical object known. There is no way to physically lay one-nanometer lengths across the atom so that they may be counted to total ten nanometers. Instead, we would have had to use some indirect method, such as performing an experiment to determine roughly how many atoms there must be along the length of a meter stick, then working out the size of each atom mathematically. In this example where atoms are ten nanometers in size, our experimental results would have indicated that there were roughly 100 million atoms along a meter stick, which would work out to an individual size of ten nanometers per atom (one meter divided by 100 million atoms).

It is important to note, however, that we haven't actually *measured* these ten tiny nanometers of space across the inside diameter of the atom, but have merely *deduced* that each atom must be ten nanometers across if there are 100 million of them along the length of a meter stick. And, in fact, we have never literally measured any distances *within* the atom at all. All sizes and distances within the atom that are quoted today have been deduced indirectly and labeled by an abstract division of our standardized units down to whatever arbitrarily small size was necessary. A simple example of this process is to now take our deduced atomic width of ten nanometers and make the further conceptual leap that the inner radius of the atom must then be five nanometers across. Although this is a perfectly logical conclusion, we have not actually *measured* this radial distance from the center of the atom to its edge. The danger here is that we have arbitrarily *assumed* that space-as-we-know-it *outside* the atom also applies *inside* the atom, without actually verifying this assumption. We merely logically deduced that the inner radius of the atom must extend across five nanometers of space since its radius must be half of its diameter – a diameter that was also only logically deduced. But neither of these stretches of space have actually been experienced or measured at all. However reasonable it may seem to make these logical spatial deductions, they still remain unsupported assumptions.

In actuality, our experience and definition of space has always been entirely in terms of the distances *between* atoms (and between objects composed of atoms), but never *within* them. The seemingly innocent logical deduction of the inner dimensions of the atom leads to the often-unquestioned assumption that the space we are familiar with *outside* atoms also applies *within* the atom. This leads to a conceptualization of space-as-we-know-it as a pre-existing volume within which electrons, protons, and neutrons congregate to form atoms, leaving the same empty space within the atom as outside the atom. Yet, space-as-we-know-it has only truly been experienced *outside* the atom. Therefore, in a very real sense, *atoms do not exist within pre-existing space, but rather, space-as-we-know-it exists between pre-existing atoms*. These two very different concepts of space are illustrated in Figure 4-4.

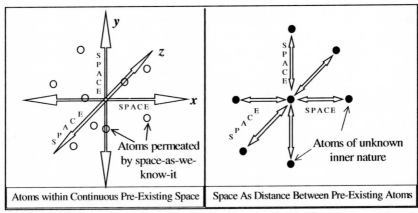

Fig. 4-4 All-Permeating Space vs. Space Only External to Atoms

The left frame of Figure 4-4 shows today's conceptualization of space. Empty space is shown as a continuous volume in all directions, with atoms existing within this space and permeated by it – seen here as hollow spheres whose insides contain the same space as that outside their edges. The protons, neutrons, and electrons within these atoms exist and move about within regular space-as-we-know-it, forming into atoms with nuclei and orbiting electrons within regular space.

However, the right frame of the diagram shows a very different concept of space. The atoms are shown as obscured spheres whose inner realm is of unknown nature, and with space-as-we-know-it only existing *between* these atoms. As beings composed of atoms, we have never known a time when atoms did not exist, and since the universe can essentially be described as atomic matter separated by empty space, space-as-we-know-it today can only be accurately described as *the space between atoms*. We have created models of the atom that show regular, familiar space within – such as in Figure 4-1 – but these are only hypothetical models that are inferred or deduced from experiments. Although it is a natural assumption that the realm within atoms would be no different than the world outside them, this has not been definitively proven to be the case. So, what if the inner atomic realm is actually very different from space-as-we-know-it here on the outside? Why might this be so? *How* could this be so? And why would we even contemplate such a possibility?

One reason to contemplate this possibility is precisely because it *is* a possibility. As just mentioned, there is no definitive reason to claim that the inner atomic realm must be just like the world outside it. It is widely acknowledged that today's atomic theories, which make this assumption, are quite mysterious and bizarre. These models may accurately agree with experimental results, but it could hardly be otherwise after undergoing nearly a century of refinements to ensure this is the case. This process of continual refinement toward agreement with experiment is typically represented as evidence that our models correctly capture the bizarre nature of our universe – but in actuality they have been deliberately engineered for experimental agreement while becoming increasingly mysterious and providing little in the way of clear physical understanding. So then, let's explore the possibility of a very different inner atomic realm.

NEW IDEA

The space within the atom has its own independent nature, and is not a mere extension of space-as-we-know-it outside the atom.

Returning to Figure 4-4, recall the right-hand frame where our experience of space is actually only the space *between* atoms, saying nothing about the inner atomic realm. In this case, which actually *is* the reality of our spatial experience, the inner atomic dimension could theoretically have any nature imaginable – or even unimaginable. It could be a realm composed of brilliant swirling colors, or new and exotic forms of energy, or additional dimensions beyond our comprehension – or expanding subatomic particles bursting outward at tremendous rates within an entirely separate inner atomic space. All of these possibilities describe only *internal* mechanisms that might underlie the existence of the atom, whose existence then gives definition to our regular experience of space outside the atom – *regardless of the atom's internal details*.

Although we can imagine a universe of completely empty space without atoms, this can only be a figment of our imagination since a universe filled with atomic matter is a prerequisite for *us* to exist so that we can imagine such a thing. Also, the nature of this empty universe in our imagination can only be based on our prior experience of the empty space between atoms, which again requires that atoms already exist. It is natural to attempt to frame our early investigations of the *inner* atomic realm in terms of this familiar *external* experience of space, but we must also be prepared for all such attempts to fall short of a true understanding of the atom. Such straightforward atomic models that simply mirror our experiences outside the atom are a logical first step; however, so far, they have resulted in precariously delicate (Rutherford-Bohr), unexplainably mysterious (quantum-mechanical), and physically law-violating (electric charge) descriptions of the atom. *Expansion Theory* now suggests that we consider an entirely separate inner atomic space where charge-less subatomic particles continually expand outward at a tremendous rate simply to support the stable atomic structure we know from the outside. This view gives the rather paradoxical result that the seemingly passive atoms composing all known objects are actually based on an inner core of enormous ongoing outward expansion – a paradox that is explained by the separate natures of internal and external space.

This separation between the inner and outer dimensions of the atom means that even though *Expansion Theory* states that the atom is supported by a tremendous outward expansion from within, *this does not cause the overall atom itself to expand.* As mentioned earlier, the inner dynamics supporting the mere *existence* of the atom as a simple sphere in space could have been any phenomenon imaginable (or unimaginable) – including the outward expansion just described – without imparting any particular external quality to the atom. The internal expansion of the atom does not occur within space-as-we-know-it, and therefore, does not *consume* space; instead, it merely supports the overall structure of the atom, which then *defines* space-as-we-know-it *outside* the atom (Fig. 4-5). Even with its enormously expanding inner realm, the atom would still only be a simple, non-expanding sphere in space from our outside perspective – were it not for an additional mechanism that will be discussed shortly.

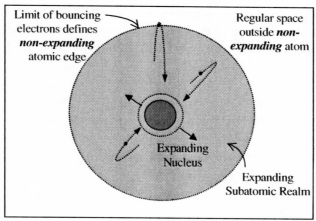

Fig. 4-5 Paradox: Inner Expansion without External Expansion

This discussion of inner and outer atomic dimensions now allows us to complete the new atomic model. Recall that we had redefined subatomic particles as charge-less expanding entities, electron orbits as electrons bouncing rapidly off the expanding nucleus, and were only left with the disparity between the enormous subatomic expansion rate and the tiny expansion rate of the overall atom. This last point now has a resolution in the fact that the inner atomic space can be an entirely separate realm from space-as-we-know-it outside the atom. This allows the apparent paradox of a rapidly expanding inner core supporting an atomic structure that has only a minute external expansion (which, again, will be explained shortly).

It may seem that the ongoing inner expansion of the atom would mean that it could have no particular fixed size; yet, atoms *are* known to have a fixed microscopic size. The reason for this is actually very straightforward. Although nature does not give the atom any particular absolute or fixed size, it *appears* as if atoms are of a particular microscopic size simply because it takes so many of them to compose the human body. Since many trillions of atoms are required to compose our bodies, it follows that individual atoms will always be trillions of times smaller than we are – hence their apparently fixed tiny (relative) size. As mentioned earlier, the atom has no *absolute* size in the sense of having a true measured inner diameter across regular space; it only has an *externally deduced* size, obtained by indirect methods of comparison or theoretical calculation. We naturally mistake our familiarity with atomic objects and the regular space between them for a familiarity with the atom itself and the space within, but since the true nature of the atom has been sizably misunderstood, it is actually quite foreign to us today. However, *Expansion Theory* now allows us to truly begin to understand the atom and its subatomic particles. The only remaining element of the new atomic model to be explained is the tiny external atomic expansion rate – the phenomenon that we know as gravity.

The Birth of Gravity

Although the atom's inner expansion is not the direct cause of the tiny external atomic expansion rate, it *is* indirectly related by way of a side effect that essentially "leaks out" at the edge of the atom. To see how this occurs, recall (from the development of the *Atomic Expansion Equation* in Chapter 2) that in *Expansion Theory* the geometry of two expanding objects is defined by their outward radial expansion toward each other from their centers. Therefore, in the new bouncing-electron model of the atom, the atomic radius is defined by the expansion dynamics between the *center* of the expanding nucleus and the *center* of the outermost bouncing electrons. However, this center-to-center expansionary dynamic between nucleus and electrons leaves the *outer half* of the bouncing electrons to lie *outside the definition of the atom*. That is, as far as atomic structure is concerned, the geometry of the atom is complete once the expansionary dynamics between the center of the nucleus and the center of the electrons are taken into account. As such, the *outward*-facing expansion of the outer half of these electrons does not contribute to the inner geometry of the atom, but *does* add a half-electron *external expansion* to the otherwise passive atom, turning it into an *actively expanding atom* (Fig. 4-6).

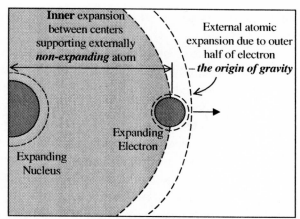

Fig. 4-6 Atomic Expansion (Gravity) due to Outer Half of Electron

This expansion of only the outer half of the atom's electrons into the outside world is the *sole reason* atoms exhibit the tiny external atomic expansion rate, X_A, introduced in Chapter 2. This tiny external atomic expansion, accumulated over the countless atoms composing our planet, is ultimately the cause of the *Earth's* expansion, which in turn creates the effect we call gravity. Recall that we calculated the value of this atomic expansion rate in Chapter 2 by dividing the distance all objects "fall" in one second by the overall size of the planet, since this actually calculates the fractional expansion of the planet, which must match the expansion rate of its component atoms.

However, such a straightforward relationship between an object's expansion and its size reaches the limit of its usefulness when brought down to the level of an individual atom. Each atom, or element, in the periodic table has a different number of protons, neutrons, and electrons, and can vary greatly in overall size. The atoms of some elements are many times the size of others, yet the size and expansion rates of their electrons do not vary – all electrons are identical, as is the universal subatomic expansion rate of all subatomic particles, X_S. Therefore, simply dividing the expansion of the outermost electron by the overall size of the atom would seem to give a different expansion rate for each element in the periodic table. Of course, this is impossible, since all atoms must expand at the same rate – otherwise, some would grow enormously and others would shrink into oblivion relative to each other over time.

The resolution to this issue once again lies in the fact that the inner atomic realm is not simply an extension of regular space outside the atom. Recall that the logical assumption that a ten-nanometer-wide atom must have a five-nanometer radius is not the same as actually getting inside the atom and measuring its radius. There are not necessarily five nanometers of *regular space* stretching between the center of the atom and its edge. Instead, our external concepts of space, size, and distance must stop at the atomic boundary and give us pause before being applied within the atom. This leaves us with the rather paradoxical fact that we can *calculate* the effective size of an atom from the *outside* based on the location of its outer edges in space, but we cannot actually *measure* the distance across the atom *internally*. The atom does effectively take up a certain amount of space from our outside perspective, yet it does not have an actual size across its diameter in the sense that regular objects do. Therefore, we cannot simply divide the expansion of an electron by the apparent size of any given atom in the periodic table to arrive at the universal atomic expansion rate. All we can say is that the outer surfaces of all atoms have the same external half-electron expansion effect, giving all atoms the same universal expansion rate.

This discussion shows that gravity is not a mysterious energy – or "gravitational force" – emanating from atoms and subatomic particles, as believed today, but is simply a rather mechanical secondary effect of the expansionary structure of the atom as a whole. This is why it is entirely appropriate to state that gravity *is* atomic expansion, since it is purely a result of the mechanics of the overall expanding atom – not a mysterious force somehow emanating from both atomic and subatomic matter.

Chemical Bonding

Chemical bonds between atoms function as the "glue" that holds our universe – and everything in it – together. The very existence of all solids, liquids, and gasses depends on chemical bonds, as does life itself via organic molecules made up of atoms held together in various configurations by chemical bonds. Although there are several varieties of chemical bonds, the basic idea is that if an atom loses one or more of its negatively charged electrons it no longer has equal numbers of electrons and protons, tipping its usual neutral charge balance toward an overall positive charge from the nuclear protons. Such an atom with missing electrons is known as a positively charged *ion*. This net positive charge on the atom is said to attract the negatively charged electrons of other atoms, creating a lasting bond between the atoms.

This electric charge model of chemical bonds has served us well – as a model, but introduces sizable mysteries and law violations when considered to be the literal description of atomic bonding. Several immediate problems can be seen with the core concept of positive and negative charge balance within the atom, since it was shown earlier that the concept of charged subatomic particles is an unexplained abstraction that also endlessly expends energy with no known power source. To see just how pervasive this core problem is, we need look no further than common objects in the world around us.

The Mysterious Force Within

Solid objects are perhaps the most familiar and easily understood form of matter. Unlike the complexities of liquids and gasses, solid objects are merely hard, inanimate lumps of matter. When we push on a solid object we feel its substance as a pressure or force pushing back against us. In classical physics this effect often conjures up the phrase "For every action there is an equal and opposite reaction." That is, in a very real sense, when we push on a solid object it resists and pushes back with an equal force from within. We are so accustomed to this effect from countless interactions with solid objects that we typically overlook the fact that it actually presents a deep mystery that violates the laws of physics.

To see this, we merely need to consider a world where atoms are all simple, charge-less spheres. Solids made up of such atoms would have no chemical bonds, and would be extremely delicate – the slightest touch would cause them to crumble into a fine atomic powder. However, there is no reason why the atoms in this hypothetical world *must* be charge-less, as long as they have an identifiable power source within that produces their charge and also drains as it expends energy to hold atoms together. In this case, it would be perfectly understandable if objects offered resistance from within when an external force is applied, maintaining their solid form by drawing on their internal power source to fortify their atomic bonds as needed. This would be the "equal and opposite reaction" to the external force.

VIOLATION

Solid Objects Violate the Laws of Physics

However, these tiny power sources within each atom must weaken over time as they continually battle the external force – in accordance with the *Law of Conservation Of Energy* – eventually resulting in the object deforming or crumbling apart entirely. Yet, neither the scenario of delicate objects composed of charge-less spheres nor solid objects with internal atomic power sources describe a typical solid object in our world. The world we are familiar with has solid objects with *endless* energy supporting their atomic bonds, and *no known power source* to explain the origin of this unending energy. We know of no power source supporting the bonds between the atoms of a table, for example, yet we expect these bonds to work endlessly to keep the table together – even if heavy objects rest on top of it indefinitely. We can hold a block of wood and squeeze it with all our might, but our muscles will tire while the ability of the wood to push back from within is not the slightest bit diminished. This is an impossible "energy-for-free" scenario that violates the laws of physics. This apparently impossible situation is not truly a violation of natural laws of course, but appears as such because of the equally impossible notion of "electric charge" underlying chemical bonds in today's atomic theory. Once again, as an abstract model this concept is very useful, but if we want to truly understand the physics of the atom and of chemical bonds we must seek a deeper truth. And, as the following section shows, similar mysteries can be found in liquids as well.

The Mystery of Freezing Water

The *Law of Conservation Of Energy* is essentially a law of energy balances – energy *in* must equal energy *out*. Therefore, to stay within the laws of physics, any system that does work or expends energy must be supported by an energy input that drives this process. For example, a balloon left sitting in the sun will heat up internally and expand. The balloon's expansion is an example of a system doing work – according to the proper application of the *Work Function*, which states that *work* equals *force* times *distance*. That is, the pressure of the heated gas within pushes the skin of the balloon outward by a certain distance, doing the work of stretching the balloon as well as pushing outward on the surrounding atmosphere and on anything that may be in contact with the balloon. This work and energy expenditure is balanced and driven by the input of energy from the sun. Alternatively, when night falls and the surrounding air cools, energy is *removed* from the gas in the balloon, its internal pressure and energy content drop, and the surrounding atmosphere now does work on the balloon instead, shrinking it to a much smaller size. It would be completely unexplainable if the balloon instead continued to expand once the sun set, growing even larger and doing even more work while internal heat energy was continually *drained* from it. This would be the very opposite of an energy balance, with only energy output and no input, and would violate the *Law of Conservation Of Energy*.

VIOLATION

     Freezing Water Violates the Laws of Physics

We know, of course, that balloons do not violate the laws of physics in this manner, yet *freezing water does*. It is well known that water expands when it freezes, but since this is a very common experience we tend to overlook the deep mystery this raises when checked against our laws of physics. Cooling water means its energy is being continually lost to the surrounding environment as its atoms move about more and more slowly. This decreasing internal energy balances with the energy lost to the surroundings – a situation that continues until the water reaches zero degrees Celsius, at which point tremendous, unstoppable energy emerges from within the newly forming ice with a force that can easily burst metal pipes. This is a tremendous amount of work and energy expenditure from within the cooling water, yet not only is there no energy input to account for this, but energy is actually being *removed* from the water all the while. There are no explanations for this mystery in our science today that stand up to logical scrutiny. Both solid objects and freezing water are completely unexplainable today due to our flawed belief that the electric charge *model* of chemical bonds also describes a literal "electric charge" *property* that exists in nature.

Solving the Mysteries – Rethinking Chemical Bonding

The new atomic model according to *Expansion Theory* is not based on "charged" electrons and protons, but on their continual expansion as the electrons bounce off of the nucleus. And when two "neutral" atoms with all their electrons meet, their bouncing electrons make no distinction between either atom. Any given electron at the top of its bounce might find itself briefly located between both atoms, and would be just as likely to "fall back" toward the nucleus of *either* atom. Although, on average, each atom would remain essentially separate and unchanged by this exchanging of electrons, this does represent a degree of overlap or ambiguity in their separate identities, which could explain weak associations or bonds between some types of "neutral" atoms, known as *van der Waals forces*. That is, the randomly bouncing electrons would rarely exchange places between atoms at exactly the same time, so at any given moment – and for the briefest period of time – either atom could have an extra electron while the other atom lacked an electron. This would cause a very weak "ionizing" effect that would allow the subatomic realms of the two atoms to essentially be weakly "attracted" to each other in a greatly diminished version of the typical atomic bonding process, to be described momentarily.

Further, if these atoms did not simply meet gently, but collided forcefully, they would continue to move closer, causing their exchanging electrons to bounce ever more vigorously within the decreasing distance between the two nuclei. This increasingly vigorous bouncing would act as a repelling force that would either keep the atoms apart and separate or cause them to bounce off one another entirely. This explains the stability of solid objects, solving the mystery of the endless force that pushes back from within. What we are actually feeling is the expansion force of countless electrons bouncing vigorously between atomic nuclei – a process that is not driven by "energy." The various forms of "energy" discussed so far – the "gravitational force," "electric charge," and the "Strong Nuclear Force" – are merely misunderstandings of the fact that we do not live in a world of passive matter pushed or pulled about by unexplainable forces, but simply a world of *actively expanding matter*. Matter and expansion are one and the same, while the term "energy" is an abstraction invented in an attempt to model expanding matter without having any knowledge of it.

As a result, laws such as the *Law of Conservation Of Energy* are not true laws of nature at all, but merely man-made rules-of-thumb that attempt to turn our artificial abstractions into universal truths. Once we realize this, we can see that the apparent law violation of freezing water is also not actually a violation of a law of nature, but merely one of many perfectly natural exceptions to our artificial rules-of-thumb. The standard explanation for the *mechanics* of freezing water – a forceful realignment of water molecules into crystals that take up more space – does not change, but this process is driven by subatomic expansion, not by an unexplained "energy" from within. The geometry of individual water molecules is such that subatomic expansion effects occur between them when they near each other, tending to pull them together and lock them into the particular geometric alignment that forms solid ice. At room temperature, the internal atomic vibrations within water are too vigorous to allow this natural crystallization to occur, keeping ice from forming. The forceful crystallization of water into ice due to subatomic expansion once the temperature drops is similar to the manner in which static electricity causes objects to attract (or repel) one another – an effect that will be explained in depth when it is introduced shortly as the *crossover effect*.

Now, as mentioned earlier, the typical chemical bond between atoms occurs when an atom is *missing* an electron, becoming a "positively charged" ion. However, *Expansion Theory* shows that it is not the exposed "positively charged" nucleus that attracts "negatively charged" electrons from another atom; rather, there are now simply fewer bouncing electrons to repel other atoms. As a result, a second atom can approach much more closely, which also allows one of its electrons to readily bounce across to the first (ionized) atom. This creates a situation where the atoms are not merely *exchanging* one electron for another, as in the above discussion of two "neutral" atoms, but are now *sharing* an electron between them to complete both of their structures. As a result, the atoms no longer experience the double bouncing force of two electrons merely changing places as they rebound back and forth between the nuclei, but only the singular repelling force of one shared bouncing electron. As such, two separate spherical atoms with their own separate internal realms have now moved closer and partially merged to become one extended atomic structure with a common internal realm, sharing a common bouncing electron (Fig. 4-7).

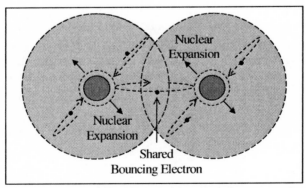

Fig. 4-7 Chemical Bond: Electron Separates Common Inner Realm

The two atoms in the diagram are not merely resting against each other, but have now formed a single, stable atomic structure – a powerful chemical bond. The strength of this bond lies in the fact that the nuclei are still expanding rapidly outward and toward one another, and would soon expand into one another if they weren't kept apart by the vigorously bouncing electron between them. Recall that such tremendous ongoing expansion occurs within the inner atomic realm completely independently from the tiny overall atomic expansion (gravity) caused by the outer half of the bouncing electrons. Therefore, there is a tremendous effective "attraction" between these atoms since the missing electron provides an opportunity for their nuclei to rapidly expand toward each other within a common inner atomic realm, with their shared bouncing electron keeping them apart. This effect is somewhat like two boxers who have been allowed to get too close to each other, with a referee struggling to keep them apart.

Electricity

Although electricity is considered to be a form of energy known as *electrical energy*, we also know that electrical circuits do not actually have pure "electrical energy" flowing through them, but rather, the physical electrons *themselves* flow through circuits. It is thought that electrons are negatively charged and that this flow of electrons carries a river of charge along with it through the circuit to power the various circuit components. However, once again, closer examination shows that this is only a superficial abstraction that does not explain the physics involved at all. For example, all electrons are considered to have an identical negative charge upon them that *never* changes – this is considered to be one of the key properties that defines or identifies an electron. So then, if this flow of electrons simply moves through the circuit without slowing down and without the electrons losing any charge, how can energy possibly be imparted to the circuit components? How can simply moving charges through a circuit power its components when charge simply moves along with the electrons through the circuit and out the other end, unchanged and intact along with the electrons? This question is never addressed today since there *is no answer* – it is a great mystery that is overlooked in favor of a set of abstract mathematical models known collectively as *Circuit Theory*. This self-consistent system of abstractions is designed to help us *model* electric circuits to create useful devices, but should not be mistaken for an explanation of the *physics* involved.

Since, even today, we recognize that current flow involves the flow of physical electrons, we can see that electricity is actually a manifestation of raw subatomic particle behavior – i.e. freely flowing electrons outside the atom. And it was shown earlier that, according to *Expansion Theory*, the subatomic particles within the atom are not somehow "charged," but rather, they are *expanding*. Therefore, it is logical to assume that these particles would not suddenly lose their core expanding natures and become non-expanding "charged" particles once they leave the atom. This means that protons and electrons are *always* charge-less expanding particles whether they are found inside or outside the atom, which further means that *the concept of charged particles – and indeed electric charge itself – does not exist.* This very important point deserves special note:

NOTE

 Neither charged particles nor electric charge actually exists anywhere in nature.

The above statement means that we actually have no explanation for electrical circuits today, despite the appearance of understanding created by our invention of *Circuit Theory* as a higher level abstract overlay. As shown in the previous chapters on gravity, as long as we only function at the level of our abstract models and do not question them too deeply, it is possible to overlook our lack of *physical* understanding. However, *Expansion Theory* allows us to look beyond our abstractions and to truly understand our physical world. This deeper understanding of electricity and electric circuits will be explored shortly; however, first it is important to return to the origin of the electric charge concept itself to see how it arose and what *Expansion Theory* has to say about this issue.

Electric Charge (Static Electricity)

As mentioned earlier, electric charge is considered to be a force that emanates from subatomic particles and comes in two varieties – positive and negative. This charge is said to hold atoms together, as well as holding molecules and objects together via chemical bonds. Charge is also thought to build up on the surface of objects in certain circumstances, surrounding them with static electric fields – a phenomenon known as static electricity. We currently have no scientific explanation for precisely what charge is, what makes a positive charge "positive" and a negative charge "negative," and how electric charge continues to do its work without weakening or even drawing on a known power source. Perhaps we can find some clues by investigating the origin of our current beliefs in this mysterious, law-violating phenomenon. To do this, we turn to the work of Benjamin Franklin.

Benjamin Franklin's Error – Violation of the Laws of Physics

It has long been known that certain objects can be influenced to take on qualities that cause them to either attract one another or repel each other. This is a very common observable phenomenon in nature, but one that did not have a clear explanation until Benjamin Franklin (1706-1790) performed a series of experiments to investigate this issue.

EXPERIMENT

 Benjamin Franklin's "Electric Charge" Experiments

Several centuries ago, Franklin showed that rubbing two suspended glass rods with silk caused them to forcefully repel each other. He concluded that the glass rods now each possessed a quality he called *charge*, which somehow repels other charged objects. Franklin then rubbed two rods made of wax (usually plastic today) with fur and found the same mutual repulsion between them, but he further noted that the wax and glass rods were attracted to each other (Fig. 4-8). From this he concluded that the wax rods must have a different type of charge from the glass rods, which, for some reason, results in an attraction between them. A series of such experiments solidified the concept in Franklin's mind that there must be two types of charge in nature, which can either cause attraction or repulsion between objects. Since the effects of attraction and repulsion are opposites, he considered these two types of charge to also be opposites, borrowing the terms "positive" and "negative" – which are easily recognized opposites – and inventing the concept of *positive and negative charge*. Franklin's results can be summed up in the well-known statement *"Like charges repel and opposite charges attract."*

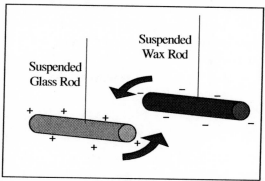

Fig. 4-8 Franklin's Charge-Based Interpretation of Attracting Rods

At the time of Franklin's experiments the concept of expanding subatomic particles was unknown, and the mysteries and violations associated with the concept of charged particles were not widely recognized, nor of primary concern. It was more important to develop a useful explanation for his observations, upon which technological advancement could be built. As a result, all experiments and theories related to atoms and subatomic particles today are unquestioningly framed within the interpretation made by Benjamin Franklin and his glass and wax rod experiments centuries ago.

VIOLATION

    Electric Charge Violates the Laws of Physics

Yet, as simple and useful as Franklin's conclusions are, the concept of "positive and negative charge" can only be considered as an hypothesized abstraction or model unless it lies within the laws of physics – and preferably also has a clear physical explanation for its nature. But Franklin's "charged" rods would have continued to repel or attract each other hour after hour, day after day, and week after week without these forces showing even the slightest sign of weakening (assuming no other influences affected the rods). Further, this energy expenditure would have had no recognized power source to explain its origin. In fact, Franklin never explained *why* like-charged rods should repel each other; he merely gave the same circular reasoning that we use even today – that charged rods repel each other because they are charged. Yet, this is like saying dropped objects fall because they are dropped. Neither statement truly explains why charged rods repel each other or why dropped objects fall. While it is understandable that Franklin's breakthrough idea has become a widely accepted and extremely useful model for this phenomenon, we still have no physical explanation for the nature of "positive and negative charge," and it still remains a clear violation of the *Law of Conservation Of Energy*.

Rethinking Electric Charge

As mentioned in the discussion of the new atomic model, *Expansion Theory* shows that subatomic particles are not "charged," but rather, they are *expanding* entities. This not only applies within the atom, but outside the atom as well. This means that every observation currently attributed to electric charge is actually solely a manifestation of subatomic expansion. With this in mind, let's revisit Benjamin Franklin's experiments.

EXPERIMENT

Expansion and the Franklin Experiments

The *electric charge* interpretation of the repelling wax rods states that the rods have a surplus of negatively charged electrons on their surfaces from which a negative electric field emanates, somehow repelling the negative electric field from the other rod. And, as already mentioned, this effect is a completely unexplained hypothesis even today. *Expansion Theory*, on the other hand, states that the rods are objects that have a surplus of *expanding* electrons on their surfaces, whose tremendous subatomic expansion rate causes them to expand rapidly outward in a growing electron cloud. That is, there is no mysterious "negative electric field" emanating from "charged" particles, but rather, the *expanding* particles themselves move outward into space. The repelling force between two such objects is merely the force of these two expanding electron clouds pushing against one another (Fig. 4-9).

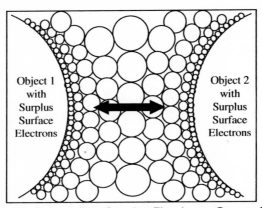

Fig. 4-9 Freely Expanding Surplus Electrons Cause Repulsion

Figure 4-9 shows a straightforward physical explanation for the repelling force between the objects, as well as a new concept that can only be properly understood by recalling the earlier discussion of the different inner and outer atomic realms. This new concept – the *crossover effect* – is discussed below, while the straightforward element in Figure 4-9 is simply that two clouds of expanding electrons meet in the middle and push against each other, causing the repelling force that keeps the objects apart indefinitely. As mentioned in previous discussions, electrons expand by the very nature of their existence, and are not driven by our current abstraction called "energy." Therefore, the phenomenon currently believed to be "electric field energy" endlessly pushing the objects apart in violation of the laws of physics is simply a misunderstanding of the expanding nature of electrons, which push between the objects in freely expanding clouds when outside the atom.

This growth of electrons in the atomic realm has the effect of creating a barrier that keeps them from simply entering back into the microscopic atomic structure of the two objects, since they have lost much of their subatomic definition to their freely growing behavior in the atomic realm. The resulting repelling force that occurs when the two electron clouds meet to push on each other represents the end of the free growth of the electrons at the atomic level and a return to their subatomic definition as identical particles undergoing mutual subatomic expansion against one another. This redefinition occurs because the two clouds now form a continuous "matter bridge" of subatomic particles stretching between the subatomic realms within each of the objects from which they radiate. This results in the two electron clouds essentially becoming one continuous bridge of electrons all expanding against one another and repelling the two objects. This is what is currently thought of as a repelling force due to a "negatively charged" electric field emanating from the surplus electrons at the surface of the objects.

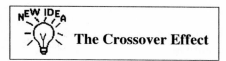

The new concept referred to in the above discussion, which can also be seen in Figure 4-9, is the *crossover effect* between the *subatomic realm* within the atom and the *atomic realm* outside the atom. The earlier discussion of the inner and outer atomic realms addressed the separate natures of these two realms, where the subatomic realm could have tremendous inner expansion while paradoxically supporting a largely *non*-expanding atomic sphere externally. In fact, support for this concept can be seen in the above explanation of the repelling objects, which requires that the externalized electrons have a far greater growth rate than the tiny expansion of the atom. If this were not the case and the subatomic expansion rate of electrons, X_S, was actually the same tiny rate as for the overall atom, X_A, the externalized electrons would merely sit atop the atom in the same manner that static piles of rocks do on Earth. A pile of rocks expands at the same rate as the Earth, and so, effectively just sits there as a static feature and does not expand rapidly out into space. Therefore, the tremendous subatomic expansion rate of electrons is observable by the fact that they do expand rapidly out into atomic space, repelling the two objects.

This also shows that the separate atomic and subatomic realms do not always remain separate, but can interact via the crossover effect when subatomic particles are liberated from the subatomic realm. The surplus expanding electrons on the surface of the two objects in Figure 4-9 are an example of this effect, where the electrons initially expand rapidly outward but soon meet between the objects and strike a balance between their expansion force and the effort of pushing the objects apart. At this point, it seems the electrons somehow cease their natural expansion and settle as particles of varying fixed sizes within the overall electron cloud between the objects. Yet, if this were truly the case, then all electrons would not be identical – especially considering the tiny electrons within the atom with their ongoing expansion – and it would seem that a deep principle underlying our universe (subatomic expansion) is easily brought to a halt.

Of course, electrons *are* all identical, and we *cannot* so easily tinker with the core machinery of our universe. Electrons cannot truly vary in size and expansion rate, but must all be identical in these respects in order for there to be order and stability in our universe. Instead, the answer lies in the difference between inner and outer atomic realms. That is, it only appears that the electrons in the electron cloud have stopped expanding when we view the electron cloud as an *atomic* object. In actuality, the electron cloud is a *subatomic* object (composed of electrons which are always *exclusively* subatomic particles) that is stretching across atomic space between the subatomic realms within the atoms of each object. As such, the apparently static electrons within the electron cloud are actually still expanding at the same universal subatomic expansion rate as those within the atom, and they are also the same size as well. This apparent paradox is a manifestation of a new concept called the *crossover effect*, where the identical ongoing expansion rate and unchanging relative sizes of all subatomic particles can have very different manifestations in the atomic world when these particles are displaced from their usual subatomic realm.

To distinguish between subatomic and atomic behaviors, the term '*expansion*' will continue to refer to the single universal expansion rate of all subatomic particles in the subatomic realm, while the terms '*growth*' and '*shrinkage*' will be used to describe the *crossover effect* of their manifestations in the atomic realm. With this understanding, we can now say that Figure 4-9 shows free electrons that are effectively microscopic in size at the surface of the objects – where they are closely associated with the subatomic realm within the surface atoms – and grow in size the further they extend out into atomic space. We can also say that these electrons have effectively halted their *growth* and are pushing outward, creating the repelling force between the objects. As far as the electrons within the cloud are concerned though, they are still in the subatomic realm among identically sized, continually *expanding* electrons. The enormous growth of these electrons relative to those still within the atom, and the resulting repelling force between the objects, are only atomic-level side effects of the unchanging nature of these expanding subatomic particles. Also, it is worth noting that:

NOTE

 This explanation for the repelling force between "charged" objects is the *only known explanation* for this effect; today's science offers neither a clear physical mechanism nor a scientifically viable explanation for *how* or *why* "charged" objects repel each other.

Of course, "charged" objects do not only repel one another, but can also attract each other as well. Benjamin Franklin believed that attraction occurred when one object had a surplus of "negatively charged" electrons on its surface, while the other object had a deficit of electrons, giving it a "positive charge." This can only mean that the normal complement of electrons surrounding the atoms of the object has been reduced, thus ionizing the atoms and allowing the "positive charge" of the nucleus to emanate out of the atom to forcefully pull on distant objects. Yet, an atom stripped of one or more of its core structural electrons, while also continually expending its "positive nuclear energy" to pull on distant objects, would seem to be an atom in a rather perilous and unsustainable situation – a situation created merely by rubbing the object with silk. And, of course, the endless attracting effort expended by the "positively charged" nuclear protons is completely unexplained in today's scientific paradigms, as is the very reason why "positive charge" should attract "negative charge" at all. Again, Benjamin Franklin's "charge" concept has served us well as an abstract model, but it cannot truly be the physical explanation.

Expansion Theory suggests a mechanism for attracting objects that follows along similar lines to the expansion-based explanation for repelling objects. In both cases, "charging" the objects by rubbing them deposits an *excess* of electrons on their surfaces. However, objects of differing materials may have different characteristic distribution densities of the excess electrons on their surfaces. This would result in different types of electron clouds growing out into space from the objects' surfaces, with some being denser than others. And, since all electrons are actually identical subatomic particles in the subatomic realm, their free growth into space in the atomic realm must still tend toward this equality. This reigns-in the free growth of electrons since neighboring electrons can only differ in size by a tiny amount in the atomic realm without straying too far from their identical size definition in the subatomic realm. This effectively creates a tension between their freedom to grow without definition in the atomic realm and their true natures as identical subatomic particles – another manifestation of the *crossover effect*.

So, there can only be the most gradual increase in size as the electrons fan outward in a cloud from each object. This means that two electron clouds of different densities will fan out differently. A denser cloud will be more free to expand further out into space without violating the close size association between neighboring electrons since it has many more electrons to stretch across any given length – its electrons are basically on a longer leash. Also, a denser electron cloud will grow outward with more force, tending to push any less dense clouds back. Therefore, two electron clouds of differing densities will not meet at the mid-point between the objects from which they originate, but rather, the more dense cloud will extend well beyond the mid-point. This means that the electrons at the outer edge of the more dense cloud have been able to grow over a longer distance, and are, therefore, larger (in the atomic realm) than those of the less dense cloud (Fig. 4-10).

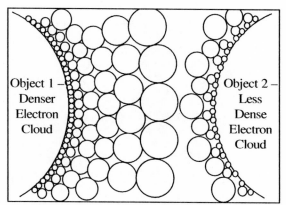

Fig. 4-10 Different-Density Electron Clouds on Charged Objects

But, as in the earlier discussion of repelling objects, when the two electron clouds meet, they form a continuous "matter bridge" of subatomic particles between the two objects – a continuous *subatomic* realm. And indeed, as in the earlier discussion and as suggested by Figure 4-9 earlier, two "positively charged" objects would also repel each other as their electron clouds of similar densities meet and push against each other. However, in the current scenario shown in Figure 4-10, the significant *atomic* realm size difference between the electrons at the fronts of the two clouds must immediately adjust to one of near equality as the electrons begin to regain their mutual *subatomic* identities. This means the electrons in the more dense cloud must shrink, and therefore, so must the overall cloud as well. And, since the less dense cloud cannot grow to fill the gap without stretching its limited number of electrons beyond their nearly equal sizes, this shrinking subatomic matter bridge pulls the two objects toward each other (Fig. 4-11). Further, as the two objects draw closer, the less dense cloud also begins to shrink to fit within the closing gap, requiring that the more dense cloud shrink even further to attain electron size equality, reinforcing the attraction. Also, the definition of the subatomic realm within this continuous "matter bridge" becomes stronger the closer the two objects draw to each other as the subatomic realms within each object close in on either side, reinforcing all of the aforementioned effects.

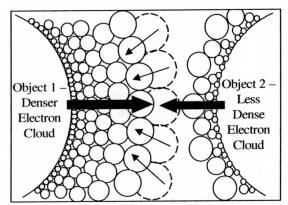

Fig. 4-11 Shrinking Electron Clouds Draw Objects Closer

If the two objects are allowed to continue drawing nearer, the electron clouds soon shrink to two microscopic clouds of subatomic particles between the two objects. As long as the two objects don't get too close or touch they can be repeatedly pulled further apart then allowed to draw toward each other again, with the electron clouds acting much like an elastic fabric of subatomic particles that can repeatedly stretch and shrink.

If the two objects draw too near, however, the smaller, less dense electron cloud becomes so tiny that it retracts almost entirely within the subatomic realm of its associated object, drawing the more dense cloud extremely close to the object's surface. At this point, the situation rapidly changes from that of two separate electron clouds in a tug-of-war between two objects, to that of a single dense electron cloud spanning the gap and drawn down into the subatomic realm within the second object. Once this occurs, there is nothing stopping the dense electron cloud from immediately and rapidly entering the subatomic realm within the second object, creating a rapid surge of electrons between the objects that we see as a spark across the gap. The fact that sparks often flash and crackle suggests that the electron surge may also be powerfully torn apart during this phase, emitting groups of electrons into the surroundings in various configurations, as explained in later discussions.

This process removes the surplus electrons from the surfaces of both objects, leaving them "neutrally charged." This does not occur when two objects with the same type of electron clouds touch, since the clouds do not pull each other across the gap and down into the subatomic realm, but rather, they push on each other, deforming and spreading out. The electrons in these clouds would retain much of their free growth tendencies as they are simply pushed aside by the approaching objects in the atomic realm rather than being drawn tighter into the gap.

Since the expanding electron clouds surrounding all "charged" objects can only be of two possible types – similar or dissimilar densities – the objects will always either attract or repel each other, seeming to support Franklin's idea of "positive or negative charge." However, the concept of electric charge is merely an abstraction in the absence of an understanding of subatomic expansion. In actuality, there are no "negatively charged" and "positively charged" objects, but merely objects with different densities of expanding electron clouds surrounding them.

Electric Circuits

Earlier, we began the discussion of electricity by touching briefly on the flow of electricity through circuits before stepping back to examine the concept of electric charge itself. We now return to discuss electric circuits in more detail. As mentioned earlier, science currently has no explanation for the physics underlying the abstractions of electric charge, and *Circuit Theory* in general, leaving mysteries such as "negatively charged" electrons powering circuit components as they flow through a circuit *without losing any of their charge in the process.* Such apparent "energy-for-free" mysteries are solved when electricity is seen as subatomic expansion instead of "electric charge."

A basic electric circuit has a power source, such as a battery, connected via wires to a circuit component, such as a resistor, as shown in Figure 4-12. *Circuit Theory* states that negatively charged electrons are pulled from the negative terminal of the battery and attracted through the circuit by the pull of the positively charged battery terminal.

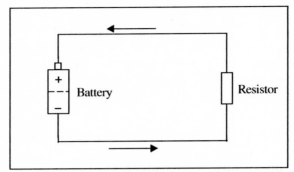

Fig. 4-12 Basic Electric Circuit: A Battery and a Resistor

 Electric circuits are driven by subatomic expansion, not "electric charge."

Expansion Theory, on the other hand, describes this as a surplus of electrons in the "negative" half of the battery, which are essentially trapped between subatomic and atomic realms. That is, they are free enough to attempt to grow out into the atomic realm, as in the earlier static electricity discussion, yet are also confined by a loose association with the atoms within the battery as well as by the walls of the battery casing. This situation presents a large expansion pressure within the battery that seeks a way out. Connecting wires to the battery provides a conductive path for the electrons to expand out along the wire and through the circuit. The "positive" battery terminal at the other end of the circuit is not actually "positively charged," but rather, it contains a material that has a deficit of electrons, essentially creating a region of very low expansion pressure that readily accepts the expanding electrons from the circuit.

Therefore, once the circuit is completed, the surplus electrons from the high-pressure ("negative") battery terminal are able to expand freely along the wire, through the circuit, and into the low-pressure ("positive") terminal in a continuous flow. Also, similar to the spark in the earlier static electricity discussion, once a continuous span of electrons extends to the other battery terminal the whole river of electrons tends to rapidly return to their subatomic definition within the far terminal – somewhat like one long, controlled spark through the circuit. As this process continues, it equalizes the high and low expansion pressures within the two sides of the battery, gradually reducing the current flow and eventually bringing it to a halt entirely, leaving a drained battery. The battery is not actually drained of "charge," however, but has simply reached a state of equalized expansion pressure on both sides.

VIOLATION

Electric Circuit Components Violate the Laws of Physics

This explanation of electric current solves the mystery of how, for example, the resistor continually heats up simply because "charged" electrons flow through it, while the electrons lose neither charge nor even kinetic energy (speed) in the process. That is, although not generally recognized today, there is no identifiable energy transfer from the flowing electrons to the circuit components, leading to the logical conclusion that the battery only powers the movement of the electrons through the circuit – but *not* the circuit components themselves. Since the circuit components do, nevertheless, consume power in order to operate, this is a deep "energy-for-free" mystery that is only solved by realizing that this whole process is not driven by "energy" at all, but by the ongoing subatomic expansion of all electrons.

Once again, the term "energy" is merely an abstraction that was invented in an attempt to explain observations in the absence of an understanding of expanding matter. With this understanding, we can see that it is not actually the *battery* that powers the circuit, but *subatomic expansion* that drives this whole process (as it drives *all* things), in this case using the battery to enable a temporary crossover effect into the atomic realm known as *current flow*. The details of precisely how subatomic expansion powers circuit components will be addressed further as the discussions continue, leading up to an analysis of a very special type of resistor – the light bulb.

As might be expected, as the electrons expand and push each other along the wire, they would experience far less difficulty expanding along the outside of the wire than through its dense center, which should result in electricity tending to flow along the outside of the wire. And, in fact, there is a well-known but little understood effect called the "Skin Effect" that describes precisely this behavior of electric current. This effect is so-named because electricity is known to flow as a thin coating or "skin" along the outside of the wire rather than flowing equally through the whole volume of the wire itself (Fig. 4-13).

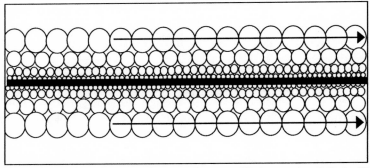

Fig. 4-13 Electrons Expand along Wire - with Important Side Effect

This tendency to expand along the outside of the wire creates a very important side effect, which can be clearly seen in the diagram. As the expanding electrons push one another along the outside of the wire they would also freely expand out into space, creating a surrounding cloud of electrons that radiates outward as it moves along with the current. This is actually a very well known and measurable effect, known as a *magnetic field*. Although today the Skin Effect is thought to occur mostly in circuits having rapid back-and-forth current oscillations rather than continual one-way current flow, *Expansion Theory* suggests this is an oversight due to a failure to recognize that the magnetic fields surrounding all wires carrying current are actually such externalized electrons. This defines a new concept for magnetic fields:

 NEW IDEA

The magnetic fields surrounding all wires that carry electric current are actually the electrons themselves expanding outward into space as they move along the wire.

 MYSTERY

The Mystery of Magnetism

The true nature of magnetic fields, revealed above, is unknown today since the theory of expanding matter is unknown. Instead, it is currently believed that the magnetic field surrounding the wire is yet another form of "energy" – known as "magnetic field energy" – which somehow emanates from moving "charged" particles, such as the electrons flowing through the wire. Yet, this is another belief that introduces sizable unexplained mysteries. One mystery is how a magnetic field emerges from moving electrons. This is a commonly stated fact in our science today, but one that has no solid physical explanation for how or why it occurs. A closely related mystery is how such "magnetic field energy" emanates from particles that do not lose any energy themselves; the flowing electrons are not considered to compensate by either slowing down or losing "charge," even if the magnetic field expends energy to continually repel another magnetized object. If a magnetic field emanates from flowing electrons and even expends energy without removing any energy from the electron stream, this is a sizable unexplained mystery.

Expansion Theory once again solves the mystery by showing that it is not "magnetic field energy" that emanates from "charged" electrons, but the electrons themselves expanding outward – much as in the earlier discussion of static electricity. Both electric current and the resulting magnetic field are merely temporary *crossover effects* as freed expanding electrons release pressure by expanding out of the battery and along the wire, returning to the subatomic realm within the low-pressure side of the battery. This explains the collapse of the magnetic field once the power is turned off, since some electrons would shrink back into the battery along the conductive wire pathway once the expansion pressure is removed, while the remainder would continue being drawn into the far battery terminal. This description of the surrounding magnetic field would also explain the well-known fact that it circulates around the wire while the current flows along. Since the magnetic field is not "pure energy," but fast flowing matter particles, it is reasonable that a spiraling action might emerge as the most efficient way to travel, just as water flowing down a drain also follows a natural spiral pattern. We will return to the topic of magnetism later in the chapter, delving into this issue in greater depth.

Radio Waves

The previous discussions of electricity flowing in a wire surrounded by a magnetic field lead naturally to the topic of *radio waves*, which today's Standard Theory considers to be waves of electromagnetic energy that radiate away from a wire when the current oscillates back and forth. This oscillation occurs when the wire is connected to a power source that supplies *alternating current* (an *AC* power-supply), turning the wire into a transmitting antenna that emits radio waves. Specifically, today's theory states that while *magnetic fields* surround wires that have a steady current flow in one direction, known as *direct current* or *DC*, *radio waves* radiate away from electrons that *accelerate* as they constantly change speed to oscillate back and forth.

 The Mystery of Radio Waves

This explanation might sound reasonable enough – until we attempt to explain precisely how this emanation occurs in real physical terms. In actuality, it is poorly understood today how a switch from magnetic fields to radio waves occurs simply by changing from a steady flow of electrons to a continual acceleration or oscillation. Further, radio waves are literally considered to be energy that is liberated from the circuit and sent off into space, supposedly emanating from the accelerating "charged" electrons. Yet, the mechanism for how energy emanates from electrons whose individual charge never varies is also largely unexplained. The radio wave energy leaving the wire can only be extracted from one source – the electrons moving within the wire. This would require that the electrons somehow transport energy from the circuit's power source and release it into space as they oscillate along the wire. However, electrons are not considered to behave in this manner – as if they were tiny rechargeable batteries. All electrons are considered to have a precise, identical negative charge that never varies, and so cannot pick up extra charge at the power source and later release it into space in the form of radio waves. The only possible remaining explanation – that the kinetic energy of the electrons is somehow converted into radiant electromagnetic energy – would also have no clear physical explanation.

NEW IDEA

Radio waves are not "electromagnetic energy," but waves of electrons expanding out into space.

The solution to these mysteries can be found in the fact that, as with the other forms of "energy" discussed so far, radio waves are not truly a manifestation of radiating "electromagnetic energy" but are a manifestation of expanding electrons. When a current oscillates along a wire it rapidly surges from standstill to a peak current flow and back to standstill again before surging again in the opposite direction. These repeated back-and-forth surges would cause the surrounding magnetic field to also surge outward into space in step with the current since the magnetic field is merely an extension of the flowing electrons expanding outward. The outward momentum of such a surge of electrons into space would counteract the tendency to shrink back toward the wire as each surge subsided, creating a separate, self-contained band of electrons in space with each successive surge of current. The separate, dissociated natures of these bands would be further accentuated by the fact that, as mentioned earlier, the surrounding electron cloud (magnetic field) is known to circle around the wire – but also changes direction each time the current does. This alternate direction of circulation between each band of electrons would create further detachment and dissociation between them, resulting in separate, self-contained electron surges sandwiched between each other and freely expanding off into space. This is what we currently call *radio waves*, or "electromagnetic radiation" (Fig. 4-14).

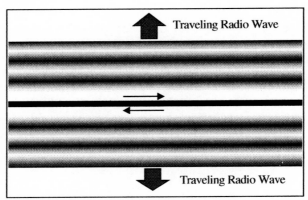

Fig. 4-14 Radio Waves: Bands of Electrons Expanding Outward

This is the first clear *physical* description of radio waves, and shows why we call this phenomenon "electromagnetic radiation." It is *electrical* in nature, since it is composed entirely of expanding electrons, and it is *magnetic* as well in the sense that each band of expanding electrons is essentially a detached magnetic field similar to that which typically surrounds a wire, as shown earlier in Figure 4-13. We can also clearly see why the frequency of radio waves matches the frequency of oscillation within the transmitting antenna, since every oscillation creates a separate band of electrons expanding off into space. And the fact that radio waves travel at the "speed of light" is simply a manifestation of the expansion rate of free electrons in space as they continue their natural expansion from within. Also, the energy mystery is solved by realizing that the oscillating electrons in the wire are not somehow emitting "energy" in a manner that is scientifically unexplainable, but rather, the electrons *themselves* are radiating off into space.

Although this discussion of radio waves suggests that it also applies as a description of electromagnetic radiation in general, the next chapter shows that this is not the case. This important point means that, although today we describe radio waves, microwaves, visible light, X-rays, etc., as simply different frequencies of the same phenomenon – namely, electromagnetic radiation, this is *incorrect*. As we will see, the *electromagnetic spectrum* is not actually a continuous range of electromagnetic wave frequencies, but rather, there are two very distinct classes of phenomena with very different natures and behaviors within this spectrum that today is painted with one broad brushstroke.

Electric Fields

The phenomenon of *static* electric fields was explored earlier, as free electrons on the outside of an object left to expand out as growing electron clouds. However, there are also *dynamic* electric fields, typically created between two parallel metal plates that leave a gap that interrupts the flow of electricity along a wire. Such a device is known as a *capacitor*. This dynamic electric field is similar to the *static* electric field on "charged" objects, but instead of surplus electrons being deposited on an object's surface, electrons push their way through an electric circuit and onto the first metal plate. This results in an electron cloud that radiates out into the gap between the plates and across to the opposite plate. Since the opposite metal plate is connected via a wire to the "positive" or low-pressure battery terminal, it readily accepts the high-pressure electron cloud crossing the gap. There is no conductive path between the plates to allow a continual flow of current, so the electron cloud merely tightens across the gap, creating what is known as an electric field (Fig. 4-15).

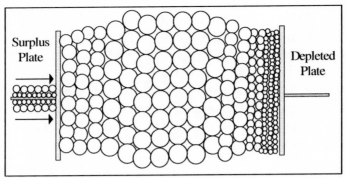

Fig. 4-15 Electric Field between Parallel Plates of a Capacitor

As just mentioned, there is no actual flow of electrons across the gap since there is no material between the plates for the electrons to flow along or through. Surplus electrons readily *expand* out into space, but generally require a dense, conductive material path to continuously flow along. However, if the current is not merely attempting to flow in one direction (DC) but continually flows back and forth (AC), the situation changes. There is still no conductive path for an actual current flow across the gap, but there is an *effective* AC current flow "through" the capacitor nonetheless because of the back-and-forth current flow in the circuit. This occurs because the electric field between the plates is a dense reservoir of electrons, and when the current reverses direction this reservoir pulls back into the half of the circuit from which it originated, while another electric field begins to stretch across the gap from the other plate. This is a continual process where each plate alternates between electron surplus and deficit, and where electric fields emerge then retract from each plate alternately, allowing the current in the circuit to slosh back and forth. Although there is technically no true current flow between the plates, the effect is just as if there were. This, according to *Expansion Theory*, is why a capacitor is said to block DC and pass AC signals through circuits.

Particle-Beam Deflection

A well-known phenomenon that has lent support to the concept of positively and negatively charged particles is the deflection of a beam of "charged" particles by electric or magnetic fields. This is a common procedure in laboratory experiments and in devices such as the cathode ray tubes of television sets, typically involving a beam of "negatively charged" electrons that are deflected as they pass between a pair of parallel electrified plates (Fig. 4-16). *Circuit Theory* states that the negatively charged plate repels the negatively charged electron beam while the positively charged plate attracts it, causing the beam to bend as it passes between them.

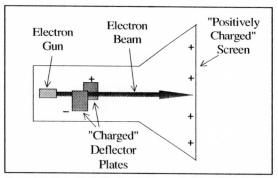

Fig. 4-16 Cathode Ray Tube Creates and Deflects Electron Beam

While the concept of electric charge is a useful abstraction to describe electron beam deflection, it cannot be the *literal* explanation for cathode ray tube operation. That is, although a power source is used to expel the "negatively charged" electrons from the electron gun and to vary the amount of "charge" on the deflector plates, this power source is not powering the core deflection itself. Instead, it is the attraction and repulsion caused by the "charge" on the plates interacting with the "negatively charged" passing electrons that focuses them into a beam and deflects them to one side or the other. To accentuate this fact, if we weren't concerned about *varying* the beam's deflection, we could just as well have replaced the electrified deflector plates with statically "charged" objects that are not connected to any power source. As long as we ensured that the passing electrons did not alter the "charge" on the objects, we would still expect the objects to endlessly focus and deflect the electron beam despite their lack of a power source. Such an endless forceful shaping and deflection of the electron beam with no power source violates the *Law of Conservation Of Energy*, which also applies to the electrified deflector plates whose power source merely *varies* their "charge" but does not cause the deflection itself.

According to *Expansion Theory*, the electrons emitted by the electron gun would expand outward in a large electron cloud. And since the entire screen at the other end of the cathode ray tube has a conductive mesh or coating connected to the "positive" terminal of a power supply, it has a deficit of electrons. Therefore, much like the electric field that tightens between the plates of a capacitor, the large expanding electron cloud would soon encounter the screen and tighten to an electric field extending from the electron gun to the screen at the far end. This electric field is confined to a narrow beam and further deflected across the screen in a controlled manner by the series of "charged" deflector plates in front of the electron gun. However, it is not actually an "electric charge" repulsion that controls the electron beam as it passes between the deflector plates, but rather, an expansionary effect caused by the tight electron cloud stretched between the deflector plates in the form of a separate electric field.

Expansion Theory shows that the electric field between the deflector plates actually functions as a subatomic reference frame for the electrons in the freely expanding electron beam. Recall that the electron cloud radiating from a surplus deflector plate encounters the depleted plate across the gap and shrinks toward it, tightening across the gap to form an electric field. If a beam of expanding electrons then passes through this electric field, these expanded electrons effectively become part of the electric field. When such a group of expanded electrons is introduced into a tightened electric field, there must be an immediate shrinkage of these new electrons toward an identity with those already composing the field. Since these shrinking electrons are effectively part of the electric field while passing through, it is as if the overall field is shrinking – similar to its original tightening across the gap, where the electron cloud from the surplus plate shrank toward the depleted plate. This continual process of a shrinking readjustment as the expanded electrons continue to arrive at the center of the electric field and progress through it results in a continual deflection of the electron beam away from the center and toward the depleted (or "positive") plate (Fig. 4-17).

This causes a narrowed, off-center electron beam to strike the screen at the far end, producing a small, persistent off-center dot on the screen.

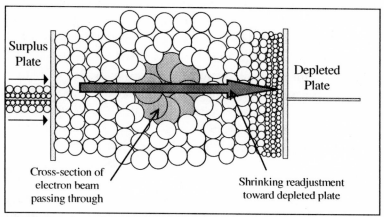

Fig. 4-17 Electron Beam Causes Shrinkage toward Depleted End

This process can give the impression that a "negatively charged" plate is somehow repelling a "negatively charged" electron beam, while a "positively charged" plate across the gap is simultaneously attracting the beam. Of course, as shown earlier, electric charge violates the laws of physics, and its attracting and repelling properties are never truly explained; today's science says little more than "opposite charges attract because they are oppositely charged." This point is worth special note:

NOTE

 The preceding explanation for the deflection of an electron beam is the *only known explanation* for this effect; today's science offers neither a clear physical mechanism nor a scientifically viable explanation for *how* or *why* an electric field deflects an electron beam.

The expansion-based explanation of the electron beam deflection suggests that a greater voltage on the deflector plates would create a denser volume of electrons in the electric field, making these more numerous electrons smaller and even more strongly associated with the subatomic realm within the dense electric field. This would increase all aspects of the deflection process just described, causing a greater deflection. In this manner, the beam can be moved back and forth as the voltage on the plates varies. This is the expansionary explanation of how the electron beam of a television screen is controlled. Note also that although electrons do not readily flow through space without a conductive medium to travel along, the series of deflector plate pairs concentrates the passing electrons into a tight beam across the cathode ray tube, rather than a diffuse electron cloud filling it. Such a tight beam acts as its own conductive "matter bridge," almost as if it were a wire running through space, allowing the electrons to proceed across the tube. Also, unlike the scenario of a capacitor, where electrons merely stretch across the gap and the current stops flowing in the circuit, the electron gun of a cathode ray tube is part of a separate circuit that continually emits electrons into the tube. This guarantees that, even without the deflector plates, the electrons will eventually either flow or spark across the cathode ray tube and out via the depleted screen coating as their numbers build within the tube. The deflector plates simply make this process as efficient and controlled as possible.

Experiments also show that some types of particle beams deflect in the opposite direction to electron beams as they pass between deflector plates. *Circuit Theory* represents this as a beam of positively charged particles that are repelled by the positively charged plate and attracted by the negatively charged plate. However, since this appeal to electric charge has been shown to be scientifically unsound, let's explore the expansionary explanation.

A typical beam of "positively charged" particles is composed of *alpha particles*, which are helium nuclei (a cluster of 2 protons and 2 neutrons) that have lost their orbiting electrons in the natural process of being expelled from certain radioactive substances. Rather than these bared nuclei presenting a "positive charge" to the world, they present a strong subatomic reference to any freely expanding electrons nearby, causing the electrons to undergo a rapid return to their definition as subatomic particles. Therefore, as these large expanding alpha particles pass through the center of the field of electrons between the deflector plates, they essentially divide the electric field into two smaller spans of electrons on either side, with the alpha particles in the middle. This powerful subatomic reference presented by the alpha particles (essentially bare atomic nuclei) across the much shorter halved distance to the plates on either side causes the electrons within the field on either side to shrink and tighten even further. The fact that alpha particles are drawn to the surplus (or "negative") plate is evidence that the electron clouds on either side do not shrink with equal pulling force. Instead, the electron cloud on the surplus side pulls more strongly, due either to more rapid shrinkage because of a size or density difference with the cloud at the deficit plate, or to a firmer anchoring within the surplus side of the circuit than the other cloud has within the deficit side. The stronger pull from the surplus side would cause the alpha particles to deflect toward the surplus plate (Fig. 4-18), creating the appearance of "positively charged" alpha particles that are somehow attracted to a "negatively charged" plate.

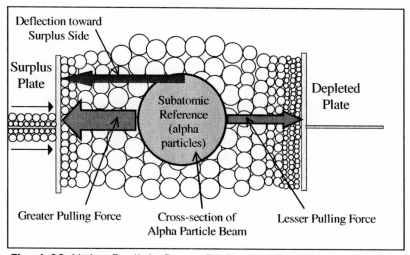

Fig. 4-18 Alpha-Particle Beam Deflected Toward Surplus Side

Magnetism

The earlier discussion of current flowing through a wire addressed the issue of the surrounding magnetic field, which is considered a rudimentary form of magnetism. Typically though, magnetism is associated with a physical bar of iron that is permanently surrounded by its own self-contained magnetic field with a north pole at one end and a south pole at the other – a *permanent magnet*. This is a very common and familiar object that can be found stuck on refrigerator doors worldwide. Yet, it is not commonly recognized that this behavior violates the laws of physics.

VIOLATION

    Permanent Magnets Violate the Laws of Physics

We know, of course, that it takes energy to hold on to the side of a cliff, hanging on against the effective pull of gravity and straining under the effort of supporting our body weight. Likewise, this is precisely what is required of a magnet if it is to hang on to a refrigerator door, holding its weight against gravity as well. And in fact, fridge magnets typically hold *more* than their own weight since they are often used to pin other objects to the fridge door. A strong magnet may even hold many times its own weight and will do so *indefinitely* – without weakening and *even without a recognizable power source*. The magnet is not mechanically glued to the fridge, but holds itself there by "magnetic energy" that mysteriously and endlessly emanates from within – no textbooks exist that show the power drain curve of such a permanent magnet. This is a very common observation and so tends to be overlooked as a clear violation of the *Law of Conservation Of Energy*.

Today's scientific explanation for why this is not a violation of the laws of physics appeals to the same flawed *Work Function* logic discussed in Chapter 1, which is used to justify why the endless work done by gravity also does not violate the laws of physics. That is, today's claim is that fridge magnets do not move, and since *work* equals *force* times *distance* there is no work done and hence no energy expended by a fridge magnet. This type of logic was shown to be a flawed use of the *Work Function* in Chapter 1, which can be seen as further flawed since *Expansion Theory* shows that gravity is not a force pulling objects down, but an expansion of the planet that accelerates objects upward.

Therefore, the fridge magnet *is* in motion – it is hanging on tightly while being accelerated 4.9 meters upward every second by the expanding planet below. No matter how we look at it, a fridge magnet is definitely expending effort to continue clinging to the refrigerator, yet it has no identifiable power source; and, unless it is adversely affected externally (such as being dropped or excessively heated) its hold on the refrigerator will never weaken.

Further evidence of this mysterious and unexplained magnetic energy can be experienced directly by attempting to push both north poles or both south poles of two separate magnets together. As we know, like poles repel each other, and we will feel a strong force pushing the magnets apart. Our muscles will soon tire from this effort, as will the muscles of any number of people who step in to replace those who tire. The magnets will endlessly fight all efforts to push them together without weakening, and again, without even drawing on a power source.

Even today's very *concept* of magnets falls apart upon examination. To see this, we simply need to ask where the magnetism of a permanent magnet comes from. The simplest answer given today is that the overall magnet results from the alignment of thousands of sub-regions within the iron bar, each of which acts like its own tiny magnet with north and south poles. So then, we must delve further and ask where each of these tiny magnets gets its magnetism. This line of investigation ultimately ends with the iron atoms themselves being the tiny magnets upon which the overall magnetism of the iron bar is based. The theory states that when the magnetic poles of these tiny iron atoms all line up, their combined magnetism gives us the familiar strong permanent magnet with its north and south poles. But this only shifts the focus from the mystery of how a magnet does endless work without a power source to the mystery of how its component atoms do this work without a power source. Also, this does not explain the phenomenon of magnetic fields themselves – what they are *physically* and why like poles repel and opposite poles attract. Magnetic fields and their behaviors are merely abstract characterizations of observations in our science today, with no clear and solid physical or scientific explanation.

In fact, looking back on the discussions so far, Standard Theory states that the atom expends "strong nuclear force energy," "electric charge energy," "gravitational energy," and now "magnetic energy" – all without any clear physical explanation, and often for billions of years with no known power source. This shows that the atom plays an extremely crucial and pivotal role in our understanding of our world and our science, which is why these first four chapters have been dedicated to an exploration of both its external behavior and its internal nature. And, now our investigations show that "magnetic energy" is also a flawed concept. Today's science cannot explain magnetism at all, only providing abstract models and equations at best. While these abstractions are useful, they leave a great mystery to be solved when we look beyond our models to the underlying physics. So then, if magnets are not objects from which "magnetic energy" mysteriously emanates, then what exactly *are* magnets?

Rethinking Magnetism

Permanent Magnets

A permanent magnet is typically created by repeatedly passing an existing magnet in the same direction across an iron bar until the bar becomes magnetized as well. As mentioned earlier, Standard Theory states that there are thousands of tiny magnetic regions throughout the iron bar, but which are all initially oriented randomly. These tiny *magnetic dipole* regions, as they are called, are thought to align when an external magnet is repeatedly passed over an iron bar, turning the whole bar into one overall magnet. The resulting magnet is now said to have "north pole magnetic energy" emanating from one end and "south pole magnetic energy" emanating from the other – phenomena whose nature and behavior have just been shown to be completely unexplained.

From the perspective of *Expansion Theory*, however, the process of repeatedly passing a magnet over the iron bar causes the electrons within the iron bar to migrate toward one end, and then further out onto the outside of the bar. These externalized electrons naturally expand out as a cloud radiating in all directions. This is similar to the objects in the earlier static electricity discussion, except that such objects are typically non-conductors whose surface becomes coated with electrons from an external source. In the case of the iron bar, however, the conductive metal of the bar allows the electrons within the volume of the bar to be pulled toward one end of the bar during creation of the magnet. This end becomes saturated with electrons that have been dragged from the other end, which is now depleted of electrons. The end of the iron bar that is now saturated with electrons is the *north pole*, and has an electron cloud radiating from it due to the surplus electrons.

This process would leave the *south-pole* end with a deficit of electrons, resulting in a low-pressure region much as in the earlier "positive" battery terminal discussion, and also likely an external electron cloud of lesser density as in the static electricity discussion. Depending on which of these two effects is dominant, as the dense electron cloud at the north pole fans out into space it either encounters the less dense south-pole electron cloud or the low pressure of the south pole directly. In either case, the dense electron cloud would be rapidly drawn into the subatomic realm within the south pole, aided by the conductive nature of the iron bar. This return to the microscopic subatomic realm at the south pole would cause a size equality readjustment to immediately ripple back through the dense electron cloud, causing it to tighten around the iron bar in the characteristic field lines of all bar magnets (Fig. 4-19). Since a sizable number of electrons are still stretched between the poles outside this newly created magnet in the form of a magnetic field, the south pole still remains significantly depleted. This description shows that, once expanding matter is considered, magnetism can be seen as an emergent *group* behavior of many electrons, and not the result of an unexplained magnetic property of *individual* electrons and atoms within the magnet, as thought today.

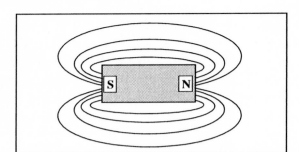

Fig. 4-19 Magnetic Field Lines Around a Permanent Bar Magnet

Whether two north poles or two south poles of separate magnets meet, they encounter tightened clouds of expanding electrons under tension, which push on each other and cause the repelling force between like poles. When a north and south pole meet, however, the north pole is closer to the depleted south pole of the other magnet than it is to its own, and the cloud begins to unwrap from around its magnet and extend across the gap to the other magnet's south pole. As the cloud tightens across the gap in the same manner that it ordinarily tightens around its own magnet, the two magnets are drawn toward each other (Fig. 4-20). This is the attracting force between north and south poles. When the magnets touch each other, the electron cloud shrinks to a microscopic cloud of subatomic particles sandwiched between the two magnets, which can be forcefully stretched back out like an elastic fabric of subatomic particles if the magnets are pulled apart.

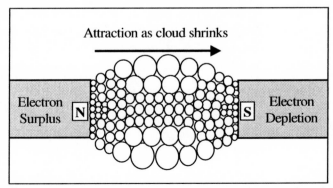

Fig. 4-20 Shrinking toward Subatomic Realm of another Magnet

Unlike "electrically charged" objects, the north-pole electron cloud does not discharge into the other magnet and neutralize, or in this case, demagnetize both magnets when they touch. The "charged" objects discussed earlier do this because of a surface coating of excess electrons isolated outside of non-conducting material, which will readily leap to another object capable of accepting these isolated surplus electrons if it is close enough. However, the surplus electrons at the north pole of a magnet are part of a continuum of surplus electrons that extends throughout the north half of the conductive iron bar. Therefore, the north-pole electron cloud is actually an extension of a densely packed electron cloud within the magnet, and does not readily separate and discharge from the surface the way the electron cloud of a "charged" object does.

Also, "charged" objects are not strongly attracted to a magnet's depleted south pole because the large and rapidly growing electrons in the freely expanding electron clouds of "charged" objects are largely isolated from the magnet's depleted south pole by the tightly wrapped electrons of the magnetic field. In a sense, the tight magnetic field acts as a "force field" that deflects the cloud of larger, freely expanding electrons. There is also no strong repelling force between a "charged" object and either pole of a magnet for essentially the same reason. That is, the tight magnetic field of the magnet does not expand outward freely to cause a repelling push, but acts more like a "force field" that mechanically deflects or brushes aside the object's electric field when it nears. It is also worth noting that:

NOTE

The preceding explanation for the repelling and attracting forces between magnets is the *only known explanation* for this effect; today's science offers neither a clear physical mechanism nor a scientifically viable explanation for *how* or *why* magnets repel or attract each other.

Electromagnets

Another type of magnet is the *electromagnet*. Although the expanding electrons surrounding an electrified wire are commonly referred to as a magnetic field, this scenario lacks the identifiable north and south poles that typify a magnet. A true electromagnet typically results from electricity passing through many turns of such a wire wound around a cylindrical core. As mentioned earlier, the expanding electrons moving along the wire also circle in space around the wire, which is now wrapped in side-by-side windings along the core. This means the electrons in every winding all end up circling in one direction side by side on the outside of the cylinder, and in the opposite direction on the inside of the cylinder (Fig. 4-21). This results in an effective coating of electrons all moving en masse up the outside of the core and back down the inside, somewhat like a large conveyor belt delivering electrons from one end of the core to the other along its outer surface and returning them along the inner surface.

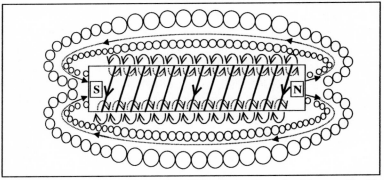

Fig. 4-21 Electrons Circling Wire Windings Create Magnetic Field

The north pole and south pole distinctions in electromagnets arise from the direction of current flow and the fact that the inner and outer surfaces of the core have opposite curvatures. The electrons traveling on the outside of the core freely expand out into space from its outwardly curving (convex) outer surface, while the inside surface of the core is inwardly curving (concave), focusing the expanding electrons into a tight central stream as they pass through the core. Therefore, the freely expanding electrons that typify a magnetic field occur mainly on the outside of the core, and travel in a particular direction from one end to the other that depends on the direction of the wire windings and the current through them. The end of the core that the electrons travel toward along the outside is equivalent to the south pole of a permanent magnet, since they wrap around and enter the core at that end. The other end where the concentrated central stream emerges from the core and out into space is the equivalent of the north pole of a permanent magnet, where surplus electrons emerge and wrap around through space toward the south pole.

This is essentially a dynamic magnet that behaves similarly to a permanent magnet, but which is supported by a power source that drives the continual current flow through the wire windings. The magnetism itself is not powered by the power source, any more than the magnetism of a permanent magnet needs such a power source. Rather, the power source only supports the continual current flow through the wire, and it only drains because the decreasing expansion pressure at its high-pressure source terminal is eventually equalized with the increasing pressure at its destination terminal. We know that a permanent magnet has no such power source driving a continual circulation of electrons in the magnetic field between north and south poles, and if it did, the power source would have to drain in a similar manner. We also know there is no such circulation of electrons in a permanent magnet because this would cause the magnet to continually heat up, as occurs with electromagnets.

The Light Bulb – Do We Really Understand It?

Finally, we return to the basic electric circuit with only a battery and a resistor to discuss a very special type of resistor – the _light bulb_. One of the most important phenomena both in human experience and in our science is that of light, and, more broadly, electromagnetic radiation in general. Scientists throughout the ages have studied light, producing a rich body of knowledge regarding its refraction into separate colors, its speed of travel, its nature as either a particle or a wave, and even its place in the mysterious theories of *Quantum Mechanics* and *Special Relativity*. Any all-encompassing physical theory would have to provide a description of light that clearly shows the physical reality behind all significant observations, theories, and experiments regarding its nature and behavior. As will be shown in the next chapter, *Expansion Theory* does just that, beginning with the following overlooked mysteries surrounding the common light bulb.

The Production of Heat and Light is Unexplained Today

One of the simplest electric circuits for producing light is the basic 'battery and resistor' circuit shown at the beginning of the electricity discussion, but with the resistor replaced by a light bulb. A light bulb is simply a specially designed resistor that readily heats up and sustains a white-hot glow as electricity flows through it. This seems to be a very simple process that is fully understood today and described by simple circuit equations which show a power drain from the battery to account for the power output from the light bulb. When viewed at this level of abstraction there seems to be nothing unusual or inexplicable going on. However, when we look beneath our abstractions and investigate the underlying physics, we see something quite different.

Although we have equations showing a transfer of the stored energy within the battery to the heat and light energy put out by the light bulb, the physical reality is that this radiating energy must actually be removed from the stream of electrons as it travels through the light bulb. That is, the light bulb doesn't somehow take energy *directly* from the distant battery, as our abstract models imply, but from the *electron stream* passing through the light bulb. The first conceptual mystery here is that our science has no explanation for how electrons passing through the resistive material of a light bulb generate radiant heat and light. We can see how the atoms of the light-bulb filament might vibrate more and more rapidly (heat up) as current passes through to jostle them about, but that is a form of heat known as *conduction*, where an object is only hot to the touch. This does not explain the warmth we feel from a light bulb at a distance – a form of heat known as *radiant* heat.

The same mystery exists for the additional phenomenon of light radiating from the filament. The leap from "charged" electrons passing through a filament of rapidly vibrating atoms to radiant heat and light energy emanating from the light bulb introduces a great mystery that has *no physical explanation in our science today*; today's explanations simply state that photons of heat and light are "given off" by excited atoms. A close look at these explanations shows that the precise answers to how and why such photons would be produced are absent, as is a clear explanation of the nature of the photons themselves, leading to a centuries-old debate about whether light is a wave or a particle.

VIOLATION

The Light Bulb Violates the Laws of Physics

A second and deeper mystery actually qualifies as a clear violation of the laws of physics – specifically, the *Law of Conservation Of Energy* once again. This follows from the fact than an electron stream enters the light bulb, causes radiant heat and light to be emitted out into space, and then leaves the light bulb *unchanged* (Fig. 4-22). From both logical and scientific standpoints, there must be a clearly identifiable difference in energy content between the stream of electrons entering the light bulb and the stream leaving it to account for the radiant heat and light energy that emanates from the light bulb. Let's see if there is any way to resolve this mystery using the scientific knowledge of today.

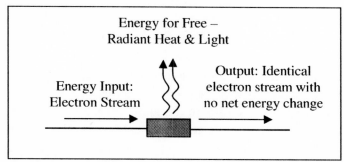

Fig. 4-22 The Light Bulb – a Mysterious Free-Energy Device

We know that the amount of electric current does not change as the electron stream passes through (i.e. the number of electrons per second passing by remains the same) since a constant current flows around the circuit. The electrons have nowhere else to go, and would pile up much like a traffic jam if they lost *kinetic* energy while passing through the light bulb and slowed down in the second half of the circuit. Therefore, the passing electrons are not slowing down and somehow transforming kinetic energy into light energy. Connecting ammeters (current meters) on either side of the light bulb to measure the current verifies this constant current flow claimed by *Circuit Theory*.

We also know that there is no possibility of the electrons losing some of their "charge" to produce heat and light as they pass through the light bulb. Electrons are always considered to have the same "negative charge" upon them, which is considered to be one of the main qualities that defines them as electrons today. They are not considered to function as tiny rechargeable batteries that can be charged up at the battery and discharged at the light bulb, and there is certainly no claim today that such a process occurs.

The passing electrons also are not dropping down to lower orbits about the nuclei of the atoms in the light bulb filament to somehow emit photons of heat or light in the process, as is often stated today. Not only is such a claim also a complete mystery in itself that has no clear physical explanation, but the electron stream through the filament does not involve some of its electrons dipping down to join atomic orbits within the filament. Even if this did occur, these same electrons would have to regain their lost energy so that they can climb back out of the atom and continue on, meaning that a photon was still emitted with no net change in energy for the electron. Even the possibility that the passing electron stream somehow excites the electrons within the atoms of the filament, which then produce a photon of light as they drop back to normal, still leaves us with the same mysteries. If the electrons in the passing stream excite electrons in the atoms then the passing electrons must be losing energy in order to impart energy to other electrons, yet science has not identified any such loss in the electron stream leaving the light bulb.

Finally, resistors and light-bulb filaments are known to have an electric field surrounding them when connected to a circuit, which can be measured as a voltage by connecting a voltmeter across them; in fact, all circuit components will give such voltmeter readings. But, in actuality, these readings are essentially only a reflection of the *battery's* electric field or voltage, the reading of which becomes increasingly weakened as more components are included in the measured path when looking back toward the battery. That is, the full voltage of the circuit can be measured by connecting a voltmeter directly across the battery before connecting it to the circuit, while the circuit components have no inherent voltage and would, of course, give a zero voltage reading before connecting the battery. So, once the battery is connected, we can only expect to read the *battery* voltage – as seen from different locations around the circuit; this is actually all a voltmeter does when it reads the voltage across a given component. And so, although the voltage across the light bulb would diminish over time as it produces light, this is merely a reflection of the *battery's* diminishing electric field or voltage as the difference in expansion pressure between its two halves equalizes. This is not a drainage of energy from the electric field of the light bulb, and if it were, this energy would have to be continually replenished by the passing electrons in the circuit to keep its electric field from immediately collapsing. And this again leaves us with the same question – what form of energy are the passing electrons losing as they replenish the electric field of the light bulb, and how is this energy transfer physically occurring?

The fact remains that today's science cannot explain the physics of the light bulb. The electron streams entering and leaving the light bulb have no clear physical or theoretical energy difference in today's Standard Theory, yet, in between, we get radiant heat and light – apparently for free. This is an impossible free-energy device from the standpoint of our current theories, and a clear violation of the *Law of Conservation Of Energy*. This is no small failure of our current scientific theories and beliefs, but is as elemental a failure as one can imagine. The discovery of electricity and the invention of the light bulb are two of our most pivotal scientific achievements as a technologically advanced species, yet we are unable to explain the underlying physics more than a century later. The answer provided by *Expansion Theory* is presented in the next chapter, which deals with energy in general, light in particular, and all the implications that follow – including the theories of *Quantum Mechanics* and *Special Relativity*.

Rethinking

Energy

Light and Electromagnetic Radiation

Solving the Light Bulb Mystery

As pointed out in the previous chapter, Standard Theory cannot explain the physics behind the operation of a light bulb. There is no way to account for the heat and light energy emanating from a light bulb by inspecting the electron stream leaving the light bulb and comparing it with the stream entering it. Both streams have identical energy content, both in theory and in measurement. Yet our laws of physics, as well as our common intuition, tell us that we can't get something for nothing. The light bulb appears to be an impossible free-energy device when viewed from the perspective of today's science. Yet, light bulbs are considered to be simple, well-understood devices that offer little challenge to our advanced science. How can this deep mystery be present in such an everyday device when science currently does not even recognize this problem?

The answer is that our current explanation of the light bulb employs the abstraction of *Circuit Theory* to create a model of an energy drain from the battery, balanced by an equivalent energy output from the light bulb. This completely bypasses the issue of energy extraction from the electron stream to supply the radiating heat and light of the light bulb, instead inventing an abstraction that essentially implies energy jumps directly from the battery to the light bulb. If we construct such mathematical equalities between the power of the battery and the power output of the light bulb, and do not investigate the underlying physics too closely, it is possible to function entirely within this self-consistent system of abstract logic, creating the *appearance* of explaining light bulb operation.

Certainly, *Circuit Theory* is a very important abstraction that allows us to design useful electrical circuits and devices, but it does not provide us with a physical and scientific understanding of the principles behind our creations. We tend to overlook the mysteries and inconsistencies in our theories today by taking more of an *engineering* approach to our world than a *scientific* one; that is, where the scientific explanation eludes us, we often install an engineering model into our science instead. But, of course, energy does not jump directly from the battery to the light bulb, but is physically transmitted along wires via the electron stream and is extracted from the stream by the light bulb in a process that modern science cannot explain. Let's now examine the operation of the light bulb from the perspective of *Expansion Theory*.

In *Expansion Theory*, as shown in the previous chapter, electricity is the flow of expanding electrons along a wire as their expansion pressure pushes them along while they are also drawn back into the subatomic realm at the other end of the circuit. As the electrons enter the filament of the light bulb they sink into the volume of the filament and cause its atoms to vibrate as they move through. So far, this is similar to the standard explanation of how the filament heats up internally, but it does not yet address the *radiant* heat and light. The next phase is where the two theories have a dramatic departure from each other. Standard Theory claims that energy is extracted from the electron stream to produce radiant heat and light energy in a conversion process that is not physically understood and which cannot even be validated by a corresponding energy loss in the stream leaving the light bulb. *Expansion Theory* states that it is not "energy" that is being extracted from the electron stream to create light, but rather, *it is the electrons themselves that are being ejected into space*:

The True Nature of Light Revealed

 Light is not composed of waves or photons of "energy," but *clusters of expanding electrons.*

According to *Expansion Theory*, the vibrating molecules of the filament impede the flow of electrons passing through while also agitating them, causing them to dissociate from the atoms and gather into pools where their rapidly building expansion pressure pushes them out into space. This describes a vast sea of expanding electron clusters of all sizes across the entire surface of the filament, pushing each other out into space – *a phenomenon we know as radiating heat and light* (Fig. 5-1). *Expansion Theory* claims that heat and light are not waves of "pure energy" mysteriously produced by electrons flowing through the resistive material of a light bulb filament, but rather, they are clusters of electrons of various sizes pushing one another through space.

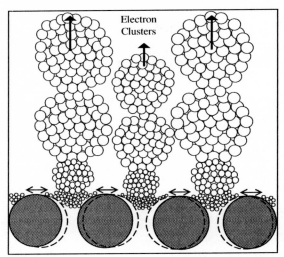

Electron
Clusters

Fig. 5-1 Vibrating Atoms eject Electron Clusters (Heat & Light)

The initial heat that radiates from the filament before it begins to glow white-hot is actually the larger electron clusters, caused by the larger pools of less agitated electrons that would gather while the atoms vibrate relatively slowly and expansion pressure builds more slowly before ejecting clusters. As the atoms vibrate faster and faster the increased agitation creates ever-smaller pools of expanding electrons that eject smaller clusters more frequently – a phenomenon known as *light*. This is the reason radiant heat is considered to have a longer wavelength than light; heat is composed of larger clusters of electrons.

At first, it may not seem possible that heat and light could be expanding clusters of matter, since our experience isn't one of tiny groups of matter particles striking us, but rather, one of the warmth, brightness, and color of "pure energy." However, we also know that the subjective experiences of warmth, brightness, and color are generated by our brains in response to stimulation of our heat and light receptors;

neurosurgeons can cause these same subjective experiences to occur in our heads merely by stimulating the appropriate brain regions directly. The receptor cells in our skin and eyes simply fire neural impulses when they detect the appropriate external stimulus, then our brains take over from there and generate our subjective experiences of warmth, brightness and color. So, although light is indeed bouncing around in the outside world, it does not necessarily follow that the light itself is literally bright, colorful waves of "pure energy." Our eyes and brain have evolved to respond to light – whatever its true nature may be – and to generate subjective visual representations of the world in our heads.

To be more precise, *brightness* is the word we use for our subjective internal experience of the *intensity* of the visual radiation striking our eyes. And, 'intensity' is merely an objective term referring to the *amount* of radiation arriving each second, whether that means the number of "energy waves" per second (classical light theory), the number of "energy photons" per second (*Quantum Theory*), or the number of electron clusters per second (*Expansion Theory*). Regardless of the true physical nature of light, our subjective experience of brightness is simply how the brain represents the intensity or amount of it that arrives each second in our conscious experience.

Likewise, *color* is a subjective term for the objective quality of *hue* within the visual radiation around us. And, the term 'hue' refers to the physical *size* of the elements composing the light that strikes our eyes, whether this is the length of its "energy waves" (wavelength), the quantity of "energy" in its "photons," or the number of electrons within the electron clusters composing the light. Regardless of the true physical nature of light, our subjective experience of color is simply how the brain represents the size of the elements composing light in our conscious experience.

As suggested by the preceding light-bulb explanation, *Expansion Theory* states that the external stimulus that we call *light* is actually a sea of electron clusters of varying size and number impacting our receptor cells. Even on a bright sunny day we are actually standing in the dark, in a sense, while a sea of electron clusters bounce around us – some of which strike our eyes to generate a scene of a bright sunny day in our heads. Figure 5-2 shows this concept of tiny matter particles bouncing around in the dark (left frame), and our resulting conscious experience and conceptualization of this as if waves of bright, colorful "light energy" literally surrounded us.

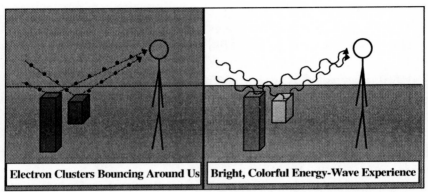

| Electron Clusters Bouncing Around Us | Bright, Colorful Energy-Wave Experience |

Fig. 5-2 Physical Electron Clusters vs. Energy-Wave Perception

This new concept of the nature of light clears up many of the mysteries in our current theories and beliefs today, beginning with the mysteries presented by the light bulb itself. The first mystery this new concept solves is how "electrical energy" flowing through the filament of a light bulb becomes transformed into radiant "heat and light energy." We can now see that no such mysterious energy transformation occurs, but rather, electricity and light are simply different forms of expanding electron behavior. A clear physical mechanism can now be seen to explain how electricity becomes heat and light. It also should be noted that:

NOTE

 This description of the nature of light and how it is produced from electricity is the *only known explanation*; today's science offers neither a clear physical description of light nor a scientifically viable explanation for *how* or *why* it is produced by electricity flowing through a resistor.

The second mystery this solves is the "energy-for-free" violation of the *Law of Conservation Of Energy*. Whereas today's theory states that "energy" radiates from the light bulb while leaving the electron stream unchanged, *Expansion Theory* states that there is a very significant change in the electron stream – *it continually loses electrons*. This may initially seem like an impossible claim that is easily falsified by simply measuring the current flow on either side of the light bulb and showing it to be unchanged. Since a current measurement indicates the number of electrons flowing by per second, identical current measurements on either side may seem to invalidate the concept that electrons are continually being ejected from the circuit. However, let's take a closer look at this issue.

In actuality, the ammeters we use to measure current do not literally count the passing electrons and report back how many electrons per second are passing by. This is a common representation of the current reading provided by an ammeter, but that is not actually what an ammeter does. An ammeter is a device that diverts the current flow in the circuit through its own inner coil of wire before allowing it to flow out and continue on through the circuit. As the current flows through the ammeter's inner wire coil, the resulting surrounding magnetic field pushes against the magnetic field of a small magnetized needle that is free to rotate; this is the needle that we see being deflected to indicate how much current is flowing through the circuit. So, although the current reading of an ammeter is commonly thought to indicate the number of electrons per second flowing in the circuit, it actually only indicates the strength of the magnetic field produced by the current flow.

This use of the ammeter as a current meter is based on the assumption that magnetic field strength is a direct indicator of the number of electrons passing per second through the wire – an assumption based on the belief that "magnetic energy" somehow emanates from the passing electrons. The logical conclusion of this belief is that more passing electrons should add up to more total magnetic energy in a one-to-one relationship, so it has never been a concern that our "electric current" readings are actually magnetic field strength readings. However, when magnetic fields are given the clear physical explanation provided by *Expansion Theory*, a somewhat different interpretation emerges, as shown by the following analogy.

EXPERIMENT

The Garden-Hose Analogy

The explanations of electricity, magnetism, and light in *Expansion Theory* allow a simple analogy to be drawn between a garden hose with a puncture and an electric circuit with a light bulb. The water running through the hose pushes itself along with a constant pressure throughout, even if water is being ejected from a small puncture somewhere along its length. Since there is less water continuing on within the section of hose after the puncture, it takes longer to fill the same volume per length of hose to keep the same pressure throughout. This describes a punctured hose with constant pressure throughout, maintained by a faster-moving volume of water (greater current) prior to the puncture and a slower-moving volume (lesser current) after the puncture. Put another way, the greater pressure that would ordinarily result from the greater current in the first half of the hose is partially released as water spouts from the puncture, equalizing the pressure in both halves.

Of course, in general, if we want to force more water through the same hose we need more water pressure to achieve this, so there is often a direct one-to-one relationship of this sort between pressure and current flow. However, if we simply measured the equal water pressure on either side of a punctured hose and assumed that water pressure and water current always have a direct one-to-one relationship, we would falsely believe that the same water current existed on both sides. But, of course, if there were truly no change in the current on both sides of the puncture, it would be impossible to explain the water spouting in the middle. This is precisely the mystery we face with the light bulb; we measure what we believe is unchanging current and no energy difference within the electron stream before and after the light bulb, yet we have radiating heat and light in between. This mystery is solved once we realize that radiating heat and light are simply electron clusters "spouting" from the circuit and out into space, while our "constant current" readings do not actually indicate unchanging current, but merely a constant magnetic field pressure before and after the light bulb.

As mentioned above, once light is understood to be clusters of expanding electrons, it follows that the various frequencies (colors) of light correspond to various cluster sizes. A higher frequency of light is typically thought of as having shorter wavelengths of "pure energy," but can now be seen as a beam of smaller electron clusters. Likewise, if the frequency is decreased, shifting toward the red end of the visible spectrum and eventually into the longest wavelengths, known as infrared radiation or heat, this actually describes a progression to larger and larger cluster sizes, with heat being the largest. As also mentioned earlier, the feeling of "heat" does not indicate the existence of "heat energy," as believed today, but is merely the form of internal representation that our minds use to notify conscious awareness of the fact that large and potentially harmful electron clusters are striking our skin.

It is important to note that "cluster size" does not refer to the *physical size* of the expanding electron clusters, but to the *number of electrons* within the clusters. As with the electron clouds in the previous chapter, electron clusters are not atomic objects and have no true size definition in the atomic realm. The expanding electron clusters composing a light beam may double in size after traveling a given distance, but this does not represent a change in the frequency of the light since the clusters maintain a constant number of electrons. An electron cluster with a particular number of electrons will stimulate the cells of the retina to produce a given color experience in the brain regardless of how much it may have grown in the atomic realm. A given electron cluster in the visible spectrum is simply accepted back into the subatomic realm among the atoms of the corresponding retinal cells based on the number of electrons composing the cluster, regardless of its apparent size in the atomic realm. The retinal cells can be seen as tools that are specially designed to respond to certain electron groupings, possibly even splitting the arriving clusters apart and directing the resulting electrons along the neural fibers toward the appropriate visual areas of the brain. This means that the arriving light itself may provide a stream of electrons that help power the operation of our visual neurons –

a process that likely also occurs in the photosynthesis of plants. The sizable heat given off by the brain can also now be seen as excess electrons escaping, but in the form of larger electron clusters.

Solar Cells – The Mysterious Conversion of Light into Electricity

---

Solar cells are wafers of nearly pure silicon that generate electricity when light shines on them. Solar cells are very familiar devices today, but how does science explain the conversion of pure "light energy" into flowing electrons? The answer is that it doesn't. Instead, today's theory of solar cell operation claims that the incident light imparts energy to the electrons in the silicon wafer, causing them to flow out of the silicon and around the circuit. This describes a closed system of electrons, where the naturally occurring electrons in the solar cell, wire, and circuit components are merely put in motion by incident light, and whose motion through the circuit somehow powers the circuit components. There are many ways to show that this belief is fatally flawed, one of which is demonstrated by the following example:

Consider a battery-powered circuit that is attached to a sphere for the purpose of depositing surplus electrons on the sphere to charge it with static electricity – using today's electric charge terminology for this example. Clearly, when the charged sphere is removed from the circuit, it carries away the surplus electrons that were deposited on it from the battery, leaving the battery drained of charge and voltage. Now, if we replace the battery with a solar cell, today's theory would state that the charged sphere now holds many of the electrons that once resided within the closed circuit of circulating electrons driven by the incident light. By this logic, the circuit is now depleted of electrons and should be less able to charge another sphere, and after a few such charges the circuit should be unable to provide any charge at all for any electrical purpose thereafter – even with continual incident light on the solar cell. This means that the solar cell should now be unable to develop any voltage across it at all in response to the incident light until the missing electrons are replaced from outside the circuit.

However, *Expansion Theory* has shown that the only viable explanation for the conversion of "electrical energy" into other forms ("radio-wave energy," "heat energy," "light energy," etc.) is that the electrons themselves are leaving the circuit. Therefore, *Expansion Theory* also holds that the ability of a solar cell to continually power transmitting antennae and heaters and light bulbs means that solar cells are devices that convert incident light into electrons that are injected into the circuit to replace those that are lost. This also means that a solar cell would also be able to continually charge sphere after sphere as long as there was incident light, and it would not cease to operate – as today's theory would require. But today we have no explanation for such a mysterious conversion of energy into matter (light into electrons) by a simple wafer of silicon; *Expansion Theory* provides the answer that this is merely a process of incident electron clusters breaking apart in the solar cell and contributing electrons to the circuit. So, a solar cell is not a device that merely drives the existing electrons in the circuit to flow in response to incident light, but rather, a device that extracts electrons from the incident electron clusters within light and injects them into the circuit.

This electron-cluster description of light is also consistent with many other observed properties of light. To begin with, even though light is currently considered to be "pure energy," it is also known to have momentum – it strikes objects with a tiny impact force. In fact, one method of futuristic spacecraft propulsion often discussed by scientists is that of a *solar sail*, which is a very large, thin sail that can be pushed along through space solely by the pressure of the light from our sun. The fact that light has momentum, which is a term from classical physics that refers to *mass* in motion, can be better understood once light itself is seen as a phenomenon of mass (i.e. clusters of electrons) and not "pure energy."

Superconductivity

Another phenomenon related to this new understanding of heat and light is that of superconductivity. Although a wire is a conductor, it is not a perfect conductor and offers a small amount of resistance to current flow, which generates waste heat. The phenomenon of superconductivity refers to the discovery that a super-cooled wire (near a temperature of absolute zero) offers no resistance and generates no waste heat. This is a poorly understood phenomenon today, made even more mysterious by the fact that some materials have been found to offer zero resistance at temperatures that, while still quite cold, are much higher than absolute zero. The ultimate goal of superconductivity research is to discover a material that offers zero resistance at regular room temperatures to eliminate waste heat in all electronics in common use, but there is no clear understanding of how to achieve this today. Further, a closer look at superconductivity shows an even deeper mystery than is currently recognized.

A wire is typically super-cooled by immersing it in an extremely cold liquid, such as liquid helium. But this process does not cause the wire to "absorb coldness" from the liquid to remain cool; rather, the liquid must continually remove the heat that would ordinarily build up in the wire due to the current flow. The heat generated by current flow begins in the form of atomic vibrations in the wire, which, if left unchecked, eventually build and begin to radiate heat into space. The cooling effect from surrounding the wire in a super-cooled liquid merely preempts this *radiating* heat loss by surrounding the wire with nearly stationary, densely packed molecules of liquid that absorb and conduct away the atomic vibrations within the wire. This is a *continual* process of the liquid conducting the internal *thermal* heat vibrations away from the wire, which merely removes the waste heat in the form of thermal conduction rather than radiation. Therefore, even with super-cooling, we still have a situation where waste heat continually escapes from the wire, yet it is claimed that there is no loss of waste heat – a claim that is backed up by measurement and that can amount to a power savings of several percent. How can this continual drainage of waste heat to maintain super-cooling also paradoxically eliminate the problem of waste heat at the same time? This is a deep mystery in today's science, which accentuates the fact that superconductivity is a phenomenon that has been discovered and exploited to some extent, but which is not at all understood.

Expansion Theory provides a much clearer picture of this phenomenon. As shown in the light-bulb discussion, radiating heat is actually large electron clusters that expand off into space when large pools of freely expanding electrons are allowed to form between vibrating atoms. Both the eventual radiation of these heat clusters into space and the initial thermal vibration of the atoms in the wire are driven by the *expansion of the electrons* as they cross over from the subatomic realm into the atomic realm. This is evidenced by the fact that, if the atomic vibrations were caused by mere classical collisions between electrons and atoms, then the flowing electrons would have to be slowed down after imparting some of their kinetic energy to the atoms. Yet, we do not see an ever-increasing buildup of electrons further along a wire as the electrons that speed out of the battery encounter a traffic jam of ever-slowing electrons ahead.

Therefore, the paradox of how the energy of atomic vibration is continually conducted away by the surrounding super-cooled liquid without removing energy from the circuit is solved. The atomic vibrations were not driven by the kinetic energy in the current flow, but by the *crossover effect* of subatomic expansion as the jostled electrons briefly emerged from the subatomic realm within the current flow, freely expanded into the atoms, then returned to the current flow. Therefore, the real waste energy – the free expansion of electron clusters (radiating heat) away from the circuit and off into space – is eliminated by dampening the atomic vibrations caused by the *crossover effect*.

This understanding of superconductivity has many important implications. For example, we can now see that superconductivity is not strictly a process of cooling, but a process of preventing the pooling of electrons between vibrating atoms. This can be achieved either by dampening atomic vibration by conducting it away (i.e. cooling), or by inventing materials that either naturally have more internal atomic dampening or whose atomic vibrations disrupt the process of electron pooling so that electron clusters cannot form. This understanding can be used to guide further development toward the ultimate goal of room temperature superconductivity.

As a further example of the benefits of this new understanding, consider superconductors in space. Outer space is generally considered to be an extremely cold place, often represented as being as cold as absolute zero in the shade. This thinking prevails because the near-perfect vacuum of space means that there are hardly any surrounding molecules to vibrate and create thermal heat. Logically then, it may seem that all wires would become superconducting in space – at least in the shade. However, a closer look shows that this is not a forgone conclusion.

It is true that there are no surrounding molecules to cause thermal heat (warmth), but there are also no surrounding molecules to conduct thermal heat away, dampening atomic vibrations and cooling the wires. Without the nearly stationary, dense surrounding molecules of a super-cooled liquid the atoms of objects are free to vibrate without dampening – provided there is a process causing thermal vibration, such as current flow in a wire, incident sunlight, or the temperature-regulating system within a space suit. As seen in videos of spacewalks, tether cords bend freely and the material of spacesuits remains flexible and mobile. If space were anywhere near absolute zero tether cords would immediately become brittle and would readily snap, while the outer material of space suits would freeze solid and shatter when an astronaut attempted to move. An object in the shade could radiate heat and eventually cool to near absolute zero in space since there are no surrounding vibrating air molecules to warm it, but the object could also be kept relatively warm with little effort since there is also nothing to conduct its warmth away.

Therefore, we should not expect current-carrying wires in space to experience a super-cooling effect as if immersed in liquid helium, but rather, they would undergo internal thermal heating that may well be typical of electrified wires at room temperature on Earth. This means that electron pooling and radiating heat (electron clusters) would occur in space – even in the shade – with very likely a similar loss of waste heat to that experienced at room temperature on Earth. Therefore, we would very likely receive an unpleasant surprise if we designed satellite circuits with the expectation that they would become superconducting in space – an expectation that might seem quite reasonable according to many of our current beliefs about heat, superconductors, and the "cold vacuum" of outer space.

 Mysteries of Lightning

Although lightning is an electrical phenomenon, it is also a very visible phenomenon as it emits light, making it pertinent to the current discussion of light. Although lightning is little more than a very large spark, it is shrouded in mystery today. While we know that electricity is the flow of electrons, some descriptions of lightning include mention of the flow of *positive* charge between the ground and the clouds. Such a characterization introduces quite a mystery, since this could be interpreted to mean that either nuclear protons or entire ionized atoms (the only agents of positive charge in today's *Electric Charge Theory*) zip between the ground and the clouds within lightning strikes. Although such an interpretation could be made of explanations involving the movement of "positive charge" within lightning strikes, if this were actually the case it would be a very strange and unexplained phenomenon indeed. Further, it has recently been discovered that there are various types of flashes that occur above the clouds during lighting strikes. These effects are being categorized, named, and studied today, but are also currently considered to be puzzling mysteries.

Expansion Theory shows that lightning must always be a flow of electrons expanding forcefully across a gap, and is never a phenomenon involving the flow of "positive charge" across a gap, removing the first mystery. Also, a lightning strike can now be seen as a forceful burst of *expanding matter* between clouds and the ground, rather than a bolt of energy. This removes the mystery of why there might be flashes of light above the clouds merely because "electrical energy" sparks between clouds and the ground. Instead of this simple point-A to point-B flash of "energy," we can now see that the burst of matter particles in a lightning bolt would likely have a backward kick, and would also be unable to stop instantaneously once it reaches its target. All matter must push backward in order to move forward and all matter has momentum that takes time and effort to stop. Therefore, the various classes of light flashes above the clouds would simply be the result of electrons (and possibly ionized atoms) being rapidly pushed upward out of the clouds, either as a kick-back effect in cloud-to-ground strikes or as an impact effect in ground-to-cloud strikes. This violent churning region of electrons and ions above the clouds would understandably form electron clouds and clusters of all sorts, generating electric fields, radio waves, and light flashes.

 **A Prism is a Tiny Mass Spectroscope**

Also, the refraction of light into a rainbow or spectrum of colors by a prism can now be better understood when light is seen as tiny clusters of matter. This process is a form of *spectroscopy*, which today is defined as the process of separating a beam of energy or particles into its component parts, known as a spectrum. Although the most well known example of spectroscopy is perhaps the separation of white light into its component colors, physics labs routinely perform *mass spectroscopy* as well. Mass spectroscopy is the deflection of a beam of speeding *particles* around a corner, typically by using electric or magnetic fields, resulting in the heavier particles being deflected the least due to their greater mass and momentum, and the lighter particles deflecting the most. This separates the *particle* beam into its components, from heaviest to lightest particles, in a manner that is analogous to the separation of "light energy" into a rainbow of colors using a prism.

However, we can now see that this is no longer just an analogy. The splitting of light into its component colors is actually not a separation of "energies" but a miniaturized version of *mass* spectroscopy, with the larger and more massive electron clusters bending through the least angles and the lighter clusters bending through the greatest angles. And indeed, red light is bent through the smallest angle, while violet light is bent through the greatest angle by a prism (Fig. 5-3).

Since the red light at the low-frequency end of the visual spectrum is known to have a longer wavelength than the violet light at the high-frequency end, this would correspond to a larger number of electrons in the clusters of red light. Therefore, it is not surprising that red light is bent the least due to its more massive electron clusters.

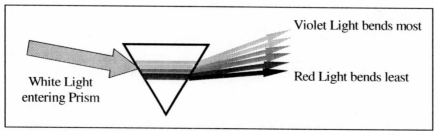

Fig. 5-3 Tiny Mass Spectrometer: White Light Separated by Prism

One of the well-known qualities of light is the fact that all frequencies of light (and, in fact, all forms of electromagnetic radiation) travel at the same speed in a vacuum – known simply as the *speed of light*. This observed behavior is given a clear physical explanation in the earlier description of the light bulb operation according to *Expansion Theory*. The production of electron clusters continually radiating off into space is purely a result of the nature of agitated, pooled electrons on the surface of the hot filament. The *size* of the clusters is determined by factors such as the current flow in the circuit and the filament material, but the production of a continual beam of light is only dependent on a continual supply of surplus, agitated electrons on the surface of the filament. As long as there is a continual presence of such electrons, there is a continual process of these electrons expanding out into space at a rate determined by their natural subatomic expansion. Therefore, it is perhaps not surprising that there is a uniform speed of production and progression of this sea of freely expanding electrons off into space – known as light –regardless of the size and number of electron clusters into which these electrons are grouped (Fig. 5-4).

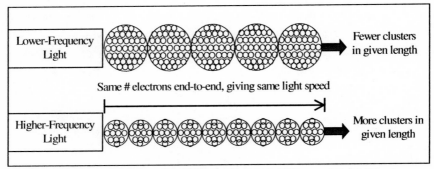

Lower-Frequency Light

Fewer clusters in given length

Same # electrons end-to-end, giving same light speed

Higher-Frequency Light

More clusters in given length

Fig. 5-4 Two Light Beams: Same Speed, Different # of Clusters

Different frequencies of light are also considered to have different energy content in Standard Theory. *Expansion Theory* suggests a number of reasons why higher frequencies are characterized as having greater "energy content" today. First, referring back to the production of light at a light bulb, a higher frequency (smaller cluster size) occurs when the atoms of the light bulb filament vibrate faster. And, heating the filament to this higher temperature requires a greater current through the circuit, which means higher-frequency light would generally be accompanied by a greater drain on the power source. Therefore, a greater amount of "energy" is put into producing higher frequencies of light.

Secondly, higher-frequency light would be more likely to penetrate materials due to its smaller cluster sizes, and also to a greater depth if smaller clusters remain intact longer and encounter fewer atomic collisions. The smaller cluster size would also mean a beam of high-frequency light would fit more clusters along a given length than a low-frequency beam, as shown earlier in Figure 5-4. Therefore, since both lengths of light beam would enter a material in the same amount of time due to their equal speeds of travel, more electron clusters would enter the material in a given time period for a higher frequency of light. Although this does not necessarily translate into a significant difference in the number of electrons striking the material since each of the more numerous smaller clusters also has fewer electrons, it does mean that whatever response a material has to an *overall* electron cluster now occurs at a greater rate. For example, as mentioned earlier, a brighter light beam in conscious experience results from a more rapid firing of visual neurons, which is caused by a greater number of electron clusters striking the eye per second (greater light intensity). So, a beam of violet light, with far more electron clusters per length (and per second) than red light, would cause a far more rapid firing of neurons, which would tend to create a brighter conscious experience. This would naturally lead to the further characterization of a higher "energy content" in the higher-frequency light.

Mysteries in Light-Transmission Theory

In the field of optics it is well known that, although light always travels at the universal *speed-of-light* in a vacuum, it varies in speed when passing through different materials. Light attains its highest speed in a complete vacuum, where there is no matter to impede its progression through space, and travels slower through material such as glass. Stated more formally, the speed of light varies with the *index of refraction* of the medium it is traveling through. This is the reason light rays bend when they pass from air into water or glass, and the degree of bending determines the index of refraction of the medium – a relationship known as *Snell's Law*. However, today's quantum-mechanical concept of light is as "photons of energy," which are considered to be separate particles or "packets of energy" that fly through space somewhat like bullets fired from a gun. This bullet-like travel through space is implied in today's *Quantum Theory* since no form of photon self-propulsion or ongoing external push is included in the photon theory of light, which even allows for a single photon to be produced and to travel continually through space at light speed.

 Unexplained Behavior as Photons pass Through a Material

This bullet-like theory of light has sizable problems, which can be seen by observing the behavior of light passing through a transparent material, such as glass. When an actual bullet passes through a material it is slowed down and emerges from the material at this slower speed; the bullet cannot speed up of its own accord to return to its previous speed once it leaves the material. Yet, light slows down when it passes from air into glass, then returns to its original speed upon exit. This behavior is only partially explainable in today's science by considering that photons would zigzag as they bounce off the atoms within the glass, causing them to take a longer, more winding path that effectively slows their forward progress within the glass. This is undoubtedly *part* of the explanation since light beams do widen or disperse while passing through a block of glass, but this means the photons must also slow down with each collision. Light is known to have momentum, and to impart some of that momentum to objects that it strikes, which means it must slow down as a result. Otherwise, if light imparted some of its energy to another object in a collision, yet kept traveling at the same speed after the collision, this would be an unexplainable creation of energy at each collision – a violation of the *Law of Conservation Of Energy*.

In fact, this energy transfer from the photons to the glass molecules is easily verified by noting that glass heats up when an intense light beam passes through it continually. Since light does not somehow carry heat along with it, today's science could only explain this heating of the glass as an energy transfer from the photons to the atoms during collision; this is how lasers cut through material or vaporize tissue in laser surgery. But, if this "energy" is not taken from the momentum – and therefore the speed – of the photons, then there is no explanation for the heating. Yet, light enters and leaves the glass with no net change in the speed or energy content of its photons in Standard Theory, despite clear evidence that it generates heat within the glass (Fig. 5-5).

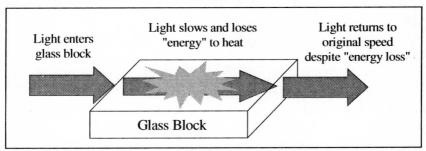

Fig. 5-5 Light Loses Energy yet No Net Speed or Energy Reduction

Solving the Light-Transmission Mysteries

Expansion Theory, on the other hand, does not consider light to be separate "energy photons" coasting through space after being fired from a light source, but rather, a continuous beam of expanding electron clusters constantly pushing one another along. The "energy" for this does not come from some initial firing of "photons" into space as if from a light gun, but from the continual inner expansion of all electrons. As light pushes its way through the glass, it does indeed collide with atoms and cause them to vibrate faster, becoming hot to the touch, but the light beam is still driven onward by the continual inner expansion of its electron clusters. The light does not "lose energy" in this process, and simply resumes its original speed upon exit, as determined by the ongoing expansion of its electron clusters once it is unimpeded by the glass atoms. There is no longer any mystery of light slowing, losing "energy" to the glass, and then speeding up again with no apparent energy difference.

Note also, that if the heated glass begins to radiate heat into space, this indicates a loss of electrons from the light beam to create the large electron clusters of radiant heat. This redistribution and loss of electrons could also contribute to the slowing of the light beam while expansion pressure builds once again within the beam to compensate. This would also explain the frequency shifts that can be caused by some materials, since the remaining cluster sizes in the light beam may become altered by this process.

A New View of the Electromagnetic Spectrum

As mentioned in the previous chapter, today's electromagnetic spectrum of "radiant energy" assumes that all forms of radiation, from low-frequency radio waves through to visible light and up to high-frequency X-rays and gamma rays, are merely different frequencies of the same electromagnetic radiation. However, the explanation of radio waves in the previous chapter shows that both the method of production and the resulting nature of radio waves are very different from the description of light just provided. Radio waves are alternating bands of freely expanding electrons (detached magnetic fields) produced by oscillating current in a wire or antenna, while light is a sea of expanding electron clusters produced by passing electricity through a resistor and essentially overheating it. Therefore, radio waves are much like true waves of compression composed of expanding electrons, somewhat analogous to the pressure waves of sound in an atmosphere of air molecules; however, light is not a wave phenomenon at all, but rather, it is a sea of expanding electron clusters.

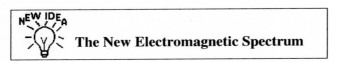

The New Electromagnetic Spectrum

This means that our current concept of the continuous electromagnetic spectrum of "energy waves" is in error; instead, there are actually two very different categories of radiation in this spectrum, neither of which are true waves of "pure energy," while one is not even a wave phenomenon at all. The lower frequencies have a nature similar to the electron bands of radio waves, while the higher frequencies are manifestations of electron clusters as in visible light. Instead of the smooth, continuous spectrum of electromagnetic radiation known today (Fig. 5-6, top), we should actually divide the electromagnetic spectrum into a lower-frequency section of *electron compression bands* and a higher-frequency section of *electron clusters* (Fig. 5-6, bottom).

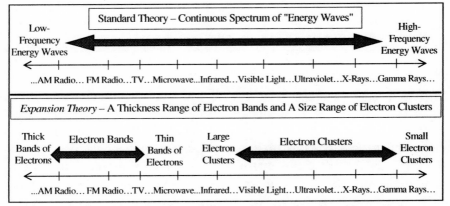

Fig. 5-6 Electromagnetic Spectrum Today vs. Expansion Theory

Interestingly, the methods used to produce the frequencies in the *electron band* section of the spectrum are quite different from those used to produce the frequencies in the *electron cluster* section. Radio waves through to microwaves are typically produced using oscillators – either wire antennas where electricity oscillates back and forth or hollow tubes designed to produce resonating or oscillating electromagnetic waves. Yet, infrared radiation through to gamma rays is essentially produced by heating a material until it "radiates energy."

No device exists where we can simply turn a knob to smoothly move along the electromagnetic spectrum, progressively producing radio waves, then microwaves, then light waves. In fact, scientists who specialize in electromagnetic radiation are well aware of a segment of the electromagnetic spectrum that they sometimes refer to as the "forbidden zone" due to their difficulties in producing and understanding it. This "forbidden zone" of terahertz or "T-Wave" radiation lies precisely where the transition from electron bands to electron clusters is shown in Fig. 5-6, according to *Expansion Theory*, and has some very unique properties that scientists are currently attempting to understand and exploit.

So, even our physical production of electromagnetic radiation today suggests the two-category spectrum found in *Expansion Theory*. Further evidence of these separate categories of radiation will be seen in the next chapter, where space missions and astronomical observations are discussed further.

Throughout the discussions so far, sizable problems have been shown with the concept of "pure energy," with it being repeatedly replaced by subatomic and atomic expansion instead. As a result, the concept of "energy" has been playing an ever-diminishing role in the description of the world around us. Yet, "energy" is a term that is used extensively today to describe both everyday experience and scientific endeavors. Therefore, we now turn to the topic of "energy" itself, taking a look at both the everyday and the scientific use of this term.

The True Nature of Energy

We are all familiar with the concept of *energy*. In today's terminology, the sun's energy warms us and causes plants to grow; objects transform gravitational energy into kinetic energy as they roll downhill; electrical energy runs the devices we all rely on daily, and chemical energy powers automobiles, rocket ships, and even our own minds and bodies. Energy can be emitted or absorbed – processes that science tells us are actually a transformation from one form of energy to another. In fact, this is the foundation of our *Law of Conservation Of Energy* – the fact that energy can never be created or destroyed but merely transformed from one form to another. Yet, on closer examination, how does today's concept of energy differ materially from the medieval concept of magic? Aside from a terminology change from "magic" to "energy" and the invention of formalized mathematical models of its observed behavior, do we truly have a deep understanding of the actual physical nature of energy and the reasons why it behaves as it does? Do we know why "gravitational energy" attracts rather than repels, or why magnets and "charged" objects attract or repel other objects?

The previous discussions have shown that today's beliefs about energy are filled with unexplained abstractions, mysteries, and even clear violations of our most cherished laws of physics – a state of affairs that is indeed reminiscent of the magical beliefs of the past. *Expansion Theory* shows that all known forms of ethereal "energy" are actually forms of *physically expanding matter* instead, which means that today's energy terms are merely euphemisms for the overlooked or misunderstood principle of expansion in nature, as shown below in Table 5-1:

Euphemism	Physical Description in *Expansion Theory*
Gravitational Energy	An effective attraction between all objects, which is actually caused by their ongoing expansion as objects composed of continually expanding atoms.
Strong Nuclear Force Energy	The natural cohesion of protons and neutrons in the nucleus of an atom due to their tremendous ongoing subatomic expansion against one another.
Electric Charge Energy	Effective attracting or repelling forces between objects; caused by the tremendous ongoing expansion of subatomic particles in a *crossover effect* between subatomic and atomic realms.
Chemical Bond Energy	A manifestation of the *crossover effect* between subatomic and atomic realms, occurring between individual atoms rather than overall atomic objects.
Magnetic Energy	Clouds of expanding electrons surrounding conductive objects, causing attracting or repelling effects via the *crossover effect* between subatomic and atomic realms.
Electromagnetic Energy	Bands or clusters of freely expanding electrons that continually push one another through space due to their ongoing inner subatomic expansion.
Kinetic Energy	The appearance that absolute energy of motion is "possessed" by individual objects, but which is actually only a *relative* motion effect between objects.

Table 5-1 Today's Energy Terms According To Expansion Theory

This table captures the essence of the discussions in all previous chapters, showing that *Expansion Theory* – and indeed the alternate *theory of everything* that it embodies – is essentially a theory about the true nature of *energy*. That is, although the energy terms and concepts listed as euphemisms in the table have proven to be very useful *models* of the various manifestations of expanding matter, their *literal* interpretation as actual "energy" phenomena in nature has sidetracked us from uncovering the much-sought-after Theory Of Everything. These energy concepts have allowed us to organize subatomic and atomic expansion into a very functional and precise system of conceptual and mathematical abstractions that have served us well – and may well continue to do so – but it is imperative that we also understand the physical reality that they embody. *Expansion Theory* suggests that it is this physical understanding underlying our current energy abstractions that we are actually seeking when we pursue the Theory Of Everything – an understanding that is summarized in Table 5-1.

Perhaps the least explored item in Table 5-1 so far is the term *"kinetic energy."* This term is unique among those listed, since it is the only candidate for consideration as a true form of energy. However, it is inappropriate and misleading to call kinetic energy a *form* of energy since the discussions so far have shown that it is a mistaken belief that a mysterious, ethereal "energy" phenomenon exists in nature, taking on various forms. Such a belief is merely a misunderstanding of expanding matter. Rather, "kinetic energy" is the *only* form of energy that could reasonably be considered to exist – in the sense of an ethereal, energetic phenomenon in nature. That is, although "kinetic energy" refers to *physical* objects in motion, the physicality is only the object itself – not its motion. Motion alone is not a material concept, and can, in fact, be quite ethereal when we consider that it is often completely arbitrary whether an object is in motion or not. As shown in the rethink of *Newton's First Law of Motion* in Chapter 3, a given object can be in motion relative to one object but completely stationary relative to another; yet, an object in relative motion can make a very real impact when it strikes another object. Therefore, since the kinetic energy of an object in relative motion is the only example of a phenomenon that truly has ethereal, energetic elements (compared to the other entirely physical descriptions of "energy" in Table 5-1), it is the only phenomenon that could satisfy the definition of energy as it is conceived today.

This singular form of energy in nature means that use of the phrase "kinetic energy" is similar to describing a "round circle." The phrase "round circle" erroneously implies that there are other forms of circles. Likewise, the phrase "kinetic energy" refers to the only phenomenon that could possibly be considered a form of ethereal energy, and therefore is as misleading and redundant as the phrase "round circle." There is only one type of circle, and only one form of energy in existence – the relative motion of objects. However, since the term "energy" is firmly established today as a mysterious form-changing phenomenon due to a misunderstanding of the various manifestations of expanding matter, it may be difficult and confusing to redefine "energy" to only mean the kinetic energy of relative motion. Instead, perhaps an entirely new term should be invented to replace "kinetic energy," highlighting the fact that the term "energy" does not apply here since it is actually a euphemism for expanding matter, while the relative motion of objects is the only phenomenon that truly qualifies as active and immaterial.

It should also be noted that *all* of the active phenomena in nature – subatomic expansion, atomic expansion, and relative object motion – are ultimately manifestations of the singular phenomenon of subatomic expansion that drives and defines our entire universe. The expanding atom is merely a structure composed of expanding subatomic particles, and all relative object motion also ultimately has its origin in subatomic expansion. For example, a speeding bullet is in relative motion with the gun that fired it as a result of an explosion of gunpowder whose "chemical bond energy" is a manifestation of the *crossover effect* of subatomic particles between individual atoms. The same is true of any relative motion of objects caused by muscular force, which is also driven by "chemical energy." All motion caused by wind is also ultimately driven by "electromagnetic energy" from the sun (a manifestation of subatomic expansion), which warms the Earth and causes pressure differences in the atmosphere that drive the wind. The nature and implications of this fundamental and ubiquitous subatomic expansion underlying our entire universe will be explored in more depth in the following chapter.

Without the physical explanations of today's energy terms provided by *Expansion Theory* we are completely without physical explanations for much of the power and technology we wield. We are, in effect, modern-day wizards with more precise control over our magical devices, creating the illusion of true knowledge and understanding of nature. This illusion is reinforced via the development of scientific terminology, abstract equations, and engineering know-how that often originates in trial-and-error or accidental discovery, joining a growing scientific legacy that is then passed on to future generations. Yet, the many mysteries and violations of the laws of physics pointed out so far in even our most basic science show that we produce and use many everyday devices whose fundamental operating principles we do not truly understand – either as individuals or as a collective society. There is a sense of this lack of understanding in many of our scientists today as the search for deeper answers continues in earnest; however, current efforts to arrive at the final Theory Of Everything remain largely confined to our existing flawed paradigms. This can be seen even in the development of our most radical theories today, such as *Quantum Mechanics* and *Special Relativity*, which will now be explored.

Around the turn of the twentieth century there was a concerted effort to shake off existing paradigms in an attempt to truly understand the underlying physics of our universe and the phenomena in the world around us. We had reached a point where both our scientific ideas and our technology had advanced far enough that we were able to both intellectually and experimentally investigate the atom, the subatomic realm, and even the very nature of energy itself. There was a great deal of discussion, disagreement, and debate amongst the researchers of the day, since their discoveries about the structure of the atom, the behavior of subatomic particles, and the nature of energy were often surprising and confusing, leading to interpretations that were often controversial and counterintuitive. Many scientists refused to accept these new theories initially, only being brought into reluctant agreement when they appeared to be validated by experiments.

Even the creators of many of these new theories have often admitted to not fully understanding them, but believed they had nonetheless managed to capture something of the apparently bizarre and mysterious nature of our universe. This sentiment is summed up in a famous quote by Neils Bohr, one of the founders of *Quantum Theory,* "If Quantum Mechanics hasn't profoundly shocked you then you haven't understood it yet"; many proponents and practitioners of such theories can be heard echoing similar sentiments even today.

The now firm installment and acceptance of such theories into our science has resulted in the elevation of several of our latest and most mysterious theories almost to the level of oracles that seem to know more about our universe than the human mind is capable of comprehending. This state of affairs in our science has left us with little choice but to simply accept what these "oracles" seem to be telling us and to marvel at the apparent strangeness and mystery of nature. However, the understanding provided by *Expansion Theory* allows us to break out of this pattern of thought and see our universe as it really is – a simple, comprehensible universe whose entire structure and operation is merely based on the single unifying principle of expanding matter. Toward this end, we move next to an exploration of one of today's most mysterious theories – the theory of *Quantum Mechanics*.

Quantum Mechanics – Is It All Just A Misunderstanding?

The theory of *Quantum Mechanics* refers to a body of knowledge that deals with the microscopic scale of the physical world, namely, subatomic particles and energy. It sits opposite *Classical Mechanics*, which deals with large objects composed of atoms and is largely embodied in Newton's three laws of motion. Since it was discovered by experimentation that the laws of *Classical Mechanics* do not seem to apply to the subatomic realm and to energy, *Quantum Mechanics* was eventually developed to complete the picture. Together, these two bodies of knowledge describe our universe on both macroscopic and microscopic scales.

Despite the widely recognized bizarre and mysterious conclusions about our world that follow from *Quantum Theory*, many of its proponents also claim that it is one of the most important, elegant, and accurate theories known to science. Such radically opposing aspects of the same theory can at least be partially attributed to unexamined assumptions and logical oversights. For example, while both the subatomic realm and energy are indeed extremely important elements of our universe and the world around us, it is a logical oversight to claim that they are important manifestations of *Quantum Mechanics*. In actuality, they are manifestations of whatever fundamental truth may underlie our universe, and *Quantum Mechanics* is simply an abstraction that attempts to describe and model the resulting behavior. Of course, *Expansion Theory* suggests that this fundamental truth is actually the phenomenon of expanding subatomic particles, which would mean that it is the physical reality of *subatomic expansion* that is of such great importance, not the abstract model of this physical reality proposed by the theory of *Quantum Mechanics*.

Also, as mentioned in the previous chapter, the accurate experimental agreement of many of our current theories is often the result of intense effort by generations of scientists to ensure that this is precisely the case. The current form of many of our theories and their associated mathematical models is often the result of trial-and-error, beginning with a variety of large-scale concepts, followed by a series of refinements to bring the most promising of these concepts into eventual accurate agreement with experiment. Therefore, the accuracy of *Quantum Mechanics* is certainly a tribute to the years of effort that have gone into ensuring accurate agreement between calculation and experiment, but this is often misconstrued as evidence that this theory must then be the literal description of the physical world.

The Misunderstanding of "Quantized Energy"

In fact, a number of observations that currently stand as "quantum-mechanical mysteries" have already been shown to have much clearer and more rational explanations in *Expansion Theory*. Recall the discussion of atomic structure in the previous chapter, where it was suggested that the mysterious statistical appearances of orbiting electrons to form "probability clouds" are simply the result of bouncing electrons sampled at random points in mid-bounce. Also, recall earlier in this chapter that the current description of light as "photons of energy" is another mysterious quantum-mechanical concept that is clearly explained by the electron cluster concept of *Expansion Theory*. Such examples suggest that the bizarre nature of *Quantum Theory* is due to the fact that it is actually an attempt to describe subatomic expansion without any knowledge of its existence. With enough effort and resourcefulness we will eventually find an abstraction to model any observation we wish to explain, no matter what lengths we must go to in order to achieve this end. It is suggested that *Quantum Theory* is one such example where our mathematical and conceptual resourcefulness has prevailed over the actual physical truths, as shown by the concept of "energy quantization".

"Energy Quantization" is merely a reference to individual electrons.

Another feature of *Quantum Theory* is its explanation of energy. Light energy, in particular, is commonly used to explore and display the apparently quantum-mechanical nature of energy, which is largely embodied in the concept of "energy photons." These photons are considered to be tiny particle-like packets of energy, which can vary in the amount of energy they contain. This photon theory of light was supported by experiments that displayed effects that scientists could not explain via *Classical Mechanics*, such as the *photoelectric effect* to be discussed shortly.

These photons were further shown to be *quantized*, meaning that any energy difference from one photon to another cannot vary by an arbitrary amount, but must occur in discrete quantized jumps or steps. The size of these steps is extremely tiny, described by a value known as *Planck's constant*, after Maxwell Planck (1858-1947). The discovery of this tiny quantum jump between allowed energy levels was a revolutionary discovery since it meant that nature is not smooth and continuous at its most elementary level, but that even pure energy comes in tiny "grains" of some sort that are smaller even than photons. This discovery is an observed phenomenon that has been given a mathematical description in *Quantum Theory*, but it has never been explained in clear physical terms.

However, once again, this unexplained quantum abstraction and mystery has a clear physical explanation in *Expansion Theory*. Recall that according to *Expansion Theory* these "energy photons" composing light are actually clusters of electrons, which clearly shows that the discrete "quantum-mechanical energy jumps" between them follow naturally from the fact that electron clusters can only vary by multiples of *whole electrons*. Small as it is, the single-electron variation between electron clusters is the likely explanation for the tiny "energy quantization" seen in experiments and described by Planck's constant in quantum-mechanical equations of "photon energy." This is yet another example showing that *Quantum Mechanics* can be a useful tool for investigating and describing previously unknown aspects of our world in the search for understanding, but it does not actually provide that physical understanding itself.

The Misunderstanding of the Nature of Light

Today's quantum-mechanical characterization of light as tiny packets of quantized energy also lies behind the well-known mystery or paradox referred to as the *wave-particle duality* of light. For centuries there has been an ongoing debate as to whether light is a wave or a particle. Today, it now appears that we have experimental evidence for light behaving like *both* a wave *and* a particle. *Quantum Theory* states that a traveling beam of light exists in a bizarre state where nature has not yet "decided" whether it will be a wave or a particle until it is detected. It is thought that the method of detection itself breaks nature's uncertainty and forces the reality of either a wave or a particle to manifest itself.

This concept does not state that the detection of light simply exposes whether it was *originally* transmitted as a wave or particle, since the same beam of transmitted light can be detected as either a wave or a particle simply based on the method chosen for its *later* detection. Instead, *Quantum Theory* states that it is only once the light is detected as either a particle or a wave that its originally transmitted nature is "decided" by the universe. That is, according to *Quantum Theory*, there is a bizarre effect in nature that reaches back in space and time instantaneously – even across billions of light years to distant stars – to define whether a wave or a particle was *originally* transmitted, based purely on the outcome of its *later* detection.

This mysterious and completely unexplainable claim of instantaneous backward time travel is the currently accepted scientific interpretation of experimental results today – a claim that is commonly held up as a key example of the bizarre and purely probabilistic nature of not only *Quantum Theory* but, presumably, of the universe itself. However, as will be shown shortly, *Expansion Theory* does not require such fanciful explanations of our experimental results; but first, it is important to clarify what we mean when we speak of light waves, and in fact, waves in general.

Waves and the "Wave Nature of Light"

The world around us has many examples of wave*like* behavior, and basic wave theory often represents such phenomena as purely conceptual, disembodied oscillations in space. Such wave theory states that when two waves line up in phase so that their peaks coincide, as well as their troughs, these peaks and troughs add together to form a single larger wave in what is known as *constructive interference*. Likewise, when two waves are out of phase so that the peaks of one wave coincide with the troughs of the other, the peaks and troughs cancel in *destructive interference* (Fig. 5-7).

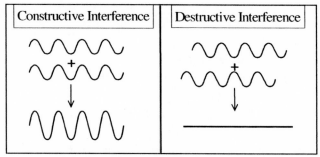

Fig. 5-7 Idealized Constructive and Destructive Interference

Although this is a common conceptualization of wave interaction, a closer look shows that it is an idealized abstraction that does not strictly apply to the real world from which it was extracted. In actuality, such a model describes idealized, disembodied waves frozen in space and time, then mathematically added together. While this may be a convenient *model* for discussing wave behavior, such idealized frozen waves are not representative of true waves in nature. The real waves around us are actually all dynamic manifestations of the wave-*like* behavior of *physical matter*, not literal idealized waves of frozen "energy" that neatly add together as shown. Both our diagrams and our mathematical descriptions of waves tend to represent them in this somewhat misleading manner. Even in situations where idealized standing waves *appear* to exist in the real world, such as a rapidly vibrating guitar string, they still result from ongoing dynamic wave-*like* vibrations of *physical matter*.

For example, sound waves are not waves of pure "sound energy," but rather, bands of alternating compressed and decompressed air molecules that are conducted along through the atmosphere in a sort of "domino effect" – a wave-*like* behavior of *matter*. Water waves are also the dynamic wavelike behavior of matter in the form of water molecules that rise and fall in an ongoing succession – again, a *matter wave* composed of water molecules. In fact, *every* waveform that we know of in nature – without exception – is actually the dynamic wavelike behavior of a large number of matter particles moving in unison. Although such wavelike behavior of matter may lend itself to representations in static diagrams of idealized waves that add mathematically, it is a conceptual oversight to assume that such pure "energy waves" literally exist in the real world and interact via such mathematical principles. The idealized frozen waves in the conceptualization of Figure 5-7 – presumably waves of pure energy – do not actually exist anywhere in nature.

It may be tempting to dispute this conclusion by referring to the example of light, which today's science tells us is composed of waves of pure energy – that is, at least some of the time. However, the true nature of light has been in dispute for centuries, and the concept of light as pure "energy waves" is merely an idealization that has been borrowed from wavelike behavior of *matter* in the world around us. A close look at this issue shows that the description of electromagnetic radiation as pure "energy waves" is an unsubstantiated human invention that does not exist anywhere else in nature, and which even leads to a violation of the laws of physics.

VIOLATION

Classical Wave Theory Violates the Laws of Physics

This is an important realization since it shows that not only do pure "energy waves" have no proven existence in nature but neither does their idealized wave *behavior* shown in Figure 5-7. We can artificially draw waves on paper and add them together mathematically such that they neatly reinforce or cancel each other, but this is merely a human conceptual artifact that does not strictly occur in this manner in the real world. And in fact, if it *did* occur, it would be a violation of the laws of physics.

To see this, consider the destructive interference shown in Figure 5-8. Instead of two idealized parallel waves, we have two parallel laser beams emitting identical pure frequencies of light such that they are also out of phase in the same manner as shown earlier in the right-hand frame of Figure 5-7. If the laser beams were brought together so that they overlapped, then, according to the pure wave theory of destructive interference, they would simply "cancel" each other out. That is, both lasers produce light energy that immediately vanishes into thin air – no heat, no other forms of radiation, but simply complete annihilation. This would be the expectation according to current wave theory regardless of the amount of energy involved, even many thousands of watts of power. This is not a transformation of energy from one form to another according to our laws of physics, but an absolute *destruction* of energy, and once again, a clear violation of the *Law of Conservation Of Energy*.

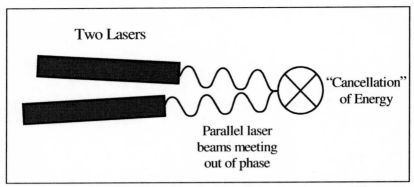

Fig. 5-8 Destructive Interference Violates the Laws of Physics

Of course, this complete disappearance of any arbitrary amount of energy without a trace simply due to a phase difference does not actually occur in reality, showing that the purely mathematical abstraction of "energy waves" and their idealized interference patterns is not a true description of the physical world. More importantly, this shows that light does *not* actually behave like a wave of pure energy even in the simplest possible experiment – that of two identical waves of a single pure frequency meeting, as shown earlier in Figure 5-7. If light *did* behave as an idealized wave then its energy would have to vanish into thin air in our laser beam example, which it clearly does not do. Yet, the idealized concept of constructive and destructive interference of light has long been held as proof of the more classical, wavelike behavior of pure "light energy." This flawed belief has persisted because selective evidence has been used to support the pure wave concept of light in our science. Despite the fact that this simple laser experiment seriously challenges the concept of pure waves of light energy, such evidence is overlooked in favor of other experiments that, on the surface, appear to fit the pure energy wave hypothesis. Once such classic experiment is the *Double-Slit Experiment*, which will be examined shortly.

This idealization of light as waves of pure energy is, in large part, responsible for the apparent paradox of the "wave-particle duality" of light in our science, yet the preceding example calls this very notion into question. How can we have a wave-particle paradox when we have yet to confirm that light ever behaves as a true energy wave? Further, the physical explanation of light in *Expansion Theory* shows that it is not sometimes a particle and sometimes a wave of pure energy, but that it is simply composed of clusters of dynamically expanding electrons – *i.e. light is entirely a manifestation of physical matter.* All evidence so far points to the fact that idealized waves, and waves of pure "energy," do not actually exist in nature, but are exclusively a human conceptual invention. Therefore, the wave-particle duality issue seems to be more of a conceptual oversight than a true paradox in nature. This possibility is examined further in the following discussions of some of the classic light experiments supporting our current quantum-mechanical beliefs about light and energy.

EXPERIMENT

Rethinking the Classic Double-Slit Experiment

The double-slit experiment, first performed by Thomas Young in 1801, has become a classic experiment in our science because it is thought to show both the wave nature of light as well as the paradox of its dual wave-particle nature. This experiment simply involves a barrier with two vertical slits, through which light is able to pass. The idea is that light passing through these slits will emerge on the other side and radiate outward as two separate cones of light that will interfere with each other in patterns of constructive and destructive interference. And indeed, with the proper selection of slit width and separation distance between the slits, the emerging light does interfere and cause light and dark bands on a far screen (Fig. 5-9).

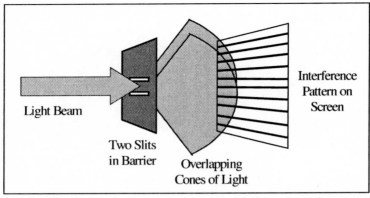

Fig. 5-9 Top View of Double-Slit Experiment

This experiment is thought to be analogous to the interference pattern that can be observed between water waves radiating from two nearby disturbances in a pond. Likewise, since light is thought to be a wave of pure energy – and, in theory, idealized waves that meet out of phase cancel each other out – the light and dark bands have traditionally been interpreted as constructive/destructive interference bands, validating the wave theory of light. However, a simple experiment with overlapping lasers, as discussed earlier, shows that light cannot be made to cancel itself out of existence in the manner idealized in abstract wave theory. In fact, it is a violation of the laws of physics to even *expect* energy to vanish in the physical world in this manner. So, although light and dark interference bands do occur within the two overlapping cones of light, the dark bands cannot be regions where waves of disembodied "light energy" cancel each other out of existence any more than waves of disembodied "water energy" cancel each other out in a pond. The analogous interference pattern with water waves results from the interaction of wavelike oscillations of *matter particles* (water molecules), and so, the logical conclusion is that the interference pattern in light is a similar manifestation of *matter particle* interaction. This would be expected if light were actually a sea of electron clusters radiating out into space, as shown in *Expansion Theory*.

NOTE

The double-slit experiment has been misinterpreted as evidence for the "wave theory" of light, but is actually evidence of an interaction between groups of *particles*.

A further reason the double-slit experiment is a classic is because it is also thought to show a deeply mysterious wave-particle paradox. The paradox supposedly arises when the intensity of the light beam is reduced to the point where only single photons of light are transmitted one-at-a-time from the light source. This means there should no longer be two cones of light interfering with each other, but rather, separate light photons traveling one at a time through one slit or the other. If these photons then proceed on and strike a photographic plate, the cumulative effect over time should develop into two bright spots on the plate – one for each slit that a photon might pass through. However, the actual result is an interference pattern much like the original experiment with the full light beam. This is thought to show that, even when light is sent toward the slits as individual particles one-at-a-time, it can still produce a wavelike interference pattern. It is completely unexplained how these individual particles seem to "know" how to land in a wavelike interference pattern on the photographic plate, doing so even though the scenario is no longer one of interference between two waves. This is the famous wave-particle duality paradox of the double-slit experiment, showing that even single particles of light mysteriously act as if they were waves passing through both slits simultaneously.

Taking a fresh look at this apparent paradox, we can now see that it is not actually a *wave-particle* paradox at all. It was just shown that even the original interference pattern in Figure 5-9 is not a proven "energy wave" phenomenon, but merely resembles known interference patterns between waves of *particles*. So, the actual mystery of the double-slit experiment is not that these particles of light somehow individually produce the interference pattern of "pure waves," but only that individual particles seem to still produce the original group *particle* interference pattern. With this clarification, the experiment simply leaves us with the question of whether this is truly a situation of separate particles fired one-at-a-time through the slits. As shown in the earlier discussion of light passing through a glass block, the current *Quantum Mechanical* theory of "energy photons" behaving like projectiles shot individually through space is *unsupported by experiment.* Despite the evidence against such an idea, this is precisely the claim that is made in the double-slit experiment; therefore, there is good reason to question even this facet of the experiment. Evidence is mounting for the possibility that the entire classic double-slit experiment may simply be a series of logical and experimental oversights regarding the nature and behavior of light. So then, what *are* we to make of the interference pattern when it is thought that individual photons are passing one-at-a-time through the slits?

Expansion Theory shows that light proceeds through space as a sea of expanding electron clusters pushing each other away from the light source. It is such an expanding beam of particles that arrives at the double slits to cause the original interference pattern on the other side – much as a wall of water molecules would behave after passing through two such openings. When the light intensity is reduced to the point where Standard Theory claims it produces single photons, *Expansion Theory* would maintain that continuous beams of electron clusters are still produced, but are very short-lived and sporadic since the light source is just on the verge of being turned off. We can picture the earlier description of agitated pools of electrons on the surface of a light-bulb filament, but rather than a continuous supply of electrons pooling and expanding off as a sea of clusters, the supply is only barely enough to produce sporadic bursts of clusters. Each burst would only extend a short distance through space before its supply of electron clusters suddenly cuts off at the source, in favor of producing another such burst a short while later.

Since a regular-intensity light beam would normally expand across the entire distance from the source to the detector, our detectors are designed to trigger based on this forceful stream of electron clusters continually arriving under their combined expansion pressure. If such a beam has its source cut off before arriving, it becomes an orphaned partial stream of electron clusters in mid-air that is free to dissipate much of its expansion pressure before reaching the detector. This is rather like a spring coiled against a wall, which shoots away from the wall and strikes a nearby target when it is allowed to uncoil. If the wall is removed part-way through the uncoiling process, the spring is unable to launch itself forward with its normal full strength and strikes the target with much less force. In fact, depending on when the wall is removed, the spring may not even make it all the way to the target. Similarly, the space between the source and the detector in the double-slit experiment may be filled with bursts of unseen partial light beams that are unable to trigger the detector but nonetheless pass through the slits and interfere with each other much as before. The occasional beam that is supplied by the source long enough to be detected (currently thought of as a single photon fired across the distance) would still be affected by interference between these unseen light beams; it would simply take longer to build up the interference pattern at the detector.

Implications of the Double-Slit Reinterpretation

This reinterpretation of the double-slit experiment carries with it some very deep implications, not only for the nature of light and energy, but also for *Quantum Theory* itself. First, it explains a long-standing experimental mystery in our science, showing that light need not be considered to have a mysterious inherent wave-particle dual nature that is fundamentally unresolved until detection.

Secondly, this reinterpretation shakes the very core of *Quantum Mechanics* since the wave-particle duality paradox of light is thought to exemplify and validate the concepts of quantum uncertainty and probability in nature that have become deeply woven into *Quantum Theory*. In fact, this wave-particle-duality concept has become so widespread in our science today that it has even been extended from a description of energy to a description of *matter* as well. In 1924, Louis de Broglie postulated that electrons, atoms, and even regular objects possess a mysterious wavelike nature as well. Via a simple mathematical equation one can calculate the theoretical wavelength of any such object, as if it truly had a wavelike nature. Even the wavelength of a truck can be calculated, and although the result is of no practical use, science today does consider such calculations to be valid applications of the wave-particle-duality principle of *Quantum Theory* – a principle whose core experimental support has just been shown to be erroneous.

Yet, as apparent proof of this seemingly mysterious wavelike behavior of matter, beams of electrons have been made to interfere with each other in a similar manner to the double-slit experiment, resulting in a similar "wavelike" interference pattern. Although this has been taken as proof of de Broglie's concept of matter having a paradoxical dual nature as a quantum "probability wave," it was just shown that such an interference pattern does not actually indicate a dual wave-particle nature at all, but merely interference between *particles*. Therefore, there is no particular reason to conclude that individual electrons have mysterious "quantum wave" natures, but merely that groups of electrons interfere in a manner much like many other known examples of interference between large groups of particles – just as we might expect.

Further, this supposed wave-particle dual nature of both energy and matter is embodied in perhaps the most central equation in *Quantum Mechanics* – the *Schroedinger Wave Equation*, named after Erwin Schroedinger (1887-1961), one of the founders of *Quantum Theory*. The *Schroedinger Wave Equation* is considered the cornerstone equation of *Quantum Theory*, and claims to capture the mysterious probabilistic "quantum wave" nature that supposedly underlies all energy and matter in the universe. Therefore, it is a sizable problem for *Quantum Theory* that the apparent experimental support for the "quantum wave" nature described by this central equation now appears to simply be a misinterpretation of straightforward particle interaction.

It appears that neither the double-slit experiment, de Broglie's matter wave concept, Planck's "quantum energy jump" concept, nor even the central *Schroedinger Wave Equation* stand any longer as the literal description of our universe. Yet, these concepts are the key support pillars for the theory of *Quantum Mechanics*, which now increasingly appears merely to be an abstract model built partly on unchecked logical oversights in experimental interpretation and partly on misunderstandings of expanding matter. This is the likely reason for repeated descriptions of *Quantum Theory* as being mysterious and bizarre, rather than it being our universe that is bizarre and incomprehensible. Once the principle of expanding matter is recognized, all of today's quantum mysteries vanish. This can be further seen in yet another classic experiment that has been taken as support for *Quantum Mechanics* – the Photoelectric Effect.

EXPERIMENT

 Rethinking Einstein's Photoelectric Effect

The *Photoelectric Effect* refers to another experiment whose results surprised scientists in the early 1900's, and which was given an interpretation by Albert Einstein that is thought to show yet another odd quantum-mechanical manifestation that supports *Quantum Theory*.

For simplicity, the experimental setup can be represented as a capacitor connected to a battery so that there is an electric field between the plates of the capacitor. The experiment involves shining a beam of light on one of the plates to knock electrons off the plate so that they are free to be pulled across the gap by the electric field and on through the circuit (Fig. 5-10). This creates a flow of electrons through the capacitor and around the circuit as long as the light shines on the plate. Although the actual apparatus is somewhat more complex, this is the basic concept.

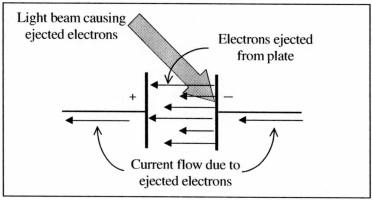

Fig. 5-10 The Photoelectric Effect

What surprised scientists about this experiment is that when they altered the light intensity they were unable to satisfactorily explain the resulting effect on the current flow using the classical wave theory of light. The three main surprises are presented below, each followed by the current *Quantum Mechanical* explanation as well as the explanation from the perspective of *Expansion Theory*:

- **Stopping-Potential Mystery:** If the battery voltage supplying the electric field between the plates is continually reduced and eventually reversed, the number of ejected electrons crossing the gap is also reduced and eventually stopped altogether as the reversed electric field is increased in strength. The minimum reverse voltage required to stop all current flow is called the *stopping potential*. So far, these results were not surprising since a reduced electric field should be less effective at promoting ejected electrons across to the other plate, while these electrons should actually be repelled by the other plate when the field is reversed. However, scientists *were* surprised to find that, once the stopping potential was reached, it didn't matter how much they increased the light intensity – *the same reverse voltage stopped all electron flow at any light intensity.* This was unexplainable, since classical wave theory states that more light energy (greater wave amplitude) should impart more energy to the ejected electrons, allowing some to overcome the stopping potential and cause current to begin to flow again.

  Quantum Explanation: An explanation for this mystery was found in the idea that perhaps light didn't behave like a wave in this experiment, but rather, like a particle. If the energy were contained in little particle-like energy packets (i.e. photons) then increasing light intensity would mean increasing the *number of photons per second* rather than increasing the *strength of the light waves*. Therefore, if the stopping potential were strong enough to overcome the electron ejection attempt of each photon individually, then it shouldn't matter how many arrived per second. Each photon would be rendered unable to eject an electron, and no current would flow at *any* intensity. This apparent validation of the "energy photon" quantization of light is considered to be a key support pillar of

Quantum Theory, and is currently the only explanation for this effect.

Expansion Explanation: The quantum-mechanical "energy photon" explanation of the stopping potential mystery can now be replaced by beams of electron clusters, whose individual force of impact on the plate is insufficient to overcome the counteracting effect of the stopping potential. In this case, it would not matter how many such clusters strike the plate (i.e. how much light intensity); if none of them has the strength to overcome the stopping potential individually, no current will flow. In addition, this solution introduces the possibility that one of the causes of the electron flow in the Photoelectric Effect may be from the electron clusters in the light breaking apart on impact and then releasing their electrons to be swept across by the electric field. Indeed, the possibility of this type of transformation from light (i.e. electron clusters) into currents of electrons could deepen our understanding of the process of photosynthesis in plants and, as shown earlier, electricity generation in solar cells.

• **Light-Frequency Mystery:** Another surprising finding arose when it was discovered that the flow of electrons could also be cut off by changing the frequency (i.e. color) of the light striking the plate. It was found that, once the light frequency dropped below a certain value, electrons were no longer ejected from the plate *no matter how much the intensity of the light beam was increased.* In standard wave theory, if light were composed of waves whose energy was reduced as their frequency was lowered, then simply increasing the intensity of the light should increase the energy (amplitude) of the waves and cause electrons to be ejected again.

Quantum Explanation: This problem is solved if lowering the light frequency means that each photon has less energy individually. Then the argument is the same as in the stopping-potential solution: if each photon is individually incapable of ejecting an electron, then it shouldn't matter how many arrive per second. No current would flow at *any* intensity below a certain frequency of light.

Expansion Explanation: Once again, *Expansion Theory* explains this effect without appealing to *Quantum Mechanics*. A lower frequency of light means the electron clusters are larger, and so, fewer would arrive each second, as shown in Figure 5-4 earlier. And, since fewer clusters means less intensity, a frequency reduction has the side effect of also lowering the intensity of the light beam. Also, larger electron clusters would have a different dynamic when striking the plate, such as being less penetrating on impact and possibly absorbing the impact force better and being less likely to break apart and eject their electrons into the gap. A more intense beam of such lower-frequency light would, simply have a greater number of these larger clusters that are less able (or unable) to eject electrons individually. Once again, there is no need to appeal to *Quantum Mechanics* to explain these results.

- **Time-Lag Mystery:** Finally, standard wave theory states that it takes time for the waves of light energy to be absorbed by the atoms in the plate, resulting in an *eventual* ejection of an electron. Even though this time lag would be extremely small it is thought to be measurable, especially if the light intensity (wave amplitude) is reduced significantly so that its energy is absorbed as slowly as possible. Despite this prediction by standard theory, no time lag has ever been detected.

Quantum Explanation: If light were composed of individual photons that must each be capable of ejecting an electron in order for current to flow, then reducing the light intensity would not affect this individual ejection mechanism. A reduced light intensity would reduce the number of photons arriving per second, but it would not increase the time lag between arrival of a given photon and ejection of an electron. Any time lag that may exist could remain immeasurably small even at the weakest light intensity.

Expansion Explanation: The concept of light as electron clusters also offers the solution that the mechanism whereby electron clusters either knock electrons off the plate or break apart themselves remains the same regardless of light intensity (i.e. number of electron clusters per second). So, here again, we have the explanation for a possibly immeasurably small time lag even at the weakest light intensity without appealing to *Quantum Mechanics*.

This discussion of the Photoelectric Effect shows that this cornerstone of experimental support for *Quantum Mechanics* is actually readily explained by *Expansion Theory* instead, making it even less necessary to seriously consider the bizarre and mysterious conclusions that must be accepted in a "quantum-mechanical" universe.

 Do We Truly Understand Polarized Light?

One well-known quality of light is that it exhibits a property called *polarization*. Current theory typically explains polarization by invoking the wave concept of light, envisioning light as idealized, two-dimensional (flat) waves of energy that may oscillate up-and-down (vertical polarization) as in Figure 5-7 earlier, or side-to-side (horizontal polarization), or any random orientation in between. Light that has no polarization is considered to be composed of waves that oscillate in all possible directions. Light is often either selected or filtered-out based on any inherent polarization that it may possess, which is the principle behind polarized sunglasses that typically block either horizontal or vertical polarization that may be created in light when it is reflected or transmitted through various materials. Natural polarized sunlight is typically created when it passes through the upper atmosphere or is reflected off of bodies of water. The filtering of light by polarized lenses typically reduces the intensity of all colors of light while causing only minimal color distortion by filtering out only one particular plane of polarization, vs. the color-distorting effect of non-polarized lenses that completely block a particular band of colors.

However, a closer look at this issue shows that perhaps our entire concept of polarization needs to be rethought. Firstly, the preceding discussions have cast serious doubt upon the concept of "pure energy" in general, and the "energy wave" concept of light in particular, suggesting that our current concept of "polarized energy waves" is highly questionable. Indeed, although clear diagrams of idealized, two-dimensional energy waves with various polarized orientations can be found in any physics textbook, the fact remains that we have never actually seen or conclusively verified the presence of such waves. This is only a hypothesized model of what is believed to be occurring based on observation and experiment. Secondly, even *Expansion Theory* does not suggest today's concept of an inherent polarization property of light, showing light to simply be a generic sea of expanding electron clusters with no particular inherent "polarization." So then, if neither Standard Theory nor *Expansion Theory* present clear cases for this inherent polarization property of light, what are we to make of the obvious *polarization effect* in the world around us?

Polarization Explained

As we have seen with the examples of gravity and magnetism, it is possible to develop a useful body of knowledge about a particular phenomenon without truly understanding the physical reality underlying it. Likewise with polarization, we know how to produce, identify, and filter various forms of polarized light, but do we truly understand it? It is largely such empirical experimentation and observation that has led to today's concept of two-dimensional light waves of varying orientations. This assumption or hypothesis has been worked into our current theories and equations of light, yet the pure "energy wave" concept of light has so far been shown to be simply a misunderstanding of various experiments. Also, while it is a useful hypothesis that light waves of a particular orientation (polarization) can be created simply by reflection or transmission through certain materials, the physical details of how and why this might occur are left largely unexplained today.

Expansion Theory suggests that polarization is not a quality somehow possessed by individual "light waves" or "energy photons," but that it is simply a straightforward optical effect caused by the reflection or transmission characteristics of overall light beams. To see this, consider one of the most common forms of polarization in nature –

the horizontal polarization of sunlight created when it reflects off of a body of water such as a lake. When a beam of light strikes the surface of the lake, it splits in two – a portion gets reflected while the remainder gets bent, or refracted, down into the water. The proportion of reflected and refracted light depends on the angle at which the original light beam strikes the surface. If the light comes in at a low angle that is nearly parallel to the surface of the lake it is mostly reflected, while higher angles become increasingly refracted down into the water.

Further, although only light rays of one particular angle bounce directly to an observer from a given spot on the lake, when we view an entire scene from a distance light from the whole scene enters our eyes. Therefore, since we see a large portion of a lake with one glance, we are simultaneously taking in a wide variation of reflection angles from many locations all across the lake. It follows then, that the overall lake scene is not uniformly bright, but varies in intensity from the top to the bottom of the scene (i.e. from distant to nearer locations on the lake) due to the reflection and refraction variations just mentioned for the various light angles. In fact, this is why the color of a lake is not uniform, but varies in brightness or color depth from the near side to the far side of the lake.

However, this variation is not the case in the horizontal direction. That is, although light striking the lake from higher or lower angles results in a vertical variation in reflected intensity, light reflected from the right or the left does not vary. Every light beam that strikes the lake at a given vertical angle is accompanied by many other beams with the same vertical angle coming from the right and left in the scene. All of these rays along a given horizontal line in the scene are reflected with the same intensity since they strike the surface at the same vertical angle (Fig. 5-11). This can be seen as a uniform brightness or color depth to the water from left to right at any given distance out on the lake. So, a lake scene actually has *inherent intensity variation* from top to bottom, but uniform intensity from left to right. This means that the majority of the intensity of the overall scene is contributed by the consistent strength of the horizontal reflections from side to side, and significantly less so by the varying strength of the vertical reflections from top to bottom.

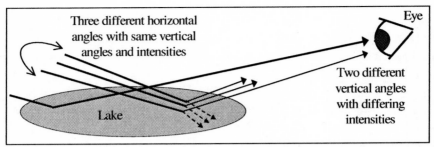

Fig. 5-11 Light Reflecting from Lake Varies only in Vertical Intensity

Now, consider a typical pair of polarized sunglasses that cuts out every other line of "photons" (i.e. electron clusters) in the scene, much like tiny Venetian window blinds. The resulting intensity reduction to the overall scene will be greater if every other horizontal line is blocked rather than every other vertical line. This follows from the fact that the unvarying intensities across each horizontal line contribute the most to the overall intensity of the scene, so reducing them will have the greatest effect on the overall brightness. This explains why a pair of polarized sunglasses worn normally will significantly reduce glare, but if we tilt our head or take the glasses off and rotate them the glare increases. This is simply the difference between blocking the *unvarying* intensity of the horizontal scene vs. blocking the *varying* intensity of the vertical scene. *Expansion Theory* suggests a straightforward optical explanation for the *polarization effect* upon overall light beams, while current theory presents an abstract and largely unexplained claim of an inherent "polarization" quality possessed by individual "energy photons" or "light waves."

The Misunderstanding of "Quantum Entanglement"

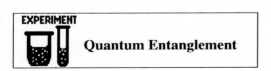

One of the most recent quantum-mechanical mysteries to be added to our science is that of "quantum entanglement." This is an experimental

observation where, when two photons are produced from the same light source, travel together through space, then are split into two separate paths of travel, they apparently maintain a mysterious link with each other. Thus, if one of the photons is somehow altered after they are separated (such as a change in polarization), the other photon also becomes *instantaneously* altered in the same fashion no matter how far apart the two photons may be at the time. This is considered to be a mysterious faster-than-light "communication" between two "entangled" photons, classified as yet another mysterious "quantum effect" between these two quantum-mechanically-defined energy particles. This is an effect that Einstein is famously quoted as calling "spooky," and which he took as a clear indication that such quantum-mechanical interpretations of observations must be fatally flawed.

Further doubt is cast on today's "quantum-mechanical" interpretation of this event by the preceding discussions showing that polarization is not likely a quality possessed by individual "energy photons," and that the very concept of such photons traveling separately through space is quite problematic. Instead, recall that *Expansion Theory* suggests light is never the result of separate "photon projectiles" but that it is always composed of continuous beams of expanding electron clusters. Therefore, a closer examination with this understanding should show that this is not an experiment involving two photons exhibiting a mysterious "quantum entanglement," but merely two separate beams of continuous expanding electron clusters that are *physically connected* back to where they were split from one initial beam.

Once this interpretation is considered, the more likely explanation of the "entanglement" effect is that any influence that alters one beam could potentially be conducted along this continuous span of physically connected electron clusters (or through a sea of unseen clusters between them) to affect the other beam. Since vibrations within solid objects travel faster the more dense the material, the speed of conduction through the extremely dense span of subatomic particles (electron clusters) in light may well be extremely rapid – perhaps even far exceeding the speed of light. Although this explanation is yet to be confirmed, it is suggested that this is the most likely explanation for an otherwise mysterious and completely unexplained phenomenon.

Faster-Than-Light Communication?

As this discussion suggests, there appears to be preliminary lab evidence of the possibility of conducting signals along beams of light at speeds that so far appear to be instantaneous, providing a practical possibility for faster-than-light communication. The possibility of faster-than-light communication would be unexplainable in science today since it violates the speed-of-light limit in Einstein's *Special Relativity Theory* – a theory that also must be rethought in light of *Expansion Theory*, which will be done shortly.

If such a faster-than-light communication method is possible, it is likely that advanced species would use this method of communication along existing beams of starlight rather than generating light or radio waves and waiting for them to physically move through space at the relatively slow speed of light. An analogy for the difference between these two methods of signal transmission can be seen in the common desktop toy made from a line of hanging metal spheres all suspended next to one another, often called Newton's Cradle. When one sphere is pulled back and allowed to swing to strike the others, a sphere at the far end is immediately ejected. A long line of such spheres would allow transmission of such a signal to the far end in this manner far faster than it would take for a single sphere to swing that same distance. Today's method of communication waits for newly-generated light or radio waves to physically move across a distance rather than attempting to conduct signals along existing light beams instead. Although this possibility is only a conjecture at this point, it is a conjecture that would not be possible in today's theories of light, which describe light as discontinuous packets of "quantum-mechanical energy photons" rather than a continuous span of expanding matter (electron clusters).

Another theory that is equally as mysterious and pervasive in our science as *Quantum Theory*, is the *Theory Of Special Relativity*. Once again, *Expansion Theory* shows that significant misunderstandings, as well as clear errors, have led to the creation and acceptance of this theory in our science today.

Special Relativity – Is It All Just A Mistake?

Special Relativity Theory, for which Albert Einstein is perhaps most well known, is actually a special case of the broader *Relativity Theory*, which was put forth by Galileo and further developed by Henri Poincaré (1854-1912) and Hendrik Lorentz (1853-1928). The work of Poincaré and Lorentz on *Relativity Theory* was the precursor to Einstein's *Special Relativity Theory*, published in 1905. *Relativity Theory* is essentially a formal mathematical description of the fact that objects do not possess absolute motion themselves, but only have motion relative to each other. The relative motion of objects was explored by Galileo near the turn of the 17th century and was also discussed in Chapter 2, showing that *Newton's First Law of Motion* overlooks this fact. Near the turn of the twentieth century there was still much debate about the nature of motion since issues such as the propagation of light through space were still unresolved. Einstein decided that neither *Newton's First Law of Motion* nor existing *Relativity Theory* could explain how light travels through space, and so, he proposed a modification of *Relativity Theory*, which became known as *Special Relativity Theory*. Let's examine how this theory – which is also widely recognized as mysterious and bizarre – came to be accepted as part of our scientific beliefs today.

Rethinking the Michelson-Morley Experiment

Since every form of wave known at the time required a physical medium for its transmission (water, air, etc.), it was widely believed that light must be waves that travel through space within an invisible, and as yet, undetected physical medium known as the ether, which must fill the universe. However, the existence of the ether had never been scientifically verified and was becoming an increasing point of debate among the scientists of the day. Finally, in 1887, A.A. Michelson and E.W. Morley devised an experiment in an attempt to resolve this debate. The results of this experiment not only ended the ether debate, but were also given an interpretation by Einstein that radically changed our view of light, time, space, and matter – embodied in his *Special Theory of Relativity*. Let's now return to this crucial turning point in our scientific legacy and take another look at this issue from our perspective a century later – and now also with the knowledge of *Expansion Theory.*

EXPERIMENT

The Michelson-Morley Experiment

The Michelson-Morley experiment was an attempt to determine whether the proposed ether actually existed. The basic premise of this experiment was that, if the universe were filled with a stationary, invisible ether which light must use in order to move through space (much as sound uses air), then light would travel the same speed through this medium in all directions. So, for example, a light beam should then take the same amount of time to propagate through the proposed ether between any two points on Earth whether the light travels in a North-South direction or an East-West direction.

However, since the Earth moves rapidly through space (and thus through the ether) as it speeds along in its orbit around the sun while also spinning rapidly on its axis in the same plane, this motion through the ether should affect the measured speed of light. Any Earthbound experiment to measure the speed of light would itself be hurtling through space along with our orbiting, spinning planet, giving the whole apparatus a sizable relative speed compared to the stationary ether all around. And, since light was hypothesized to travel through this stationary ether, the experimental apparatus would also be moving relative to a beam of light traveling within the ether.

So then, according to the ether theory, measurements of the speed of a beam of light shone in the East-West direction of the Earth's orbit and spin should be affected by the Earth's relative motion, while no such effect should be seen for light shone in the North-South direction. That is, if the either exists, it seems reasonable that we should measure the expected speed of light plus or minus the relative speed of the Earth for measurements in the East-West direction, with no such variation in the North-South direction.

Michelson and Morley reasoned that this effect, if it existed, could be measured by timing a beam of light as it traveled first in the direction of the Earth's motion through space, then perpendicular to this motion. Since light would be confined to travel at a constant speed within the immovable ether, when it is shone in the direction of the Earth's motion through the ether the Earth would move slightly further away during the light beam's travel. Therefore, instead of merely traveling from *A* to *B* on the Earth, the light beam would actually have to propagate through the ether from *A*, past the original location of *B*, then a bit further since *B* (and *A*) would have moved ahead slightly as the Earth moved through the ether (Fig. 5-12, left frame). However, if *A* and *B* were positioned perpendicular to the Earth's motion through the ether, then *A* and *B* would now move *sideways* through the ether, and the distance the light beam had to propagate would remain constant (Fig. 5-12, right pane). Therefore, if a longer travel time were measured in the first case, it would confirm the hypothesis that light propagated at a constant speed through an unseen ether that permeated the universe, acting as an absolute, stationary reference that light propagates through.

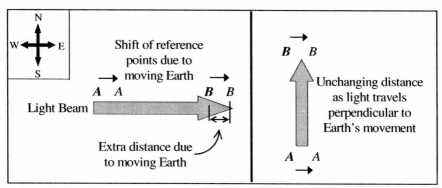

Fig. 5-12 Increased and Unchanged Distances as Earth Moves

Michelson and Morley found that there was no measurable difference in the travel time of the light beam in either direction, which meant that light did not travel the extra distance shown in the left frame of Figure 5-12. This meant that light did not travel through a medium that served as a stationary reference point throughout the universe, ending the debate over the ether and showing that it did not exist. Therefore, light did not travel through a medium, but moved independently through empty space somehow.

However, with this issue resolved, another one came to the forefront. If water waves move at the characteristic speed of wave propagation in water, and sound waves travel at the characteristic speed of vibrations in air, then what is the speed of light relative to if it does not have a medium such as the ether for reference? After all, since existing *Relativity Theory* in classical physics stated that all motion is relative, the speed of light had been considered to be relative to the stationary ether that was thought to permeate the universe. But now that the ether did not exist, what did it mean when the speed of light was quoted? If it were always quoted as an absolute speed with no agreed-upon stationary reference point (such as the atmosphere for sound or a lake for water waves), then it would seem to violate *Relativity Theory* and its well-established equations of relative motion. All observers would then presumably measure the same speed for a given beam of light whether they were stationary or speeding through space, which clearly violates common sense.

Einstein's Postulate – a Solution without a Problem

Einstein offered a solution to this conundrum, although his solution was not to *eliminate* this violation of common sense, but to *accept* it as a mysterious new behavior in our universe. He postulated that light was somehow special and outside of all known laws of motion, suggesting that the speed of light in a vacuum (empty space) was indeed somehow a universal constant that was the same for all observers regardless of their relative motion through space. This meant that an observer speeding toward a light source would, paradoxically, measure the same speed of light as an observer speeding away from it.

When Einstein worked his new constant-speed-of-light postulate into the existing equations of *Relativity Theory*, the resulting equations of motion were largely unchanged except for a term that only became significant at speeds approaching light-speed. The implications that followed from these revised equations were very mysterious and bizarre, but at the speeds of regular daily experience the new term introduced by

Einstein essentially vanished, leaving the original *Relativity Theory* equations. This suggested that the familiar world we experience in our daily activities is a special case in a universe that is actually far more bizarre, but that this bizarreness is only exhibited to a significant degree at speeds that approach light-speed. This modification to *Relativity Theory* provided a neat mathematical solution to the issues raised by the Michelson-Morley experiment, allowing light to fit in with established *Relativity Theory* by making Einstein's small modification that became known as *Special Relativity Theory*.

Not only did this bold new version of *Relativity Theory* fit neatly into current established theory, but it also agreed with the best-known experiment into the behavior of light at the time – the Michelson-Morley experiment just described. After all, this experiment showed that the same speed of light was measured whether the light-beam detector was traveling away from the light source (East-West direction) or not (North-South direction). However, this neat solution seemed to introduce sizable mysteries that violated common sense. If one simply followed the logic embodied in Einstein's proposed new version of the *Relativity Theory* equations, it became obvious that science was being asked to accept that, at near-light speeds, time, space, and matter behave in bizarre new ways that run counter to our everyday experience and understanding. And indeed, during the century that has passed since Einstein's original postulate, these ideas have become firmly integrated into our science as deeply mysterious truths of nature. But was this really necessary – and if not – why has this happened?

The following examination of *Special Relativity Theory* shows that the bizarre conclusions following from Einstein's proposal were *not* necessary or correct, yet they have become accepted into our science, nonetheless, due to an odd combination of logical oversights, experimental coincidence, and unchecked mathematical errors. As we will see, the original logic (above) that led to *Special Relativity Theory* follows one very narrow line of reasoning that overlooks a much simpler commonsense interpretation of the Michelson-Morley experiment. This oversight has gone uncorrected for nearly a century due to apparent support provided by various thought experiments that have been invented over the decades; however, even these thought experiments have clear logical flaws that have either been repeatedly overlooked or ignored. And, a major reason why the errors in these thought experiments continue to be overlooked or ignored is that even actual physical experiments appear to support *Special Relativity Theory*; however, once again, even these physical experiments have simply been misinterpreted. Finally, even the core mathematics that transform the equations of *Relativity Theory* into Einstein's *Special Relativity Theory* have numerous mathematical errors that have been both overlooked and, for the more severe errors, even hidden from view entirely by omitting the erroneous lines in published versions of the derivation. Each of these issues will now be clearly examined against the backdrop of *Expansion Theory*.

ERROR

  Oversight in Original Postulate

Einstein's original postulate of a constant speed of light for all observers – whether moving or stationary – is unlike anything we have ever experienced. If a stone were tossed toward an observer, the observer would not expect to measure the relative speed of the approaching stone to be the same whether running toward it or away from it. Running toward the stone would increase the relative speed between the observer and the stone, and running away would decrease this relative speed. Yet, Einstein's postulate says an observer would measure the *same speed* for an approaching light beam whether that observer were stationary, moving toward, or moving away from the light source. This idea runs completely counter to our common sense and to every experience we know of in the physical world. So then, why did Einstein introduce this concept to begin with?

As mentioned earlier, the results of the Michelson-Morley experiment removed the possibility of the ether existing, seemingly leaving light with no stationary reference point. We could measure its speed, but now that there was no ether, there was no medium to serve as a reference for this speed the way the stationary atmosphere is a reference for the speed of sound or a pond serves as reference for the speed of ripples upon it. *However*, this does not necessarily introduce a serious problem requiring a whole new outlook on the universe. A bullet moves through empty space with no medium as reference and without introducing any new mysteries, so why can't light do the same? For the scientists a century ago to have considered that light might move through empty space without a medium for reference would not have violated existing *Relativity Theory* any more than would tossed stones or speeding bullets. The speed of a bullet is simply relative to the gun that fired it – an obvious fact that requires no special theory to describe its speed or motion. Likewise, there is no fundamental reason why the speed of light cannot simply be relative to the source that produced it, without introducing a special theory as well. This simply means that, just as a bullet would have different speeds for observers traveling at different rates relative to the gun, so light would have different speeds for observers traveling at various speeds relative to the light source.

Indeed, no controlled Earthbound experiments have been done with moving observers (or light sources) to invalidate this straightforward conclusion – a conclusion that is also consistent with the results of the Michelson-Morley experiment. That is, a bullet would be expected to have the same speed on Earth regardless of the direction in which it was fired since the momentum of the gun (and bullet) matches that of the planet – and light also gives this same result in the Michelson-Morley experiment. According to *Expansion Theory*, this would simply mean that all beams of expanding electron clusters (i.e. light) push themselves away from the light source at the same speed relative to the light source. So, just as we would expect a bullet to travel at the same speed in either East-West or North-South directions since its momentum while inside the gun already matched that of the Earth even before being fired, we should expect as much from light. The electrons within the subatomic realm of a battery move along with the Earth prior to being ejected into space as expanding electron clusters of light.

It is only today's concept of pure "light energy" that allows us to treat light as if it were some mysterious, other-worldly phenomenon that somehow appears as if from nowhere and follows its own independent rules of motion. The results of the Michelson-Morley experiment are not at all mysterious, and do not necessitate Einstein's postulate that light has the same relative speed for all observers – *a postulate that forms the foundation of Special Relativity Theory*. However, if *Special Relativity Theory* is not only an unnecessary mystery but is indeed even incorrect, then why is it part of our science today? This question will be answered in the following discussions of the many beliefs that have sprung up in apparent support of this theory since its inception a century ago.

Errors and Misunderstandings in Supporting Evidence

There are a number of physical experiments and thought experiments that are commonly cited as proof of Einstein's *Special Relativity Theory*; however, on closer examination all of this apparent evidence can be readily shown to have fatal flaws of logic or interpretation.

Misinterpreted "Mass Increase" Experiments

One of the mysteries following from the equations of *Special Relativity* is the concept that objects gain "relativistic mass" as they increase in speed relative to an observer. The faster an object moves relative to an observer the more "relativistic mass" it gains and, therefore, the more energy it takes to continue increasing its speed. Eventually, as the object approaches the speed of light, its mass approaches infinity, according to the equations of *Special Relativity Theory*, and even an infinite amount of energy would barely increase its speed any further. This is the reasoning behind the well-known speed-of-light limit in our universe, where nothing can travel faster than light-speed.

ERROR
✗ **Misinterpreted Particle-Accelerator Experiments**

This claim would simply be a fanciful abstraction if it weren't for the fact that experiments appear to validate it. Usually cited as proof are the results from particle accelerators, where physicists can show that as particles approach light-speed it becomes more and more difficult to continue increasing their speed; and, regardless of the amount of energy used, the particles *never* exceed light-speed. Although this is widely believed to be proof of the dramatic increase in the "relativistic mass" of the particles, a much more straightforward explanation can be found in *Expansion Theory*.

The "energy" used to accelerate the particles is supplied by electromagnets that are timed to pulse as the particles pass by, giving them a boost. However, as shown in the previous chapter, a magnetic field is actually a cloud of expanding electrons. If each pulse sends a cloud of electrons expanding toward the passing particles at the speed of light to give them a push from behind, it is not surprising that the particles are accelerated less and less as they speed by faster and faster. As the particles approach light-speed they would be shooting past almost as rapidly as the electron clouds would be expanding behind them. The expanding electron clouds would barely be able to catch up with the particles as they pass, and so, would only be able to give them a tiny additional boost that gets smaller still as the particles get even closer to light speed.

Increasing the energy input simply means producing denser electron clouds from the pulsing magnets, which improves the *efficiency* of each boost (makes each push more solid) but does not increase the *speed* of expansion of these electron clouds. Therefore, as ever more energy is put into the system, the passing particles receive ever more solid nudges toward light-speed but still can never exceed the speed of light since *this is the limit of the speed boosts*. This need not be considered as validation of the mysterious "relativistic mass increase" concept of *Special Relativity Theory*, but merely as an expected result once the true nature of magnetic fields is understood.

Erroneous "Time Dilation" Evidence

A further mystery introduced by *Special Relativity Theory* is the claim that time slows down for a speeding observer – an effect known as *"time dilation."* This literally means that a speeding astronaut would age slower than a stationary observer would, and that a clock on board the speeding spaceship would run slower than a stationary clock. This line of thought lies behind a commonly cited example known as the *Twin Paradox*.

EXPERIMENT

The Twin Paradox

The *Twin Paradox* thought experiment states that, if one of a pair of identical twins embarks on an extended space mission at near light-speeds for many years according to our Earthbound timeframe, the mission would only seem to have lasted perhaps a few hours for the astronaut twin. This is because *Special Relativity Theory* states that time slows down dramatically for anyone traveling at near light-speed relative to a stationary observer, yet runs at the same unchanging rate for the observer; so, upon returning to Earth, the astronaut would be much younger than the twin who stayed home.

ERROR

Logical Oversight in Twin "Paradox"

This thought experiment is considered to show a concrete example of the "time dilation" effect that follows from the equations of *Special Relativity*. However, on closer examination, the very paradox introduced by *Special Relativity Theory* is also undone by the same theory. Since "everything is relative" in *Special Relativity Theory*, it is just as valid to consider the *astronaut* to be stationary while the *Earth* speeds away at near-light-speed. There would be an initial difference between these two views since the astronaut would feel an absolute initial acceleration as the spaceship fired its rockets to gain speed, but thereafter this completely relative view of who is traveling and who is stationary is not only supported but also *demanded* by *Special Relativity Theory*. Therefore, as the *Earth* now coasts away at near-light-speed it would be the *astronaut* who ages while sitting in a stationary spaceship, while only a few hours pass on the speeding Earth. But how can two completely different physical outcomes result from the same space mission simply because of how we think about it? Clearly this is not a true physical paradox, but merely a logical oversight in an attempt to lend validity to the fanciful claim of "time dilation."

This further means that claims that this paradox was conclusively demonstrated decades ago, by comparing a stationary atomic clock with one flown in a jet, must be erroneous since any positive "time dilation" conclusions from such an experiment would verify a theory that states such results should *not* occur. That is, since "everything is relative," the observer on the ground should see the atomic clock on the jet run slower, while the clock on the jet could be considered stationary and should then see the effectively-moving ground-based clock run slower. There cannot be an *absolute* time difference afterward, as historically reported in this decades-old experiment, if everything is *relative*, as Einstein and his special theory claim. Clearly, there must be other explanations for such claims, such as experimental or reporting errors, influences such as turbulence and acceleration upon the delicate operation of atomic clocks (this experiment was performed across a number of regular, non-direct commercial airline trips), or simple coincidence. None of these sources of error can be reasonably dismissed without an appropriate number of repeated, carefully performed experiments by independent research teams, especially considering the extremely tiny time difference that is being measured in this experiment and the enormous and mysterious implications of a positive result.

Likewise, while claims that software exists onboard satellites to make relativistic corrections due to their motion may be true, it is unlikely that this software is truly playing the role it is believed to be playing – based on the doubts cast by *Special Relativity Theory* itself, as just shown. Instead, it is more likely that either such relativistic calculations are insignificant in the overall operation of the satellite, or the potentially counterproductive results of these calculations are largely counteracted by other corrective onboard software. It is also possible that the relativistic calculations coincidentally model some aspects of a very different phenomenon, such as in the earlier example of relativistic calculations predicting the apparent "relativistic mass increase" in particle accelerators.

And, perhaps most important to note, this aspect of *Special Relativity Theory* not only invalidates the Twin Paradox thought experiment and the claimed atomic clock evidence, but it also invalidates *itself.* It is a theory whose mathematics lead its practitioners to state that there must be an absolute "time dilation" effect in nature, which this same theory also clearly shows should not occur if "everything is relative." As will be shown shortly, one of the reasons this logical conundrum exists is because the equations of *Special Relativity* can only stand if we are willing to overlook a number of improper mathematical techniques and even clear mathematical errors in their derivation. As such, it should not be surprising if logical paradoxes follow, even to the point where they invalidate the very theory that produces them.

EXPERIMENT

Misunderstanding of Half-Life Experiments

Despite these problems with the "time dilation" claim of *Special Relativity*, there are yet further claims of experimental proof of this phenomenon. Scientists have observed that unstable subatomic particles take far longer to decay when they are accelerated close to light-speed in particle accelerators than when they are stationary. This is taken as proof that time somehow slows down for the speeding particles. However, according to *Expansion Theory*, these unstable subatomic particles have a composition and nature that should remain stable for longer periods of time if they are compressed by external forces. The details of this particle composition will be explained in the next chapter; however, it can be seen that the speeding subatomic particles in particle accelerators are surrounded by highly pressurized clouds of expanding electrons (the magnetic fields) as they are accelerated. Further, the tremendous acceleration would also impose crushing "G-forces" upon these particles.

It should not be surprising that these tremendous external pressures may well result in significantly extended particle lifetimes if indeed such particles can be held together longer by external pressure. Such experiments need not be interpreted as exhibiting a mysterious "time dilation" effect upon the particles, but merely as providing simple mechanical compression which holds these unstable particles together longer.

EXPERIMENT

Re-evaluating Cosmic Ray Evidence

Another example of apparent "time dilation" effects is thought to exist in speeding cosmic rays. Particles are detected on Earth that should not live long enough to travel from the upper atmosphere, where they originate, to the ground. The current explanation for this unusually extended lifetime is that their tremendous relative speed compared to stationary detectors on the ground means time is drastically slowed down for the speeding particles, allowing them to live long enough to be detected on the ground.

However, the normal lifetimes of such unstable particles cannot be known *individually*, and are only expressed in terms of their *half-lives*. The quoted half-life of a particle is the time it takes for half of a given population of such particles to decay – not each particle individually. So, after one half-life period, there will be half the original population remaining; after twice the half-life time there will be one quarter left (one half of one half), one eighth remaining after three half-life periods, etc.

This means that, even after a time period ten times longer than the stated half-life of a given type of particle, there should still be one thousandth of the population remaining. That is, one in a thousand particles would be expected to live ten times longer than the quoted half-life of such a particle even under normal conditions. Therefore, it might not be surprising that some cosmic ray particles from the upper atmosphere live far longer than their stated half-life period – long enough to be detected on Earth. Characterizations of these results as "time dilation" effects may be a case of seeing what one expects to see. If one is looking for evidence that the mysteries of *Special Relativity* are actually true, these cosmic ray results might be misconstrued as such evidence in lieu of the possibility that a large enough initial population of particles may have existed to account for the number that survived.

In fact, once again, this belief is undone by the very theory that spawned it. Since "everything is relative" in *Special Relativity Theory*, it is just as valid to consider the *Earth* to be speeding toward *stationary* particles in the upper atmosphere. In that case, time slows down for Earthbound observers, meaning that the particles decay at their usual half-life pace in their stationary reference frame while only a tiny fraction of this half-life time passes for the speeding observers on Earth. Therefore, just as the speeding astronaut in the Twin Paradox would return to find a much older twin, the speeding Earthbound observers would encounter an extremely old population of cosmic ray particles, which means that they should all have long since decayed and should not be detected.

Even more circular logic can be brought into this discussion, since the equations of *Special Relativity* also state that lengths shorten for speeding observers (another mysterious result of the *Special Relativity* equations). Therefore, it could be claimed that the distance to the cosmic ray particles shortens, allowing the speeding observers to detect them before they all decay. However, if such reasoning invalidates the conclusion that the particles should have long-since aged and decayed, then it also invalidates the aged twin in the Twin Paradox thought experiment. Such endless circular logic permeates much of *Special Relativity Theory* today. The fact that this mysterious theory simultaneously predicts two completely opposite and irreconcilable physical outcomes is often overshadowed by the intrigue generated when experimental results seem to validate the more mysterious of these predictions. It is this type of logical oversight and consideration of only selective evidence that has caused *Special Relativity Theory* to gain such acceptance in our science today.

EXPERIMENT

 Flaw in Traveling-Light-Beam Thought Experiment

Despite the problems just shown with the time dilation concept, it is considered to have support from still other logical thought experiments. One that is commonly cited considers a light beam shone from a light source on the floor to a mirror on the ceiling above, where it is reflected back again, with the added twist that this is occurring aboard a train moving at a sizable fraction of light-speed. While the light beam would simply travel straight up to the ceiling and back down again according to an observer on the train, a stationary observer watching the passing train would see the bouncing light beam moving along with the train. And, a beam of light that bounces up and down while also moving forward obviously traces out a diagonal path as seen by the stationary observer outside the train. However, since this diagonal path is longer than the straight up-and-down path seen by the observer on the train, the light beam seems to be paradoxically traveling two different distances in the same amount of time.

Special Relativity Theory resolves this paradox by stating that, relative to the stationary observer outside the train, time on a train traveling at such high speeds will slow down significantly. Therefore, the results of both observers would be consistent – the light beam on the train travels a lesser distance but in a shorter time period, while the stationary observer sees a longer distance in a longer time period. That is, the shorter distance divided by the lesser time gives the same velocity for the light beam as the longer distance divided by the longer travel time. In fact, this type of logic is considered essential in order to uphold Einstein's original postulate of a constant speed of light for all observers that led to his *Special Relativity Theory* in the first place.

If all of this logic is beginning to seem a bit odd, there is good reason. In actuality, there is no paradox at all in this "time dilation" thought experiment, but simply flawed reasoning. This can be seen more clearly if the light beam and the train are replaced by a bouncing ball on a cart that is slowly rolling across a room. A stationary observer sitting in the room would not see the ball follow a short up-and-down path (as would be seen by an observer on the moving cart), but rather, a much longer path as it rolls past while bouncing. However, at such slow speeds the equations of **Special Relativity Theory** revert back to the everyday equations of mere **Relativity Theory**, and there are no "time dilation" effects whatsoever. So, now it appears we have the same paradox as with the light beam on the speeding train – with the ball traveling two very different paths in the same amount of time – but the speeds are far too slow to appeal to "time dilation" to resolve this apparent conundrum. So, before even analyzing this situation further, we already know that this apparently mysterious physical paradox has to be due to a simple logical oversight since such deep physical paradoxes certainly don't exist in such simple everyday events. Therefore, this already shows that a similar oversight only created the *appearance* of a paradox with the light beam on the speeding train, which further means that the appeal to *Special Relativity* to solve this "deep physical paradox" was just a meaningless logical exercise.

So then, what is the logical oversight in these two thought experiments? It is simply the erroneous expectation that the distance traveled in one frame of reference must be the same when seen from another. We are all standing on a planet that is rotating and orbiting at tremendous speeds from the perspective of someone who is observing from out in deep space, yet, as far as we are concerned, we are standing still. This is not a mysterious paradox but merely simple relative motion from different points of view, or frames of reference. These are two very different conclusions, but that is to be expected from two very different motion perspectives, or inertial reference frames. Therefore, in this often-quoted traveling-light-beam thought experiment, *Special Relativity* is commonly used to resolve a paradox that *does not exist*, just as Einstein's original postulate of a constant speed of light for all observers in response to the Michelson-Morley experiment was a solution to a problem that did not exist.

Despite the fact that both the thought experiments and the physical experiments commonly used to support *Special Relativity Theory* are either highly questionable or clearly flawed on closer examination, it might still seem that this theory at least has solid mathematical support. After all, the equations of *Special Relativity Theory* were derived and presented in the logic of mathematics for all to see. Indeed, this mathematical support is available and open for inspection, as will be done now with Einstein's own derivation from the appendix of his book, *Relativity: The Special and General Theory,* published in 1961:

ERROR

Fatal Flaws in Einstein's Special Relativity Derivation

In an attempt to form a credible foundation for his theory of *Special Relativity*, it was imperative that Einstein presented viable mathematical and logical support for his ideas and claims. Yet, upon examination of Einstein's own account of this supporting logic, numerous fatal flaws can be readily identified. Of these flaws, perhaps the most critical of all is an improper mathematical manipulation that can be clearly seen in the following simplified example:

Line 1: $x = a + b$ – original expression w/o speed of light

Line 2: $x = a + b * (c^2/c^2)$ – "harmless" multiplication by 1

Now, let the symbol y stand for the expression $(b* c^2)$

Line 3: $x = a + y / c^2$

Here, we begin with a line that has nothing to do with the speed of light, either because the speed-of-light term was never present or because it dropped out of the derivation by this point – both reasons are functionally equivalent. Next, we arbitrarily choose to multiply one of the terms in Line 1 by the expression c^2/c^2. The justification for this manipulation is that it is merely a harmless multiplication by 1 since any expression divided by itself is 1. Then, to keep the expression, c^2, from immediately canceling itself out again – top and bottom – we group all of the top symbols together and hide them from view inside a new variable, y. This hides the upper c^2 expression and leaves the lower one alone in plain view, transforming the original expression in Line 1 into one that now appears to be intimately connected with the speed of light since it now has a term that is divided by c^2.

Of course, this is merely a contrived sleight-of-hand that can be easily exposed in a variety of ways. For example, why was the multiplication by 1 done in the particular form of c^2/c^2 ? Since this rather odd way to represent the value 1 was completely arbitrary, why not e^3/e^3 or $\sqrt{f}/\sqrt{f}$? For that matter, why even perform such an odd, arbitrary manipulation at all, especially since it introduces the very real danger of confusing this arbitrary symbol, *c*, with the symbol *c* that *did* represent the speed of light earlier on, but which naturally *dropped out of the derivation*? This raises the very important point that the arbitrarily introduced symbol, *c*, is as meaningless as the symbols *e* or *f* would have been if we had chosen them. The fact that *c* is often used to represent the speed of light (and that it *did* earlier in the derivation) does not mean that it always does so whenever and however it appears. Pythagoras' famous theorem for the hypotenuse of a right-angled triangle, $a^2 + b^2 = c^2$, has nothing to do with the speed of light, of course, but only with the sides of a triangle. The symbol, *c*, only represents the speed of light if that representation follows from the logical structure and continual flow of a derivation; otherwise, it is a meaningless symbol – nothing other than an arbitrary letter chosen at random from the alphabet.

Yet, this is precisely the logic used by Einstein to ensure that the "speed of light" was re-introduced into his derivation after the *true* speed-of-light term *dropped out entirely*. One of the reasons that this fatal flaw has gone unnoticed by the scientific community is because Einstein omitted the two key lines showing the speed-of-light term dropping out and the subsequent improper operation to artificially add it back. As a result, on the surface it appears as if the same speed-of-light term continues seamlessly throughout the derivation, though this is actually far from the case. In actuality, the "speed-of-light" term found in Einstein's widely accepted equations of *Special Relativity* is merely a random, meaningless letter from the alphabet – and nothing more. For those interested in the mathematical details, the first key section of Einstein's derivation is presented below in a simplified form, surrounded by analysis that exposes not only the above fatal error, but numerous other critical errors and improper operations leading up to it as well.

OPTIONAL MATH

(x, y) **Detailed Analysis of the Flaws in Einstein's Derivation**

A simplified summary of only the salient points in Einstein's derivation is supplied and discussed below, with the full derivation available in his book, as mentioned above. The derivation begins with the classic equation of motion, *distance* equals *time* times *velocity*:

$$d = t\,v$$

This equation is presented twice, once for a stationary reference frame (subscripted *s*) and once for a moving reference frame (subscripted *m*); and, in both cases, the speed of light, *c*, is substituted into the velocity parameter:

$$d_s = t_s\,c$$
$$d_m = t_m\,c$$

These two equations are meant to represent two different perspectives, one stationary and the other in motion, much like the earlier thought experiment with the light beam and the speeding train. Since Einstein's *Special Relativity Theory* allows for distances to shorten and for time to slow down, time and distance are also subscripted to show that they may vary from the *stationary* to the *moving* reference frame. However, since Einstein further postulated that the speed of light never varies in *any* reference frame, the constant, *c*, has no subscripted difference.

Although this is a reasonable enough beginning, the logic soon becomes derailed by a series of arbitrary value assignments to different variables, while only partially following through on the effect of these assignments. This creates a situation where some new expressions resulting from these assignments are mixed in with other old expressions whose variables should also have changed, but were neglected and left unchanged.

The invalid operations that follow from this are further distorted by later arbitrary *reassignments* of new values to some of these same variables, ignoring the fact that, in many cases, these reassignments now alter the already questionable logic and equations that have been derived thus far. That is, entire expressions that would now have to be changed before being used further are instead left unchanged. Some examples that can be found in the full derivation are:

- Setting $d_m = 0$ but ignoring that, according to the earlier equation, $d_m = t_m\,c$, this means $t_m = 0$ also.
- Setting $t_s = 0$ but ignoring that, according to the earlier equation, $d_s = t_s\,c$, this means $d_s = 0$ as well.
- Setting $d_m = 1$ but ignoring the earlier assignment of $d_m = 0$, as well as the fact that this earlier assignment led to the development of certain expressions that are no longer valid if d_m is now arbitrarily changed yet a second time.

These errors result in the creation of a mixture of variables that are only partially updated to reflect arbitrary value assignments, with further distortion due to a mixture of entire expressions that are also only partially updated after further value *reassignments* are made.

Yet, despite these sizable problems, the most significant error is yet to come. However, this error is not readily seen in Einstein's published derivation because the two key lines that would clearly show the improper manipulation have been omitted. Nonetheless, it is straightforward to recreate these two omitted lines. We begin by pausing at a key line that appears a little further along in the derivation, but which arises from an odd leap of logic that is difficult to follow from the preceding line. This key line (in simplified form) is:

$$d_m = \left(\frac{v^2}{c^2} \right) d_s \quad \text{– Key line representing large leap of logic}$$

This is a sizable and largely unexplained leap of logic from the line that precedes it. This is also a crucial line since the expression, v^2/c^2, is the key term that ends up in the final equations as the only difference between the pre-existing *Relativity Theory* and Einstein's new *Special Relativity Theory*. Although Einstein points out that he made this leap of logic by substituting an expression from earlier in the derivation, he does not actually show his work. Below is this same leap, but with the two omitted lines now shown (again, in simplified form):

$$d_m = x \, d_s \quad \text{– Omitted line where speed-of-light, } c, \textit{ drops out entirely}$$

$$d_m = \left(\frac{xc^2}{c^2} \right) d_s \quad \text{– Omitted line showing improper attempt to}$$

re-introduce the speed-of-light

And, since $xc^2 = v^2$ (from earlier in the derivation), this gives:

$$d_m = \left(\frac{v^2}{c^2} \right) d_s \quad \text{– Key line shown earlier, but now with omitted}$$

lines shown above

What might be the reason that the two omitted lines above were not shown? It is very significant that these two lines show the speed-of-light term dropping out entirely, then a completely arbitrary multiplication by c^2/c^2. Although it could be argued that this is merely a harmless multiplication by the value 1, the important point is that this is an *arbitrary attempt to reintroduce the speed of light*. Prior to this, the speed-of-light constant, c, which was substituted into the motion equations at the start of the derivation, had dropped out of the derivation *entirely*. This means the derivation had stopped being one involving the speed of light in any fashion. Yet, the steps taken in the omitted lines are a deliberate (and erroneous) attempt to arbitrarily add it back in. However, since the symbol, c, was essentially merely drawn in and did not follow from the original flow of the derivation, it cannot be considered as anything other than an undefined symbol – merely the letter 'c' in the alphabet – and nothing more.

⸰ᕯ⸰ The Speed of Light is Not a Limit – Warp Speed has arrived

As the preceding analysis shows, there are numerous improper mathematical operations, as well as a fundamental fatal flaw, at the very heart of Einstein's own derivation of his *Special Relativity Theory*. We have grown so accustomed to hearing about the thought experiments, paradoxes, and mysterious experimental evidence supporting *Special Relativity Theory* that it has all become accepted and commonplace – almost passing as commonsense itself. As a result, it can be difficult to imagine how there might *not* be a universal speed-of-light limit on objects, forgetting that there is actually no clear reason *for* such a limit. There was no clear need to introduce this concept in the first place a century ago, and we have struggled to maintain support for it – and all the mysteries following from it – ever since.

In actuality, there is nothing stopping objects from traveling well beyond light speed – to any arbitrary speed at all in fact. We won't achieve this in our current generation of particle accelerators that push particles along using a method that has an inherent speed-of-light limit itself, and there are no other processes on the planet that would cause objects to attain such speeds. So far, our spaceships have carried a limited amount of chemical rocket fuel and have used the "accelerate-and-coast" approach to traveling great distances – not even attempting to reach such tremendous speeds. But then, if there is no such speed limit in the universe, why don't we see objects of such tremendous speed in the heavens?

In answer to this question, there is no particular reason why such speeding objects *cannot* exist in the heavens, although there is also no particular reason to *expect* to encounter objects of such tremendous speed relative to us. Since our solar system likely formed from a single swirling disk of gas and particles, all early matter in our solar system would have swirled about more or less in unison. As time progressed, this matter congealed into planets of different orbital periods, and random collisions sent chunks of matter off on collision courses with still other objects, but there is no reason to expect this process to result in relative speeds that exceed or even approach light-speed. Any object having such a rapid speed relative to us would likely have to originate outside our solar system, and perhaps even outside our galaxy since our galaxy also may have formed from an enormous disk of gas and particles rotating in unison.

So, while there is no physical law prohibiting an asteroid traveling at 10 times the speed of light from hitting the Earth without warning, there is no reasonable expectation of such an event occurring either. Also, even if such an object *did* career through our solar system, it would be extremely unlikely to hit a planet. This is because, at such tremendous speeds, an asteroid would essentially fly through our solar system in a straight line, perhaps slightly deflected but essentially unaffected by the gravity (i.e. expansion) of planets. And, the planets are so tiny and widely spaced relative to the overall solar system that our solar system is essentially empty space from the perspective of such an object speeding through in a straight line. It is only the familiar, slower-moving asteroids within our solar system that are effectively attracted to planets.

This also means that the dream of warp-speed space travel – multiples of light-speed – is not science fiction and does not require some exotic or futuristic new technology. Warp speed has been within our grasp ever since the early days of the space program. We simply have not achieved it because we haven't *tried*, and we haven't tried because *Special Relativity Theory* said we couldn't – and we believed it. All that is required is continuous acceleration from an extended fuel burn; a spaceship would accelerate faster and faster as its fuel burns, just as common sense tells us it should. The spaceship will not undergo a mysterious "relativistic mass increase" as it increases in speed, it will not need to burn more and more fuel to compensate for such "mass increase," and it will not have any special difficulty approaching or exceeding the light-speed "barrier." The only question about our ability to achieve or exceed light-speed – relative to our solar system for example – is whether the spaceship can carry enough fuel to reach such speeds before exhausting its supply.

It can now be seen that the absence of understanding of expanding matter has led to a wide variety of well-intentioned yet fatally flawed theories, most requiring that we suspend disbelief and accept logically, scientifically, and even mathematically impossible evidence as solid support. Since we have managed to invent models and theories that provide useful, accurate results, we have been willing to overlook their obvious failings and accept that they may be the true description of the physics of our universe – a universe that would then seem to be truly bizarre and mysterious. However, as *Expansion Theory* shows, this state of affairs is no longer necessary. The universe is indeed a simple, comprehensible place where human common sense can be trusted, and where our intuition of a single unifying principle underlying all observations is a reality. We may still choose to use our existing models, but *only* as models and not as literal descriptions of our physical world.

And finally, if *Expansion Theory* truly is the much-sought-after Theory Of Everything, then it must also speak to the branch of science that is considered perhaps the most promising frontier for arriving at this theory – particle physics. The search for the Theory Of Everything has sparked a tremendous amount of research into the subatomic realm over the past century, spawning a sizable new branch of science wholly dedicated to characterizing and understanding this realm. Here again, there are many theories and numerous scientific experiments supporting our current beliefs about the subatomic realm – embodied in a core body of work known as the *Standard Model*. The next chapter will investigate the Standard Model of particle physics, and the experiments and beliefs behind it from the perspective of *Expansion Theory*.

- 6 -

The Big Questions

This journey toward rethinking our scientific legacy began by discussing gravity in Chapters 1 through 3, showing that the concept of "gravitational energy" has many serious problems that are resolved by the notion of expanding atoms. Chapter 4 explored the likely nature of the atom from this new perspective, showing that the atom is in turn composed of expanding protons, neutrons, and electrons. This discussion proceeded to show how phenomena such as electricity and magnetism could be solidly explained for the first time – as manifestations of expanding electrons that have crossed over from the subatomic to the atomic realm. Finally, Chapter 5 provided the first clear physical description of light, showing it to also be a manifestation of externalized expanding electrons, and further showing that there is no such thing as disembodied "pure energy" at all in our universe. Ours is a universe composed entirely of expanding matter, with expanding electrons accounting for many "energy" phenomena, while expanding atoms give us "gravitational energy" and all of the dynamics that follow from it.

This final chapter applies *Expansion Theory* to some of the big questions in science, while also delving deeper to consider the nature and origin of expanding subatomic particles themselves. Chapter 4 showed that electrons and protons do not possess a mysterious energy phenomenon known as "charge," but that such a notion arose because expansion was overlooked and misrepresented. But, is it possible to say anything more about the nature of expanding electrons, protons, and neutrons? How is it that these three specific fundamental particles came to be? And, if all energy and matter in our universe follows entirely from the dynamics of expanding protons, neutrons, or electrons, then how are we to think of the even smaller *quarks* that are said to compose them? From the perspective of *Expansion Theory* so far, smaller particles that *compose* protons, neutrons, and electrons would not qualify either as matter or energy since all matter and energy has now been entirely defined in terms of *whole* protons, neutrons, and electrons. The existence of yet a deeper family of underlying particles known as *quarks* would then lie outside the definition of atomic matter, subatomic matter, and even energy, representing a truly mysterious new phenomenon in nature.

Further, if only three basic subatomic particles are required to explain our entire universe of matter and energy, as has been demonstrated thus far, then why does today's Standard Model of particle physics include entire families of additional subatomic particles? We regularly hear of a wide variety of particles emerging from high-speed collisions in particle accelerators – particles that are often said to appear spontaneously out of "pure energy" or that apparently meet and annihilate each other back into "pure energy." The ultimate form of such "pure energy" creations and annihilations are interactions that are said to involve pairs of matter and *antimatter* particles. Often these experiments even involve something known as *virtual particles* – particles that have never actually been detected but whose existence is assumed nonetheless to help explain experimental results. What does *Expansion Theory* say about such exotic entities as *quarks*, *antimatter*, and *virtual particles*, not to mention the families of additional regular subatomic particles that are said to appear in particle accelerator experiments?

What Are Subatomic Particles?

In order to answer these questions it is necessary to take a closer look at the three basic particles in *Expansion Theory* – expanding protons, neutrons, and electrons. We can gain insight into the nature of these particles by considering a thought experiment that represents these three particles as expanding ripples in a pond.

EXPERIMENT

 Subatomic Particles as Ripples in a Pond

Consider a pond of circular, spreading ripples representing expanding subatomic particles. Protons and neutrons are known to be hundreds of times larger than electrons, yet it would not be possible for expanding ripples on a pond to maintain such a large and persistent size difference. As shown in Figure 6-1, although the two original ripples started off with a nearly five-fold size difference (left frame), as their growth continued (right frame) this relative size difference was rapidly reduced to less than two-fold. This resulted from simply projecting what would occur as all ripples traveled at the same speed across the pond. That is, the edge of each circular ripple was advanced by an identical amount, resulting in a far greater percentage growth for the smaller ripple and leaving the same absolute size difference between them as there was originally. This gives the result that the smaller ripple rapidly catches up with the larger ripple in relative size, while the fixed, absolute size difference between them becomes less significant over time. In other words, all ripples eventually approach the same relative size and overall expansion rate, as all ripples propagate at a constant speed through the water. This is true both in this thought experiment as well as in a real pond.

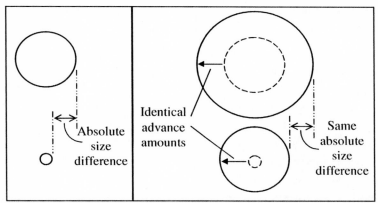

Fig. 6-1 All Ripples soon Approach Same Size and Expansion Rate

This is a natural conclusion for all subatomic particles, requiring only one simple underlying law that gives the same speed for all "ripples," just like familiar ripples in a real pond. However, if ripples in a pond are analogous to the ocean of expanding subatomic particles in our universe, this also means there cannot be several fundamental subatomic particle types of differing sizes, but only one singular particle type permeating the universe. This must be the case since all expanding circular ripples will soon grow to become indistinguishable from each other. In this case, the only way to have different-sized particles is if this single fundamental particle-type acts as the "atom" of the subatomic realm. That is, just as expanding atoms can group together to form different-sized objects that always have the same relative size and expansion rate, so a number of these fundamental expanding *subatomic* particles could group together into larger expanding *subatomic objects* in the same manner. This suggests that protons and neutrons would be just such subatomic objects composed of the one true fundamental particle – the expanding electron.

NEW IDEA

According to *Expansion Theory*, there is only one fundamental subatomic particle in our universe – the expanding electron, while protons and neutrons are actually *subatomic objects* composed of many hundreds of electrons.

This quite literally means that our entire universe is fundamentally a universe of expanding electrons – and nothing else. The two other subatomic particles that make up the atom (protons and neutrons) can now be seen simply as groups of expanding electrons, while all the various forms of energy have also been shown to be entirely manifestations of expanding electrons as well. The earlier analogy of ripples on a pond explains why all electrons are of identical size and expansion rate (currently interpreted as identical "charge"), and how the much larger protons and neutrons manage to maintain their constant relative size differential and the same subatomic expansion rate, X_S, as the electron. Just as all atomic objects expand at the same universal atomic expansion rate as individual atoms, maintaining a constant relative size, so all subatomic "particles" (actually subatomic *objects*) composed of electrons would expand at the same universal *subatomic* expansion rate as individual electrons.

This conjecture represents an important new understanding of subatomic particles; but is there any further evidence to support this possibility? One clue comes from particle accelerator collision experiments. Both protons and neutrons can be split into other particles in collisions of enough force – even if only for the tiniest fraction of a second. The electron, however, has never been split, no matter how powerful the collision. If particle physicists had to select only one stable particle that may be truly fundamental in nature, based on the experimental evidence so far, doubtless the electron would be a strong contender.

A second clue comes from the decay mode of a neutron. A free neutron is actually an unstable particle that spontaneously decays when outside the atom, *emitting an electron and turning into a proton in the process*. This fact alone makes a strong case for the neutron (and likely the proton) being a grouping of electrons. Currently, this mysterious transformation from a neutron to a proton, while somehow *creating* an electron in the process, is a completely unexplained mystery in particle physics. Also, a neutron is known to be only slightly more massive than a proton – by roughly the mass of an electron or two. These two facts together suggest that protons and neutrons are simply groups of electrons, with the neutron having one or two electrons more than the proton.

Further, since it is currently believed that free protons do not decay spontaneously, while free neutrons do, this may imply that protons are very efficiently-packed, well-balanced groups of expanding electrons, while the additional electron(s) in neutrons may upset this balance and make them unstable. Since electrons are rapidly expanding entities, a group of them could be very active, agitated, and unstable unless they are in an optimal configuration that dampens this effect. Assuming that decay experiments have not been misinterpreted, it would appear that the proton grouping is an optimal configuration, while the additional electron(s) in the neutron grouping creates enough agitation to eventually eject an electron and become a proton (Fig. 6-2).

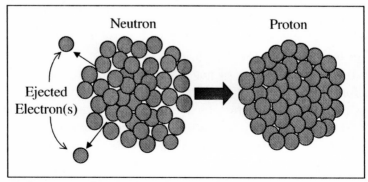

Fig. 6-2 Agitated Unstable Neutron Decaying to Stable Proton

This explanation also speaks to one of the "four fundamental forces of nature," known as the *Weak Nuclear Force*, as mentioned in Chapter 1. While this "weak force" was introduced to help explain spontaneous radioactive decay of particles, we can now see that the notion of such a mysterious, law-violating force is unnecessary once *Expansion Theory* is applied to subatomic particles.

With this perspective on subatomic particles we can begin to answer some deeper questions. First, we can see that if protons and neutrons are actually groups of electrons, it should not be surprising that they can be broken up into various sub-groups in a powerful enough collision. Since the continual, rapid outward expansion of electrons against each other within protons and neutrons would cause a powerful internal compression or "binding force" holding them together, it would be difficult to break these groupings apart, and they would tend to recombine rapidly. In fact, our belief that protons and neutrons are composed of a handful of supposedly fundamental particles known as *quarks* is based on just this type of finding. Powerful collision experiments do seem to briefly break protons and neutrons into a few sub-particles, which have been given the name "quarks," and which immediately recombine back into the original proton or neutron. *Expansion Theory* would not describe these "quarks" as new fundamental particles, however, but merely as smaller sub-groupings of the only true fundamental particle in nature – expanding electrons.

It might also not be surprising if particle accelerators managed to break protons and neutrons into a wide array of short-lived sub-groups with varying characteristics (Fig. 6-3), which would tend to recombine almost immediately into either their original groupings or some other configuration of electrons. Indeed, as particle physicists increase the energy of their collisions with ever more powerful particle accelerators, they do find a wider variety of exotic new particles emerging, typically lasting for only the tiniest fraction of a second. *Expansion Theory* provides the likely explanation for the increasing number of subatomic particles that have been discovered in recent decades and added to our Standard Model of particle physics as ever-more-powerful particle accelerators have been built.

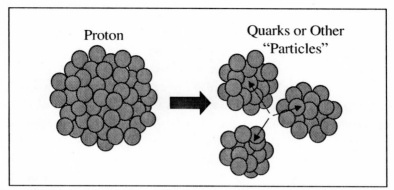

Fig. 6-3 Proton Breaking Into Smaller Sub-Groups (or "Particles")

We can also now see how the idea of particles being created out of "pure energy" might arise from particle-accelerator experiments. Physicists do claim to see subatomic matter particles appearing out of the pure energy of these accelerator collisions, but such statements are made today without a true understanding of either subatomic particles or energy. However, now that *Expansion Theory* shows the energy in these experiments to be simply electron clouds (electric or magnetic fields) or electron clusters (photons of light or other electromagnetic radiation), it is not difficult to envision this transformation from "energy" into matter.

The "high-energy" environments where this type of conversion experiment occurs would actually indicate the presence of a surrounding cloud of electrons (i.e. the magnetic or electric field "energy") that is extremely dense and highly pressurized. This surrounding electron cloud may also be very turbulent and agitated as well, especially when particles are fired at high speed into its midst, colliding with each other. It is not difficult to imagine occasional particle-like groupings of electrons forming or reconfiguring themselves in such an environment during collision experiments, especially if only for the briefest fraction of a second. In fact, one might imagine that, with enough experiments and observations, just about any conceivable combination of electrons into various exotic "particles" or various photons of radiant "energy" is possible. Indeed, millions of such collisions often occur in each of these experiments, with the potential to produce all manner of exotic new "particles" or any range of "photons" of various radiation frequencies.

These experiments need not be considered to exhibit a mysterious conversion of energy into matter, as claimed by Einstein's energy-matter conversion equation $E=mc^2$, but rather, merely a straightforward and perhaps even expected rearrangement of electron groupings when enough effort is made to create the proper environment and conditions. The equation, $E=mc^2$, may serve as a useful abstract or conceptual model for this process, but it should be recognized as such, and not as a literal description of some mysterious energy-matter conversion process. Einstein's energy-matter equation itself will be discussed in more depth in a later discussion, showing that its true origin and meaning is often overlooked and misunderstood.

Further evidence that lends support to the description of subatomic particles as various groupings of electrons appears to exist in a contribution made by Isaac Newton centuries ago – long before subatomic particles and radioactive decay were even known. Consider that, if a population of unstable, radioactively decaying subatomic particles is actually many tiny groups of expanding electrons dispersing, then we might expect the "radioactive decay" of such "particles" to be similar to the classical behavior of an ordinary cloud of atoms as it cools and disperses. Newton characterized this cooling process of gasses into a mathematical model known as *Newton's Law of Cooling*. This law has the form:

$$T(t) = T(0) \cdot e^{-kt} \qquad - Newton's\ Law\ of\ Cooling$$

This equation simply expresses how the temperature of the gas, T, changes over time, t, from its initial temperature at time $t = 0$, expressed by convention as $T(0)$. Although the details of Newton's equation are not particularly important here, it is interesting to note that, centuries later, subatomic particles and their radioactive decay were discovered and found to follow the form:

$$N(t) = N(0) \cdot e^{-Kt} \qquad - Radioactive\ decay$$

That is, the number of particles that have not yet decayed, N, changes over time, t, from the initial population at time $t = 0$, expressed by convention as $N(0)$. Note that the form of these two equations is *identical*, yet one describes the dispersion of a group of cooling atoms in a gas in classical 17th century physics, and the other describes the radioactive decay of subatomic particles in 20th century particle physics. This is not surprising from the perspective of *Expansion Theory*, which in fact, suggests that this similarity should exist when it states that decaying subatomic "particles" are not actually particles at all, but are groups of expanding electrons. It is reasonable to expect groups of active, agitated electrons (subatomic "particles") to disperse ("decay") in a manner similar to a cooling, dispersing cloud of agitated gas molecules.

What Is Antimatter?

Today's concept of antimatter actually arose 75 years ago as a mathematical aberration. In 1928, Paul Dirac (1902-1984) was attempting to develop an equation to better describe the nature of electrons, and although many problems were eventually found with his basic assumptions and resulting equation, his work, nevertheless, drew sizable attention and interest in the physics community at the time. Dirac's final equation, known today as the *Dirac Equation*, suggested such oddities as atoms mysteriously and spontaneously vanishing in a burst of energy, and electrons possessing negative kinetic energy. Although problems such as these brought sizable disfavor upon Dirac's work from many of the notable physicists of the day, these same problematic issues eventually turned into a serious scientific belief in the existence of *antimatter*.

This belief in the existence of antimatter began as a conjecture by Dirac to help explain the odd implications of his equation, and eventually became serious science when experimental results were interpreted as proof of its existence. Today, the concept of antimatter has grown to become a mainstay of both science and science-fiction, often considered to be a mysterious "opposite" form of matter that annihilates regular matter on contact and provides the ultimate power source for such things as futuristic warp-drive engines. However, *Expansion*

Theory shows that the reality is more-than-likely far removed from these beliefs.

Firstly, the concept of pure energy does not exist in *Expansion Theory*. Instead, our universe is exclusively one of regular expanding subatomic and atomic matter, eliminating the possibility that regular matter could vanish into "pure energy" by coming into contact with "antimatter." Such concepts are pure fiction, or at least a misunderstanding, according to *Expansion Theory*. In fact, the foundation for explaining our current belief in the existence of antimatter was laid in the preceding discussion of subatomic particles and particle accelerators.

Although it is tempting to think of antimatter as entire atomic objects that are just like regular objects but somehow made of some exotic type of annihilating "antimatter atoms," this is far removed from the strict definition of antimatter in our science today. In actuality, antimatter is merely defined as a subatomic particle of regular matter that has the same mass as a known subatomic particle, but with opposite charge. An antimatter electron – a "positron" – would then be a particle with the same mass as an electron but with a positive charge. In theory then, an entire antimatter atom would be composed of negatively charged protons and, presumably, regular uncharged neutrons in its nucleus, surrounded by positively charged orbiting electrons. Such an entity, however, has never been witnessed.

Instead, the antimatter that is said to be created in particle accelerators is inferred from traces left behind by subatomic particles whose trajectory suggests that they had the mass of a known particle but the opposite charge. That is, the degree of curvature in the arcing particle traces (due to the influence of the surrounding electric or magnetic field) suggests a particular mass equal to that of a known particle, but the direction of the arc is opposite that expected of the known particle. From this, it is surmised that the particle leaving the trace must be the "anti-particle" or "antimatter" version of the known one – a particle of the same mass but opposite charge. Since such traces are very rare, often occurring in one out of millions of particle traces, antimatter is considered to be extremely scarce. This rarity of particle traces that might qualify as antimatter in the midst of the enormous energy expenditure of particle accelerators is also why antimatter is said to be extremely expensive to produce. Let's now take a closer look at this situation.

Perhaps one of the strangest beliefs about antimatter is that it annihilates its regular-matter opposite particle, creating a burst of pure energy. This claim is particularly odd since antimatter is, strictly speaking, simply regular matter with opposite charge. As we know, oppositely charged objects frequently come into contact in everyday experience – whenever any two objects stick together due to static cling, for example – yet there is no concern of mutual annihilation. And, even if two oppositely charged subatomic particles such as a proton and an electron were to come into contact, there is no theory or experiment that suggests any such dramatic reaction. If there were, we would only need to cross a beam of positively charged alpha particles, which emanate from certain types of radioactive objects, with a beam of electrons to create enormous amounts of energy from their mutual annihilation.

Yet, if two such oppositely charged particles of regular matter should also happen to have the same mass, it is claimed that they would suddenly and mysteriously vanish in a flash of energy from a matter-antimatter annihilation. Precisely how or why this should occur in only this special case of equal mass and opposite charge is entirely unexplained in today's science, but is said to have the backing of experimental evidence, nonetheless. Specifically, every once in a while, the traces from a particle accelerator collision experiment appear to show odd results such as a regular subatomic particle suddenly vanishing and producing a photon of electromagnetic radiation – pure energy. From this type of evidence it is surmised that there must have been a collision with an unseen exotic particle – presumably the regular particle's "antimatter opposite" – and that both particles annihilated each other when they met, creating pure energy (Fig. 6-4).

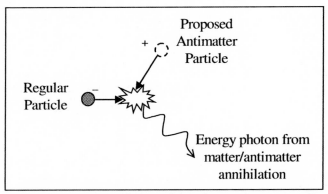

Fig. 6-4 Matter/Antimatter Annihilation in Standard Theory

 Antimatter is not a mysterious "opposite" form of matter, but merely an unstable, transitory grouping of electrons much like regular, stable subatomic particles.

However, far less mysterious explanations follow from the perspective of *Expansion Theory*, which shows that both subatomic particles and energy photons are actually similar groupings of electrons in either atomic or subatomic realms. So, when a single subatomic particle, such as a proton, is released and truly isolated in the atomic realm it may well take very little for it to become an "energy photon"; in fact, it may well even take sizable effort to *prevent* it from doing so *spontaneously*. The immense pressure from the extremely powerful surrounding magnetic or electric fields used in particle accelerators could well postpone the inevitable transformation of a truly free subatomic particle into a freely expanding energy photon (i.e. electron cluster). These fields could even *contribute* to such a transformation in certain circumstances, perhaps adding a number of electrons to (or removing them from) a free subatomic particle amidst the turbulence and chaos within these dense surrounding electron clouds, thereby triggering the transformation.

   An example of these proposed mechanisms is shown in Figure 6-5, where the surrounding fields may serve to postpone the spontaneous transformation of free protons into free electron clusters (energy photons), or may trigger such a transformation by adding undetected electron clusters ("antimatter" or "virtual particles") to the reaction.

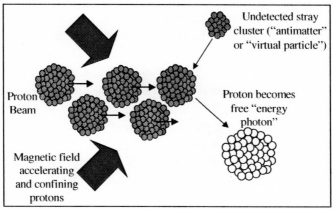

Fig. 6-5 Possible 'Antimatter Annihilation' in Expansion Theory

The key point is that today's picture of isolated matter and antimatter particles cleanly meeting and disappearing into pure energy, as shown earlier in Figure 6-4, is largely an idealized interpretation of what actually occurs in particle accelerators. In actuality, often the trace of only one of the particles is seen, while the presence of the other particle or even the resulting energy photon may be largely inferred, creating a composite built partly from theory and partly from elements of many other collision runs. It is this type of inference that has led to the concept of *virtual particles* in particle physics – particles that are not actually seen or detected, but are believed or theorized to exist in order to explain or complete a given type of collision event.

And, of course, such idealized collision diagrams exclude and ignore one of the most important components of the entire experiment – the surrounding clouds of dense electrons. Since the surrounding fields are currently only thought of as energy, this separate ethereal phenomenon is merely thought to accelerate and direct the particles, but not to make a *material* contribution to the particle reactions themselves. Therefore, a key element of all reactions – the external pressure and possibly even material contributions from the surrounding electron clouds – is *entirely overlooked* in today's collision diagrams, calculations, and theory. And, it is likely such an omission that has given physicists the impression that enormous energy is released by matter-antimatter reactions.

In actuality, it is more probable that the "antimatter particles," as well as the "energy photons" that result from these reactions, are merely electron cluster configurations supplied from the environment of surrounding electrons created by the physicists and *fed from the city power grid*. It is only by completely ignoring the immense (and almost entirely wasted) power drawn from external sources to *artificially support* these reactions that physicists are able to conclude that these "matter-antimatter" reactions offer enormous promise as future power sources. In actuality, according to *Expansion Theory*, these reactions do not foreshadow a tremendous new "antimatter power source," but merely represent arbitrary electron reconfigurations within lab experiments that will only ever *drain* power from our existing power sources.

We *could* perhaps think of today's "matter-antimatter" experiments as a new type of storage battery that releases some of the power it drew from the external power grid when the rare "annihilation collision" occurs; however, as it stands currently, this is an extremely inefficient, cumbersome, and short-lived battery technology. We may someday find some practical use for this phenomenon, but it would appear that the enormously powerful, efficient, self-contained matter-antimatter engines of science fiction are, unfortunately, likely to remain in the realm of fiction.

This description of subatomic particles also explains a phenomenon noted in the section on *Special Relativity* in the previous chapter. It was suggested that the greatly extended half-lives of the speeding particles in particle accelerators are not the result of a mysterious "time dilation" effect, but are merely the result of being forcefully held together from the outside. We can now see how this might be if these "particles" are actually groups of expanding electrons that naturally push the whole group apart very rapidly (causing spontaneous "particle decay"). And, since the magnetic field used to accelerate these "particles" is actually a highly dense, pressurized cloud of expanding electrons surrounding them, it would not be surprising if such a tremendous force managed to hold them together much longer than usual, as shown earlier in Figure 6-5. Further, the compression resulting from the tremendous G-forces caused by the immense acceleration would also contribute to the extended particle lifetimes noted in particle accelerators. Today's science has no clear understanding of the actual physical nature of either magnetic fields or subatomic particles, and so, has little choice but to accept the mysterious claims of *Special Relativity*, such as time dilation. This is especially true since experiments not only appear to validate such beliefs, but many experimental results would be a complete mystery without such explanations – that is, in the absence of *Expansion Theory*.

At this point, after showing that every known form of matter and energy can be fully explained by various configurations of expanding electrons, it may seem curious that two very different manifestations – subatomic particles and energy photons – arise from the same configuration of electrons. That is, in *Expansion Theory*, photons of radiant "energy" such as light are described as clusters of expanding electrons, but subatomic matter particles are also described as groups of expanding electrons. How can two such different manifestations have the same description? And how can it be that the energy unleashed from subatomic matter particles in an atomic bomb detonation is so explosive and radioactive, while the apparently similar electron clusters composing light are relatively weak and harmless? The answers to these questions can be found by considering the nature of an atomic bomb.

What is an Atom Bomb?

In today's science, the atom bomb is thought of as a chain reaction of atomic nuclei splitting apart and somehow converting some or all of their mass into pure energy in accordance with Einstein's energy-matter conversion equation, $E=mc^2$. Although this is a useful model, and one that may well have served as an important guiding principle for the conceptualization and engineering of such a device, it does not provide a true understanding of the physics at work. How does matter convert into energy? Why does splitting of the nucleus cause this conversion to occur? And what exactly *is* this radioactive energy released by such a device?

Science has *"know-how"* in the form of models, techniques, and engineering, but no definitive *knowledge* and *understanding* of the underlying physics to clearly answer these questions. Nuclear radioactivity was not predicted, understood, or engineered from a prior deep understanding of the physics involved, but was accidentally discovered and experimentally furthered by scientists like Henri Becquerel (1852-1908) and Marie Curie (1867-1934). The "know-how" that led to the first atomic bomb emerged more from a progression of accidental discoveries, educated guesses, abstract models, and trial-and-error than from a clear understanding of the physics at work. However, *Expansion Theory* presents a clear physical picture of the nuclear fission process that takes place within an atom bomb.

The similarity between the electron groups composing subatomic matter "particles" and the electron clusters in radiant "energy" provides the first clue. How can these two things be so similar in composition, yet as different as matter and energy? The answer lies not in any significant physical differences between them, but in a difference in the *realms* in which they exist and gain their definition. The electron clusters of "energy photons" expand freely in the external *atomic* realm, while the electron groups of subatomic "particles" are essentially confined as they expand within the *subatomic* realm. As long as an electron group contains the proper number of electrons to qualify as a proton or neutron, it has the potential to function as one of these core subatomic particles composing an atomic nucleus. If such a group of electrons exists as part of the nucleus of an atom, it functions as a subatomic particle (proton or neutron) and supports the stable atomic structure upon which our existence depends. Recall from Chapter 4 that the tremendous expansion of these particles in the subatomic realm within the atom is, paradoxically, entirely contained within the stable atomic structure with which we are familiar.

NEW IDEA

"Splitting the atom" is actually the process of destabilizing the atom such that it effectively turns inside out, releasing its subatomic particles into the atomic realm as "energy."

However, when an atom is structurally destabilized by an external force, it essentially turns inside-out. That is, when the stable atomic structure or organization disintegrates, the expanding protons and neutrons in the nucleus of the atom suddenly become redefined or bared as freely expanding electron clusters in the external atomic realm. This essentially transforms them from expanding *matter particles* contained within the subatomic realm, to freely expanding electron clusters or *"photons of energy"* in the atomic realm (Fig. 6-6). A rapid chain reaction of countless disintegrating atoms would release tremendous numbers of such freely expanding electron clusters, smashing them apart into all manner of cluster sizes corresponding to all frequencies of radiation. This is precisely what is observed in an atomic explosion – the matter contained in the core of the bomb becomes rapidly transformed into all manner of radiant energy exploding outward.

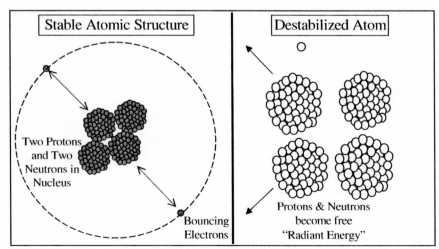

Fig. 6-6 Nuclear Fission in Atom Bomb from Expansion Theory

In essence, these newly freed electron groups from within the disintegrating atom (i.e. subatomic particles) are no different from the electron clusters in regular light described in the previous chapter. The reason for the tremendous power difference is merely that an atom bomb *instantly* releases countless *pre-existing* expanding electron clusters within a very compact space, while standard light sources slowly *construct* such clusters from electrons that are already externalized in the atomic realm as electricity flowing along a wire. The light-bulb discussion in the previous chapter, for example, shows a current of already-externalized electrons pushing each other along the outside of the wire, eventually becoming grouped into clusters in a continual "assembly line" to form light. This is a far slower and far less concentrated form of cluster generation, or radiation. Still, it is possible to design very powerful lasers that can also present significant radiation hazards, showing that the electron clusters in light are not necessarily weak or harmless – it all depends on the speed and concentration of production or release. In the atom bomb there are countless *pre-fabricated* electron clusters awaiting simultaneous release from a very compact, concentrated form – the nuclei of many very large, unstable atoms such as uranium or plutonium.

E=mc²: What is Energy-Matter Conversion?

The previous discussion dealt with the conceptual conversion of matter into "energy" in an atomic bomb explosion. This section deals with this same issue from the mathematical perspective. Einstein's famous equation, $E = mc^2$ (i.e. *energy* equals *mass* time the *speed of light squared*), is often represented as a unique energy-matter conversion equation that describes the energy released when matter is rather mysteriously converted into pure energy. In actuality, although this is a reasonably accurate characterization of this event, the phenomenon it models is not quite as unique and mysterious as we are often led to believe (as shown in the previous discussion of the atom bomb).

　　　　Consider, for example, the classical equation for the kinetic energy of a moving object, $E_k = \frac{1}{2} mv^2$, which states that the object's *kinetic energy*, E_k, is equal to one-half of its *mass* times its *velocity squared*. Although this is also an equality between mass and energy that is nearly identical in form to Einstein's equation, it is not considered a mysterious "energy-matter conversion" equation, but merely a straightforward expression of the energy of motion of an object as described by classical physics. In fact, as shown below, it is no coincidence that this equation is so similar to Einstein's equation.

A number of different derivations can be found for Einstein's $E = mc^2$ equation; and, while the more esoteric among them can become quite involved, the fact is that this equation can be easily derived in a few lines of simple classical equations of motion, as shown in the following four-line derivation:

1 -- $p = E/c$ – Momentum of light, p, equals its energy content, E, divided by its speed, c

2 -- $p = mc$ – Momentum of light, now stated in terms of its mass, m, times its speed

3 -- $E/c = mc$ – Equating lines 1 and 2

4 -- $E = mc^2$ – Rearranging line 3 gives Einstein's well-known equation

The first two lines in the above derivation are simply two common equations for the momentum of light from classical physics (physicists traditionally denote momentum by the letter p). The first line shows the momentum that light imparts to materials that it strikes, stated in terms of its energy content divided by its speed, and confirmed by experiments in which light is shone onto material from which it does not reflect back. It is worth noting, for reasons that will become clear shortly, that the momentum expression in Line 1 increases for materials that are more reflective, eventually doubling to the expression $p = 2E/c$ when the incident light is fully reflected from a mirrored surface.

The second line in the above derivation shows the momentum that light would have according to the classical momentum equality for moving objects – namely, *momentum* equals *mass* times *velocity*, or $p = mv$, with the velocity, of course, being the speed of light, c, in this case. It is interesting to note that, today, Line 1 is considered to be the proper physical description of light's momentum, while Line 2 is considered as the more abstract momentum expression since light is believed to be speeding *energy* – not mass – in today's science. In contrast, *Expansion Theory* shows that it is actually Line 1 that is the abstraction and Line 2 that is the reality since light is actually a beam of physical *matter* particles (electron clusters) and not "pure energy." Regardless, since these two lines describe the same property, they can be equated to each other (Line 3) and rearranged (Line 4) to give Einstein's famous equation, $E = mc^2$.

NEW IDEA

Einstein's famous $E = mc^2$ equation does not refer to a mysterious "energy-matter conversion" process, but simply the classical kinetic energy of free electron clusters.

The first important point to note is that Einstein's equation was derived above without reference to the splitting of an atom or any other mysterious "energy-matter conversion" process, but merely in reference to the classical momentum of speeding light (i.e. speeding electron clusters). As such, Einstein's equation can be seen merely as a *kinetic energy* equation of matter traveling at the speed of light, and so, would be more accurately written as $E_k = mc^2$. Secondly, it was mentioned earlier that the expression for the momentum of light, $p = E/c$, in Line 1 is only a statement of the zero-reflection scenario, which actually increases to $p = 2E/c$ for full reflection. That is, the impact felt when the momentum of the incident light is fully absorbed by the target would be felt doubly hard by the target if the light instead bounced fully off again at the same speed as its initial arrival.

So, repeating the above derivation of Einstein's equation for both of these scenarios – first without reflection as in Line 1 above and then with full reflection – shows that Einstein's mysterious "energy-matter conversion" equation is more correctly stated as the full and proper *kinetic energy* expression:

$$\text{Full Reflection} \rightarrow \quad \tfrac{1}{2}mc^2 \ \leq \ E_k \ \leq \ mc^2 \quad \leftarrow \text{No Reflection}$$

This expression arises from the fact that, although the momentum expression in Line 1 varies by a factor of two between reflecting and non-reflecting scenarios, the equivalent expression in Line 2 does not. Therefore, the only way that Line 1 and Line 2 can be equated (Line 3) is if the energy content of the light is considered to vary. That is, the impact-doubling effect noted in experiments involving full reflection means that the energy of reflected light must be considered as half that of non-reflected light if the resulting impact is forced to be equivalent in Line 3. Although this logic is somewhat of an aberration due to today's misunderstanding of the true nature of light, as represented in Line 1 and as evidenced by the forced equality in Line 3, it does technically result in the full kinetic energy expression above, and not merely in Einstein's $E = mc^2$ equation. The full expression above clearly shows the true nature, origin, and meaning of Einstein's famous equation – that it is merely a classical kinetic energy expression for the special case of non-reflection. But then, why is Einstein's simplification of this kinetic energy expression for light used to describe the apparent conversion of mass into energy in an atomic bomb explosion?

The answer lies in the fact that, as discussed earlier, the nuclear chain reaction of an atomic bomb is actually a releasing of the electron clusters known as protons and neutrons from the subatomic realm and into the atomic realm. This process immediately produces freely expanding electron clusters of all sizes – which we know as all forms of radiation from heat through to light and on to X-rays and gamma rays. And, since these electron clusters expand outward at the speed of light, they would each carry an amount of kinetic energy somewhere within the range from $E_k = \frac{1}{2} mc^2$ to $E_k = mc^2$, as shown in the earlier full expression. And also, since all of the bomb's atoms *instantaneously* decompose into these free electron clusters, the impact of all of this kinetic energy is just as if a regular non-explosive object of the same mass as the bomb slammed into the planet at the speed of light. That is, the damage done by the (non-reflecting) impact of such an object would give its original kinetic energy as $E_k = mc^2$, where the kinetic energy of an object causing the same amount of damage while fully rebounding off the planet would be given by the equation $E_k = \frac{1}{2} mc^2$.

Regardless of the factor-of-two difference between full-reflection and zero-reflection scenarios, we can see that exploding an atomic bomb is physically equivalent to accelerating a regular, non-explosive object to the speed of light and slamming it into a target. The atom bomb simply makes this enormously difficult task quite easy and convenient. An atomic bomb can be delivered to the target area and detonated with a relatively small amount of effort, rather than accelerating a regular, non-explosive object to light-speed and hitting the desired target, but regardless, the destructive power is the same in either case. This is the true meaning behind Einstein's $E = mc^2$ equation, and the reason why it does indeed accurately represent the amount of energy released in an atomic bomb blast. In a sense, it does describe a nuclear conversion process of matter into energy – using today's terminology – though this still-mysterious occurrence today is actually a straightforward freeing of expanding matter from the inner subatomic realm to the outside world.

Incidentally, this discussion raises the point that the force of an impact has been experimentally shown to *double* for reflection scenarios – and for no other reason than the fact that the objects involved happen to bounce off of each other. A closer look at this issue brings the odd realization that this represents an apparent *creation of energy* – up to a *doubling* of impact force merely because the materials involved happen to cause reflection. In fact, this is indeed yet another example from everyday occurrences that violate our current laws of physics. And, this is not simply an oddity of reflected light, but can be clearly seen even in the larger objects of everyday life.

Consider the scenario of a lump of clay dropped to the floor vs. a dropped tennis ball. Both objects become squashed and deformed by the impact, at which point both objects have fully expended their momentum and kinetic energy. However, the tennis ball *additionally* springs back and propels itself off the floor to nearly the original height from which it was dropped. We are so accustomed to bouncing balls that we see this as quite reasonable, yet we would be rather shocked if the lump of clay spontaneously reformed itself and leaped into the air; there would be no reasonable explanation for this surprising burst of energy. Yet, in actuality, bouncing balls are just as shocking and just as inexplicable. Their additional leap back off the floor to their original height would also have to impart an additional force against the ground of equal strength to their original force of impact with the ground. Here is the *impact-doubling* effect noted in experiments with light bouncing off of mirrors. Both light beams (i.e. beams of electron clusters) and larger material objects can spring back from an impact, and in so doing, draw on the additional hidden power source of the subatomic realm.

Such violations of our energy-based laws of physics in the atomic realm occur all around us. This is because our world is actually based on *expanding matter* – not "energy" – and involves both the *atomic realm* as well as the *subatomic realm* upon which everything rests. We should not be surprised when effects appear from the subatomic realm since the subatomic realm lies just beneath the surface of every event that occurs in the atomic realm, and in fact, supports our very existence moment by moment.

This principle can even be found in athletic performance. Athletes derive the majority of their power from their muscles, but they also owe a portion of their performance to the springiness of their bones. We cannot leap into the air by relying on muscles alone, but require the rigid support of our bones as well. An athlete who has springy bones will have an edge over those who do not. Yet, this extra bouncing-back force is only possible because the molecular bonds within the bones forcefully resist permanent deformation by pushing back – a force that can only be explained in terms of a contribution from the subatomic realm.

This is not an endless power source from beyond, but merely a *crossover effect* from the subatomic realm that effectively returns a portion of our expended power – up to a (theoretical) 100% "refund" of our power expenditure. This is how trampolines work, for example; trampolines have no visible power source, yet they allow a person to leap far higher than the slight springiness of their bones would normally provide. This is an effect that we have superficial characterizations of in our science today (standard equations for springs, elasticity of materials, etc.), but which is actually a complete energy-for-free mystery according to our current laws of physics, upon closer examination. Objects should not be able to mysteriously "push back," "spring back," or "bounce back" from within – according to today's science.

What is the Origin of Expanding Matter?

The earlier discussion of subatomic particles began with an ocean of expanding particles representing an early universe of expanding electrons – similar to expanding ripples in a pond. However, this says nothing about the nature of the universe prior to the appearance of electrons, nor why expanding electrons might spontaneously appear in the first place. It doesn't explain what these elementary matter particles *themselves* are made of, or even why they might be expanding. It also doesn't explain why they would push against each other when they meet (rather than passing right through one another) – a quality that makes matter solid and makes possible the fundamental quality of *inertia*. These are deep issues indeed, representing some of the most elementary questions in science. While it may always be difficult to answer such fundamental questions with absolute certainty, *Expansion Theory* provides a powerful new tool for exploring possibilities that have not yet been considered. One such possibility follows by considering a simulated universe of expanding electrons within a computer.

EXPERIMENT

 A Universe in a Computer

Consider a simple computer program that populates the computer screen with circles at random, representing the initial population of electrons throughout this "universe." If these circles are then made to increase in size at a given rate, they could be considered to represent *expanding* electrons, except that there would be nothing stopping them from simply expanding right through each other at this point. If additional logic were added to the program to prevent them from passing through each other, then we would have a reasonable simulation of an early universe of expanding electrons (Fig. 6-7). This additional logic would need to give all electrons a pre-determined amount of elasticity or "bounce" when they impact one another, which determines to what degree they bounce off each other and to what degree they tend to clump together when they meet. One final step toward realism would be to give these circles three-dimensional representations as expanding spheres instead of simply two-dimensional expanding circles, though this is not critical to the basic concepts of this thought experiment.

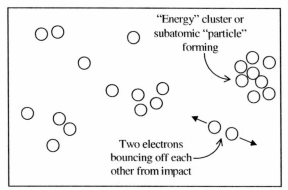

Fig. 6-7 Expanding Circles in Early-Universe Simulation

This fairly simple program now gives us a simulation of the early universe according to *Expansion Theory*. The random placement of these expanding circles means there will be varying regions of high and low population densities, giving differing dynamics. Regions of high density might be very active and chaotic as their electrons expand and bounce off each other in close quarters, while electrons in lower density areas may manage to settle into stable groupings and structures as they expand. Depending on the chosen parameters of expansion rate and elasticity of these simulated electrons, various degrees of chaos and order might be expected – anything from freely expanding clouds or clusters of electrons to long-lived "atomic" structures could emerge. As the program continues to run, and if conditions are right for certain types of stable structures to appear repeatedly, our "universe" will begin to fill with copies of a limited number of such stable "atomic" structures or templates. This is much like our *real* universe, which has countless copies of a limited number of stable atoms in the Periodic Table of Elements.

Once this occurs, there is the potential for solid universal laws of physics and chemistry to arise. This is because there would only be a finite number of ways that the limited variety of "atoms" could interact. Therefore, rather than there being complete chaos and unpredictability between an infinity of completely random structures, there would be repeatable, predictable interactions between the limited types of "atoms" that appear repeatedly throughout the universe. This amounts to recognizable chemical substances with universal physical and chemical properties. One could imagine that this process might even continue to form simulated "stars," "planets," and perhaps eventually even "life forms" within our computer universe. If not, perhaps rerunning the program with different values of expansion rate and elasticity for our "electrons" would eventually bring reasonably favorable results.

Before discussing the possible implications of such a simulation for our real world, let's consider what it would be like for a simulated life form in our computer universe. Such beings would be able to walk about as we do since the simulated atoms composing their planet would be expanding just as described earlier in Chapter 4 – creating the effect we call gravity. The ground would be solid because we programmed the simulated electrons that compose all matter to not pass through each other. The simulated sun could also provide simulated light and heat, since "light energy" and "heat energy" are simply beams of expanding electron clusters. We do not have the burden of figuring out how to create some type of ethereal energy phenomenon in this universe since such a concept doesn't actually exist even in our own universe – literally *everything* follows from straightforward expanding electrons. These simulated life forms might eventually wonder about the nature of matter and energy, as we do, eventually discovering atoms and subatomic particles. They would likely develop theories about energy, and would find that the only truly fundamental, indivisible particle is the electron – for some unknown reason. Most importantly, this simulated existence would not *feel* like a simulation to such beings; all of it would seem entirely real and material, and yet, it all springs purely from the first cause of expanding spherical patterns in our computer.

One further point to consider is what would happen if we turned off the computer screen. As we know, the computer continues to run its programs whether or not the screen is turned on. If we turned the screen off, then back on some time later, we would find that our simulated universe had progressed quite nicely while the screen was off. Therefore, the initial expanding circles were not truly little glowing circles on the screen at all since they still continue to exist and evolve into a whole universe of complexity even if they are never displayed. So then, these "electrons" are actually something even more ethereal than circles on a screen – they are little more than mere abstract patterns of activity in computer memory. Yet, despite the completely immaterial nature of it all, such particles could support a whole universe that would be very real and quite material to beings within it – *much as our own universe is very real to us.*

Now, what is the limit of knowledge that these simulated beings could attain about their universe? They would develop various theories to explain observations, perhaps inventing theories of gravitational forces and energy waves. They may even eventually come to realize that their whole universe is composed of expanding electrons at its heart – but what then? Could they ever devise an experiment to smash an electron apart and see what's inside? Certainly not with the underlying program in this thought experiment since we have only provided for our expanding circles to either meet or bounce off one another – not to shatter. And even if these electrons *were* programmed to shatter, would these beings be able to detect the resulting fragments? That depends on what we, as the programmers, decided would happen when electrons shatter; they could simply vanish into thin air or ten new electrons could appear in their place – whatever we decided upon as the programmers.

What if these beings designed devices to perform advanced experiments to figure out how electrons first came into existence, why they expand, and why they impact one another rather than passing straight through (i.e. why matter has substance and inertia)? In order to truly answer these questions such an experiment would have to be capable of uncovering the computer program that created and supports their universe – ideally even discovering the existence and nature of us, the *programmers*. Is there any conceivable experiment that such beings could perform from within their universe to uncover such things?

All this is not to say that we are necessarily such beings in a simulated universe, supported by ethereal patterns of expansion (electrons) in some type of cosmic computer, although, if this *were* the case, our experiences would be precisely as they are today. Matter would feel solid and material to us, and we would seem to be surrounded by various forms of ethereal, and still rather poorly understood, forms of "energy" (gravity, light, magnetism, etc.). Also, our universe need not *literally* exist within a computer system, supported by an actual computer program written by a programmer. While this *could* be the case, the "cosmic computer" we inhabit could also be an underlying physical realm with a physics that may even be quite simple but completely foreign to us, causing expanding spheres (electrons) to spontaneously appear and behave as they do.

For example, imagine an infinite space filled with countless "primordial particles" that pre-date even expanding electrons, and that all simply coast along in straight lines at various speeds in random directions. We should eventually be able to "connect the dots" in a given region to make some arbitrary shape such that the dots composing it are all moving away from a central point between them at the same speed, forming an arbitrary, stable expanding shape (Fig. 6-8).

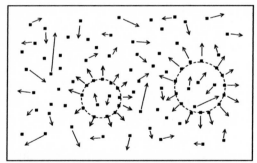

Fig. 6-8 Early Electrons: Expanding Patterns in Primordial Ocean

Also, in such a group of randomly moving primordial particles there would typically be a *normal* or *bell-curve* distribution of speeds, with the majority tending toward a certain average speed somewhere between the fastest and slowest extremes. This average speed of a population is what determines the temperature of a gas, for example, even though some molecules may be moving far faster or far slower than the overall population average. So then, we have arbitrary expanding patterns composed of primordial particles that likely tend toward a common average speed. This means that all edges of arbitrary expanding shapes would tend to proceed outward at the same speed in this ocean of primordial particles – a process that, over time, averages any shape into a smooth, uniformly expanding sphere. This tendency to approach a spherical shape over time is shown in Figure 6-9, which is why the arbitrary shapes seen earlier in Figure 6-8 were shown as spherical.

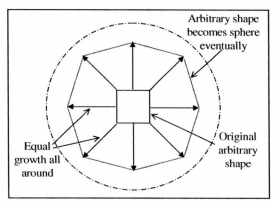

Fig. 6-9 All Shapes Become Spherical as Edges Grow Equally

Such spheres would spontaneously occur here and there within the sea of primordial particles, appearing with a random or statistical distribution. This simple process, where an infinite ocean of identical expanding electrons spontaneously appears to define our early universe, would be the fabled *creation event* – the event that we currently think of as a "Big Bang."

Since the only form and order in this early universe is this random distribution of expanding spherical patterns, if two such spheres simply passed through each other when they met, this would disrupt these spherical patterns and merely leave a formless ocean of chaotic primordial particles. Therefore, only the primordial particles that contribute to the ongoing existence of these separate spheres would play any significant role. That is, when two spherical patterns meet, the primordial particles at the edges where they meet continue traveling on (as they must), but as they do so, they leave the spherical pattern and return to the formless primordial ocean. At the same time, other primordial particles continually come and go in their place from the surrounding primordial ocean to effectively maintain a continuous stationary edge where the two spheres meet – a standing-wave pattern composed of passing primordial particles. The integrity of the overall expanding spherical patterns always remains intact (in order for the universe-as-we-know-it to exist) while the component primordial particles come and go as necessary.

This process means that the edges of the two spheres effectively stop expanding toward each other once they meet, so that they do not continue to pass through one another. However, since the remaining edges of both spheres are still free to expand, only the primordial particles that continue to define these two separate spheres as expanding away from this meeting point partake in supporting their continued existence (Fig. 6-10). This is somewhat analogous to the manner in which expanding ripples in a pond persist as continuous entities while their constituent water molecules come and go, although this is only a loose analogy between very different physical realms since water waves do pass through each other.

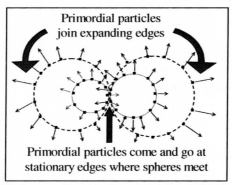

Fig. 6-10 Expanding Patterns Push on Each Other

It is also worth noting that, although the preceding discussion was intended to show a possible *physical* explanation behind expanding electrons rather than merely a simulation within a computer, it can only be considered a loose guideline for such a physical process. This is because the word "physical" can only truly be a reference to overall atoms and subatomic particles – which are all we know of a physical realm. The hypothesized primordial particles composing expanding electrons would exist in a realm that *supports* our physical universe and therefore cannot also be considered *part of* that universe and its physics. This is analogous to the pre-existing and very different computer memory in the earlier simulated-universe thought experiment, within which a universe of expanding circles (simulated electrons) existed. The computer memory *supports* these simulated electrons and the universe that emerges from them, but does not *exist within* that universe and is not *part of* that universe and its emergent physical laws.

As such, it may well be completely academic whether we consider the ocean of primordial particles to be physical or simulated, since both scenarios may well be far removed from our familiar notions of physical cause-and-effect. So, while we can conceive of the primordial realm either as a deliberate creation analogous to a designed computer simulation or as a simple and spontaneous physical process, the fact remains that the difference between these two alternatives may only be a matter of semantics. It may be that – much like the life forms in the earlier computer simulation thought experiment – we can only deduce *what* the primordial realm does by arriving at an understanding of our universe of expanding electrons, but not precisely *how* or *why* it behaves so.

It appears that we are now arriving at a point in our science where we are beginning to understand the makeup of our universe – and are perhaps reaching the limits of that understanding as well. Further understanding and advances in fundamental physics may all be up to the nature of the "program" or possible "primordial realm" that lies beneath it all. Has it provided for us to find the ultimate answers in our experiments? Is there a secret back door to this understanding built into certain types of physical or mental processes? Might there even be hidden "cracks" in the operation of our universe that provide a clue to what lies beneath it all? Are there bugs or oversights in the underlying program design or glitches in the hardware or physics of the underlying realm that we can exploit to gain insight and understanding into the nature of it all? Is there any way we could hope to ever detect the existence of a primordial realm with potentially another physics entirely? Regardless of the answers to these questions, *Expansion Theory* provides us with a powerful new perspective on our universe and our existence, and may just give us the necessary tools to begin to explore or at least consider some of these questions.

What Causes Inertia?

One of the deepest and most fundamental questions in physics is why objects have inertia – that is, why a more massive object is harder to push and harder to stop in free space than a less massive one. Although we intuitively understand this quality of objects from common experience, without a clear understanding of the principles underlying this effect we have no solid explanation for why it should be any harder to toss a bowling ball than a tiny marble. This is the case even when objects are floating in space. Objects in space may be easier to move than on Earth due to the lack of friction and the removal of the Earth's expansion against them (gravity), but more massive objects are still proportionately harder to accelerate and to stop in outer space than less massive ones. This is embodied in the equation, $F = ma$, from *Newton's Second Law of Motion*.

It was once believed that this mystery was solved by Ernst Mach (1838-1916), who proposed that the inertia of a given object resulted from the sum total of the gravitational tugs from all the matter in the universe upon each object. Yet, although Einstein also agreed with this idea, he ran into sizable problems when searching for evidence of this proposal in the equations of his own theory of gravity – *General Relativity*. Not only could Einstein find no evidence for this interpretation of inertia in his theory, but other physicists performing this same search further showed that Einstein's equations gave strange results that even *contradicted General Relativity Theory itself*, such as a universe with absolute (not relative) rotation. Although there have been many other proposed theories of inertia – many of which essentially amount to a re-introduction of the idea of an all-permeating "ether" filling the universe – the nature of inertia remains an open question in physics.

Even our current atomic models and theories about subatomic particles and energy do not provide a fundamental explanation for inertia and momentum, and many physicists believe that the answer may lie in a special subatomic particle that is yet to be discovered. Yet, although intense and elaborate searches have been ongoing for such an entity, precisely how such a question will be answered by finding evidence for the existence of yet another proposed subatomic particle has never been made clear. After all, if we are expending a great deal of effort and resources searching for evidence that such a theoretical particle exists, it would seem reasonable that its *theoretical* existence must already explain inertia in order to justify the search. Otherwise, why search for physical evidence of an entity if even its theoretical existence does not answer the inertia question that this hunt is meant to solve?

Imagine, for a moment, that we *have* verified the existence of this elusive theoretical entity. How will it answer our ultimate questions about inertia to simply fill-in the missing values of charge, mass, etc., for this theoretical particle? After all, it seems clear that the theory itself provides no revelations today regardless of the values that we plug into the equations – if it did, we would already have our answers and would merely be looking for physical proof to verify our assumptions.

From all current indications, as often occurs today, a team of researchers will likely eventually claim experimental success in a flurry of excited news releases, followed by broader confusion and disillusionment from the public when nothing further comes of such a revolutionary finding. Typically, more subdued news releases follow later to the effect that this important new discovery now seems to be the tip of the iceberg and more funding and research is required. Although this process *can* be the path to truly revolutionary discoveries, it is all too often evidence that the researchers have locked themselves into flawed beliefs and theories from the outset. *Expansion Theory* now offers us an alternative that frees us from these vague hopes for understanding, showing a clear and simple answer to this long-standing inertia question by referring to the new atomic model introduced in Chapter 4.

NEW IDEA

Inertia is explained by the nature of electrons as they continuously expand throughout our universe.

First, using the new atomic model, let's look at a scenario where objects have no inertia – no resistance to an external force. The most straightforward way this would occur is if the atom itself offered no structural opposition to external forces. This is shown in Figure 6-11.

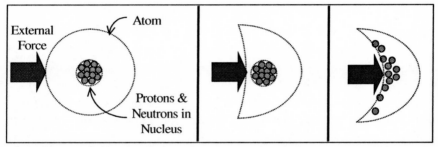

Fig. 6-11 Atom Offering No Structural Resistance to External Force

In this scenario we would feel no opposition when applying a force to an object, but the very atoms of the object would have to disintegrate in order to achieve this effect. Now, let's look at what actually occurs when we push on an object:

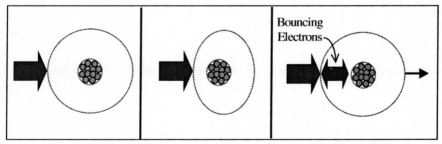

Fig. 6-12 Inner Force actually Pushes Back From Within

Figure 6-12 shows what actually happens in the real world. An external force pushes against the bouncing electrons of the atom, confining them to a smaller space and essentially distorting the spherical shape of the atom. However, this situation soon rights itself since the electrons would bounce more vigorously within their more confined space, creating a force from within that pushes back against the external force. If the external force continues, then the increased internal force of the bouncing electrons will push the nucleus of the atom further away, returning the atom to its spherical shape and causing the overall atom to move in the direction of the push given by the external force. This process of compression, internal force buildup, and rebound takes time, and the more atoms there are within an object to deform before rebounding (i.e. the more massive the object), the more time and effort this "domino effect" takes in order for the external force to move the object.

This is a very new concept that is not found in our current physics where a universe of non-expanding matter and constant energy balances – as dictated by the *Law of Conservation Of Energy* – is the prevailing paradigm. In contrast, the above explanation of inertia shows, in effect, a literal *creation* of "energy" from within the atom to push back against the external force rather than allowing the atom to simply disintegrate. As mentioned, if an external force is applied to an object long enough and with enough strength, a force from within the atom (from the subatomic realm) will cause the atoms to rebound away from the external force, eventually moving the overall object in the direction intended by the external force. However, if the external force is relatively weak or only applied for a short time, the atoms will be able to rebound back toward the direction from which the external force was applied and the object will not move as far. This opposing push from within each atom adds up significantly when there are many atoms, which is why a more massive object does not respond as readily to a brief push and requires more force for a longer period of time to accelerate it.

This "new energy" from within the rebounding atom has no identifiable power source in the conventional sense and is continuously supplied by the ocean of primordial particles from which expanding electrons arise and continuously expand. The solid, indivisible nature of continuously expanding electrons is a key element in the inertia of objects, and is determined by the physics of the subatomic and primordial realms. Whether we consider electrons to arise from the physics of an ocean of primordial particles or from the arbitrarily pre-determined behavior of expanding spheres in some type of universal computer, this fundamental nature of the electron is ultimately at the heart of inertia in our universe.

In addition to the effective *creation* of energy from within the rebounding atoms, solid objects also present a further example of how the subatomic and primordial realms break the paradigm of our current energy-based laws of physics by also exhibiting a *destruction* of energy. When we talk of energy absorption today, we are actually referring to a *transformation* of energy from one form into another – in accordance with the *Law of Conservation Of Energy*, which states that energy cannot truly be created nor destroyed but only transformed. For example, rapidly striking one object with another is a transformation of kinetic energy into sound and internal heat and vibration within the objects due to the impact. Our laws of physics require that there always be this energy balance between input and output energy.

However, this "law" is broken constantly all around us. In addition to the many examples of such violations already discussed, a further example can be found in the simple scenario of squeezing a solid object. In this scenario, a constant external force is applied to an object, yet the object continually resists this force, draining the external power supply but not *transforming* this external energy into sound or internal heat or any other form of energy. In today's terminology, this could only be described as a true *destruction* of energy as it is lost to the subatomic realm that continually pushes back from within the squeezed object.

This literal *destruction* of energy is not immediately apparent in today's thinking since it could be claimed that there *is* an energy balance as our muscles transform chemical energy into waste heat and other internal biological changes as they squeeze the object. However, the fallacy of such reasoning can be readily demonstrated by considering a heavy object sitting on a tabletop. The table continually strains under the weight of the object, yet the continual downward force from the object (i.e. "gravitational energy") effectively vanishes without a trace – causing no heating or any other transformed energy to appear within the table.

Returning, now, to the scenario of muscular effort, the transformation of chemical energy into waste heat in our muscles is a direct parallel to the transformation of a battery's chemical energy into the movement of electrons through a circuit. Yet, we do not claim that a drained battery is its own self-contained energy balance simply because its stored chemical energy was used to move electrons from one end of the circuit to the other; rather, we require an additional balancing power output from the circuit's components (heat, light, etc).

Likewise, it is expected that the energy used by our muscles results in a corresponding energy change in the external world (an object is moved, a spring is compressed, etc.). In fact, if instead, we considered the operation of our muscles to be their own internal energy balance, then moving an object or compressing a spring with muscular effort would be a completely unexplained *additional* external energy manifestation – energy for free. Therefore, muscular effort must result in some external energy manifestation, yet our muscles can continually drain while squeezing an object and produce no movement of the object and no internal heat within it. This is an effective *destruction* of energy that cannot be explained by our current laws of physics.

Expansion Theory shows that violations of the *Law of Conservation Of Energy* occur all the time – in the endless energy expended by a magnet clinging to a refrigerator, in the endless energy expended by gravity, in the endless repelling or attracting force of charged objects, etc. Such apparently mysterious generation of energy from within the subatomic realm occurs all around us and forms the foundation of our daily existence and experiences, so an apparently mysterious *disappearance* of energy into the subatomic realm should be no more surprising. A squeezed object merely has more vigorously bouncing electrons within its atoms as a result of the more confined space within its slightly deformed atoms. But these bouncing electrons are a *subatomic* phenomenon that ultimately arise from the underlying primordial realm, which need not adhere to our invented, "energy"-based laws of physics in the atomic realm. Pushed or squeezed objects literally *do* push back with an "equal and opposite force," as Newton claimed – a force that violates the laws of physics and gives matter the property of inertia, but which is also a complete mystery in the absence of *Expansion Theory*.

 NEW IDEA

Gyroscopic stability is explained – for the first time.

The preceding discussion of inertia now makes it possible to understand one of the most curious and unique phenomena in classical physics – gyroscopic stability. Gyroscopes are often used in aircraft and spacecraft navigational systems because they always maintain their orientation in space – an effect that can be demonstrated with a simple spinning top. A spinning top can be positioned so that it is tilting over, yet it will maintain this tilt and not fall over as long as it maintains its spin. Although this is a well-known effect, it is completely unexplained today. The only explanation that is to be found currently is that a spinning top has rotational inertia or momentum, which keeps it from falling over. However, this is not truly an explanation but merely an observation that says nothing about *why* rotational momentum should have this "anti-gravity" effect. *Expansion Theory*, on the other hand, offers the first true explanation of inertia, making it possible to truly explain gyroscopes for the first time as well.

First, consider a regular, non-spinning object as in Figure 6-13. The first frame shows the object standing upright, with the force of gravity effectively pushing down on it from above (of course, this downward "gravitational force" is the result of the upward expansion of the planet). As mentioned earlier, this constant effective downward force simply vanishes without a trace – it causes a constant internal compression within the object but is not transformed into another form of energy, as required by the *Law of Conservation Of Energy*. It causes no heat generation within the object, nor any other form of energy transformation. In terms of today's energy-based paradigm, this simple, everyday occurrence represents a violation of our laws of physics – a true *destruction* of energy.

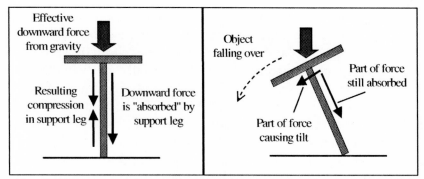

Fig. 6-13 Different Components of Downward Force during Tilt

From the perspective of *Expansion Theory*, this is not a violation of the laws of nature but a common example of a force in the atomic realm crossing into the subatomic or primordial realm – a type of *crossover effect*, as introduced in Chapter 4. There is an effective downward force, the subatomic realm pushes back to maintain atomic stability, and this struggle is evidenced by the compression within the object.

The next frame of Figure 6-13 shows the object falling over, where the downward compressing force becoming partially redistributed as a sideways tilting force once the object begins to tip over. The further the object tips, the less the compression force in the support leg, and the greater the sideways tipping force instead. So, the process of an object falling over could be thought of as a redirection of the force away from the subatomic realm, where it vanishes, and back out into the atomic realm to cause the motion of the object tipping over. With this understanding of tipping objects, we can now explain the mysterious stability of a gyroscope.

The first frame of Figure 6-14 shows an upright spinning top. This is more formally described as a scenario of centripetal acceleration, which effectively causes an outward "centrifugal force" that attempts to fling the material of the disk outward in all directions from its center. Of course, in today's scientific paradigm this "centrifugal force" is a complete mystery since it represents an unexplained force with no power source. The spinning top is merely coasting in a spin without being driven by any power source, and would continue to spin indefinitely in free space while somehow producing an endless outward "centrifugal force." However, we can now see that this rather mysterious centrifugal force comes from the subatomic realm, much as does the magnetic energy of a permanent magnet that clings endlessly to a fridge. That is, as the disk spins, its molecules would naturally continue traveling tangentially off into space, but they are forcefully constrained to travel in a circle by the attracting *crossover effect* between atoms from the subatomic realm (i.e. 'atomic bonds'). Thus, the "outward centrifugal force" that stretches and strains the material of a spinning object is actually more accurately described an inner attracting force that holds the material together. Either way, the end result is an effective force that does stretch and strain the material of the disk of a spinning top.

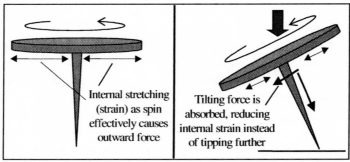

Fig. 6-14 The Forces At Work within a Spinning Top

Now, turning to the second frame of Figure 6-14, we see a spinning top that has been placed on the ground at an angle so that it is partially tilted over. Although the tiny amount of friction where it touches the ground will eventually slow its spin enough that it falls over completely, it initially remains tilted at its original angle – apparently unaffected by gravity. That is, gravity does not slowly tip the spinning top over, but rather, the top remains solidly standing at its original angle until friction eventually starts to slow its spin significantly. The top will only yield to the effective downward force of gravity and tilt further if either its spin slows or the strength of the sideways tilting force increases. But unless and until either of these events occurs, gravity is essentially cancelled out entirely – an effect that is completely unexplained today.

However, this effect can now be readily explained with *Expansion Theory*. We reintroduce the effective downward force due to gravity in Figure 6-14, and, once again, show how it separates into a compressing force in the stem of the spinning top and a sideways tilting force. And, as we now know from the previous section on inertia, a force can only begin to move an object once it has compressed all the atoms of the object and the atoms have rebounded, moving the object in the direction it was originally pushed. But, since the atoms of the spinning top are already effectively being pulled apart, the sideways tipping force simply serves to compress the atoms toward a more normal, relaxed state first. The only way this force could begin to tilt the spinning top further is if it overcame this internal stretching entirely and began to *compress* the atoms, causing them to push back with a rebounding force from within. And the only way this could occur is if either the sideways tilting force were increased (by either increasing gravity or positioning the spinning top at a greater tilting angle) or the internal stretching (or "centrifugal force") were reduced by slowing the spin. So, *Expansion Theory* explains the mystery of the gyroscope, showing how it can continually counteract gravity (and other external tipping forces such as the motion aboard an airplane or spacecraft) merely by coasting in a spin.

An important corollary to a gyroscope's ability to counteract gravity is that the term "gravity" refers, of course, to the effective downward force felt when in contact with our expanding planet – *not* to Newton's "gravitational force," which does not exist. This can be clearly demonstrated by turning a spinning top on its side and dropping it from a height, rather than merely tipping it at an angle on the ground. The dropped gyroscope will fall to the ground in exactly the same manner as any other regular, non-spinning object, which would be a mystery if gravity were caused by Newton's "gravitational force." After all, a rapidly spinning top on the ground can be placed at even the most extreme angle where it is almost tipped entirely over on its side, and yet, it can still defy gravity and not fall over any further. If this tilted spinning top were truly counteracting the "downward-pulling gravitational field" that Newton claimed surrounds the planet, then we would expect this striking anti-gravity effect to also significantly slow its fall from a height as well, which is also supposedly caused by Newton's "gravitational field." But, of course, this does not occur. A spinning top will "fall" just like any other object because it does not actually fall at all; it feels no "downward gravitational force," but floats in the air while the expanding Earth rises up to meet it. No amount of spinning can change this fact.

What Are Black Holes?

Although the subatomic realm of expanding electrons may seem far removed from the grand sweep of the cosmos, if expanding electrons are the one unifying truth in nature then they should provide insights into all the mysteries of our universe, no matter how small or how grand the scale. Indeed, one of the cosmological mysteries of our day is the phenomenon of Black Holes, and the development of *Expansion Theory* – from gravity in Chapter 1 through to particle physics – now allows a proper examination of this phenomenon.

  Black Holes – the Standard View

In today's Standard Theory, a *Black Hole* is thought to be the remnant of a large star that has collapsed inward under its own gravity once its nuclear fuel is expended, ending the outward radiation pressure that was counteracting the inward pull of its gravity. This collapse is believed to first crush the star's atoms, destroying all atomic structure and leaving only a super-dense mass of subatomic particles clumped together. This is known as a *Neutron Star* to indicate that it is now composed entirely of subatomic particles instead of atoms, and if massive enough, its gravity is believed to further crush even these subatomic particles out of existence.

What remains is said to be a *Singularity* – a microscopically-small region in space, far smaller than even a single atom, where all the matter of the original star is now compressed as pure energy, while maintaining the same gravity as the original star. The fact that the gravity of the original star was able to collapse it into this extremely dense region of energy implies that this energy is now trapped and should be unable to overcome its collapsing gravity, never managing to escape this tiny region of space in any form. This is called a *Black Hole*, since no light or any other form of identifying radiation can escape to be detected. A Black Hole is so named because its complete lack of radiating energy should make it invisible, except that its remaining gravity still means that nearby objects will either be pulled-in or will orbit – mysteriously orbiting a region of apparently empty space that nonetheless has immense gravity.

This description means that any matter nearing the Black Hole also suffers this same fate, getting pulled in and crushed into pure energy, vanishing without a trace and remaining trapped forever. While the gravity of the Black Hole is considered strong enough to overpower and contain all matter and energy within, the fact that gravity (as modeled by both Newton and Einstein) weakens with distance means that there will be a distance from the Singularity where a break-even point occurs. This break-even point is known as the *Event Horizon*, and is considered the point of no return, where even energy first begins an irreversible descent into the gravity of the Black Hole (Fig. 6-15).

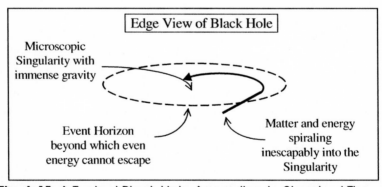

Fig. 6-15 A Typical Black Hole According to Standard Theory

However, despite this classic view of an invisible Black Hole whose existence can only be inferred by objects mysteriously orbiting empty space, this description has been increasingly modified over the years to agree with otherwise conflicting observations, and to provide for even further fanciful conjectures. Black Hole theory now states that they are often visible as objects ejecting enormous jets of radiation (since this has now been observed). This is often explained as a bizarre quantum-mechanical phenomenon that circumvents the gravitational trap of the Singularity by mysteriously creating a second energy photon out of each original one that enters the Event Horizon, leaving one to vanish into the Singularity while its duplicate escapes to be detected.

This duplication of photons, or creation of energy in violation of the *Law of Conservation Of Energy*, is said to follow naturally from *Quantum Theory*, and is considered acceptable since one of the duplicated photons is said to "disappear from the universe" down the Black Hole. Through a merging with the warped space-time idea of *General Relativity Theory*, science-fiction fantasies of "wormholes" at the Singularity that lead to other parts of the universe or even entire alternate universes have now become part of our science. Ideas such as Black Hole-mediated "time machines" involving unfathomable amounts of "negative energy" are now being considered as scientific fact. The classic description of Black Holes has now become nearly unrecognizable, serving as a safe and convenient focal point for increasing advancement of all manner of speculation and fantasy, with apparent scientific support from some of our most mysterious and bizarre established theories.

Even without these more fanciful ideas about Black Holes, there are still many unexplained mysteries even in classic Black Hole theory. For example, it has never been explained how crushed subatomic particles would become pure energy, why the pure energy of the Singularity exhibits gravity, how and why photons would duplicate themselves at the Event Horizon, how warped space-time creates portals to distant regions or even other universes, etc. And, from the perspective of *Expansion Theory*, there are even further problems with describing Black Holes in these terms since today's theories of Newtonian gravity, *General Relativity*, energy, *Quantum Mechanics*, and *Special Relativity* are merely abstractions that do not truly describe our physical world. So then, what does *Expansion Theory* say about all of this, and does it provide reasonable answers to these questions, and explanations for our observations?

Black Holes According To *Expansion Theory*

We begin with the most distinctive quality of classic Black Hole theory – its crushing gravity. Although the concept of a star being crushed by its own internal "gravitational force" is very problematic (for all the reasons that Newton's gravitational theory *itself* is problematic), this concept fits easily into *Expansion Theory*. There is no dispute as to the *effective* inward pull of stars, keeping planets in orbit and causing tremendous crushing pressure toward the star's center. However, this effective "inward pull" is not due to the unexplained attraction of a "gravitational force" from within, but rather, due to the outward expansion of all atoms, as explained in Chapters 2 and 3. Therefore, a star composed entirely of such expanding atoms would indeed experience tremendous internal pressure as the expansion force of its countless atoms pushes out from its center, bearing down on its core. So, the idea of a star collapsing inward from this crushing expansion pressure once it expends its nuclear fuel is entirely compatible with *Expansion Theory*, requiring no appeal to a mysterious "gravitational force" in this first-stage collapse.

The further mystery of the collapsed star, now a Neutron Star, further collapsing into a Singularity of pure energy that is forever trapped by gravity also deserves a second look. Since *Expansion Theory* claims that the gravitational force is merely an abstract invention of Newton (or Einstein, in the warped space-time version), it might be expected that these models would have a limit. They were invented to model everyday gravitational dynamics of large objects composed of many atoms, and, due to a lack of understanding, they have been arbitrarily extended into the subatomic realm – even to the point of being applied to pure energy. As we saw in Chapter 2, even slightly extending these gravitational models to the scenario of a tunnel through our planet stretches them to their breaking points, resulting in an impossible perpetual motion machine. So, it should not be surprising if attempts to apply them to the subatomic realm of collapsing atoms in a Black Hole yield very odd results indeed.

A true understanding of gravity in terms of expanding atoms would prevent such models from being stretched far beyond their design limits; however, modern science still lacks this understanding. As a result, today's gravitational models are stretched until they predict that all the matter of the star becomes compressed into a microscopic dot of pure energy with tremendous gravity still remaining to hold it all in. However, since *Expansion Theory* frees us from these over-extended models, it also frees us to come to far less mysterious conclusions.

Consider the belief that the tremendous gravity of a Black Hole traps even light itself. Taking a step back, wherever there is light, there must be a *light source*. And, of course, the only known light source in the heavens is *stars*. Yet, the first step said to occur in the progression from a star to a Black Hole is that the star must first expend its nuclear energy – that is, *it must burn out*. Therefore, we would not *expect* light to continue to shine from this star any more than we would expect to get light from a burnt-out light bulb. There is no need to appeal to a tremendous gravitational force holding light in, since a Black Hole only begins to form once this light source *extinguishes*. Further, the continuing progression from star to Black Hole requires that even this burnt-out star must further collapse so that even the structure of its atoms is shattered – the equivalent of not only a burnt-out light bulb, but a *smashed* one as well. So, not only is it highly questionable to extend today's gravitational models to the point where they predict "Singularities," but also, the fact that Black Holes do not shine like stars is a straightforward expectation, and lends no particular credibility to the mysterious notion of light trapped within by gravity.

Now, continuing with our analysis of the collapsing star, classic Black Hole theory presents us with three clear paths to consider for the fate of a burnt-out star in *Expansion Theory*. One is that no collapse occurs once the nuclear processes stop, a second is that the collapse stops once the atoms are crushed into a Neutron Star, and the third is that even the subatomic particles collapse into "pure energy." Since our current gravitational models are highly suspect and are not well enough understood or developed to clearly apply to these last two scenarios, let's set these models and their mysterious predictions aside and examine the situation from the perspective of expanding matter.

As mentioned earlier, the possibility that the star could collapse is feasible since there is a true crushing force within – even according to *Expansion Theory* – which might cause it to collapse once its nuclear processes cease and if the star is massive enough. The outward "radiation pressure" of "photons" that is said to counteract this collapse in shining stars can now be seen as physical electron clusters bursting outward from within, which would certainly supply such outward pressure. So, the cessation of these outward bursts of electron clusters could well result in an inward crushing collapse of the star. Also, since extremely powerful jets of radiation are now commonly observed bursting from Black Holes, such a phenomenon would be difficult to explain with an intact, burnt-out star.

Now, the description of energy provided by *Expansion Theory* shows the concept of a Singularity to be impossible. Since the concept of pure energy is merely an invented abstraction that actually refers to clouds or clusters of electrons, it is impossible to turn the enormous amount of matter found in a star into "concentrated energy" that takes up negligible space. Instead, *Expansion Theory* shows that a Black Hole is very likely a collapsed Neutron Star whose subatomic particles are now bursting forth as freely expanding electron clusters in the atomic realm.

When atomic matter disintegrates – whether due to nuclear fission in an atomic bomb or due to an immense crushing force within a star – its component electrons and electron clusters (protons and neutrons) are freed into the atomic realm. The Neutron Star would be an early stage of this collapse, where the freed subatomic particles at the center struggle to escape as an outwardly exploding mass of free electron clusters. This would explain the enormous jets of radiation observed shooting out of Black Holes as the freed electron clusters build internal pressure and seek a way out from within the immense amount of surrounding matter. In essence, such an active Black Hole is like an ongoing succession of exploding atomic bombs, at its center, in a slow, controlled release. And, since stars, like all rotating objects, tend to be somewhat flattened at the poles and thicker in the middle, it would be easier for inward pressure to escape through the poles – especially if this flattening is accentuated into more of a disk-shape as the star's atomic structure collapses. And indeed, observed Black Holes appear to be flattened rotating discs with jets of energy streaming out the side.

Another way Black Holes are detected is when stars are seen to orbit an invisible point in space. Such an observation is commonly represented as objects orbiting a sizable invisible mass, often assumed to be the massive but invisible Singularity of a Black Hole, or some type of exotic "Dark Matter" material. However, letting go of the "Singularity" concept, as well as the belief that gravity is due to *mass* rather than *size*, we are free to consider other possibilities for large bodies that do not emit light, which the stars may be orbiting. In fact, the *natural orbit effect* of passing expanding objects introduced in Chapter 3 even allows two objects to merely orbit one another in a manner determined purely by their sizes and relative speeds – with no mass or force considerations and nothing but empty space between them. There is no need to insert a mysterious "Singularity" with immense mass and zero size in the midst of this empty space. Once *Expansion Theory* is considered, it is necessary to re-evaluate all astronomical observations and interpretations based on current gravitational theory.

Did It Really Begin with a Big Bang?

Another cosmological concept that must be re-evaluated in light of *Expansion Theory* is the creation event of our universe – the so-called "*Big Bang.*" This creation event claims that the whole visible universe – billions of galaxies, each containing billions of stars – arose from a burst of pure energy from a microscopic Singularity similar to what is said to exist at the center of a Black Hole. Yet, the preceding discussion of Black Holes, and the definition of energy in *Expansion Theory*, makes such an explanation a fanciful impossibility. It can now be seen that we live in a universe of pure matter – a possibly infinite ocean of countless expanding electrons in various groupings and configurations composing atomic matter, subatomic matter, and energy. There is no evidence that electrons can be crushed, split, or converted into anything other than their current form as fundamental elementary particles of expanding matter, and there is no such ethereal phenomenon as "pure energy" anywhere in our universe. So then, what might be a viable alternative to Big Bang theory, and what are we to make of the evidence that is currently considered to support it?

One of the simplest and most straightforward beginnings for our universe can be found in the earlier discussion of an infinite ocean of primordial particles supporting countless patterns of expansion dotted about at random throughout the universe. If this ocean of primordial particles found itself completely without form for any reason at any point in the infinity of time, then these expanding spherical patterns could well emerge spontaneously from the chaos as a natural consequence of the physics of the primordial ocean, as described earlier. Alternatively, we could consider the appearance of these expanding electrons throughout the universe as a deliberate effort, much as we ourselves could cause such a "universe" to come into existence within a computer by writing a program.

Either way, a spontaneous appearance of countless expanding electrons throughout our universe could be relatively straightforward, and need not be a mysterious creation of enormous amounts of matter from an unimaginably concentrated "Singularity of energy" in some sort of explosive "Big Bang" creation event. The primordial processes that caused the early universe to appear could easily be continuing unchanged today as they underlie and support the continued existence and dynamics of our present-day universe. As mentioned earlier, it is not difficult to imagine how this early universe of countless, randomly distributed expanding electrons could result in various "subatomic particles" (electron groupings) and then stable atomic structures. And, of course, various forms of "energy" would arise as well – the "gravitational energy" of expanding atoms, the electron clusters of light and other forms of electromagnetic radiation, and the electron clouds of electric and magnetic fields. From these beginnings our current universe could well mature to its present state of complexity and order – again, by evolutionary processes that are becoming increasingly well known and understood.

This process of order crystallizing from a sea of randomly distributed expanding electrons could well explain the structure of our universe today. It is currently thought that the distribution of galaxies throughout the universe today is determined by tiny variations in the distribution of energy within the initial "Big Bang" event – variations that can presumably still be detected in an extremely faint microwave background radiation coming from all directions in space. Although this is generally taken as evidence of the Big Bang theory, it is also considered to be a sizable cosmological mystery since astronomers generally feel that the "Big Bang" should have had a perfectly uniform energy distribution, leading to a perfectly uniform distribution of galaxies today. The fact that galaxies are far from uniformly distributed mystifies many cosmologists, though this poses no particular mystery for *Expansion Theory*, where a random or statistical distribution of countless electrons is a very reasonable view of the early universe.

Is Our Universe Truly Expanding?

A major reason for belief in the Big Bang creation theory today is the apparent evidence that our universe is expanding. This does not refer to the new concept of expanding matter in *Expansion Theory*, but rather, to the fact that most of the galaxies in our universe appear to be flying apart from one another as if from a great explosion – hence the term "Big Bang." Support for this belief comes primarily from the fact that the light arriving from distant galaxies is generally shifted to lower frequencies than expected. This is thought to indicate that all galaxies are speeding away from each other, much as the sound of a motorcycle engine drops in pitch once it passes and speeds away. This familiar drop in pitch due to the *Doppler Effect* that occurs with sound waves has been adopted by astronomers and applied to light in order to explain the drop in frequency of light waves from distant galaxies. So then, a lower-than-expected light frequency from a galaxy is taken to mean that it is speeding away from us. Although this does seem to be a reasonable conclusion on the surface, based on the assumption that "sound waves" and "light waves" are similar in form and function, this assumption certainly deserves a closer look since it shapes our very understanding of our universe.

The Doppler Effect, the Red Shift, and the "Big Bang"

The *Doppler Effect* is named after Christian Doppler (1803-1853), who described the shift of sound waves to a higher or lower pitch when a sound source and a listener are in relative motion toward or away from each other respectively. Doppler also proposed that, in a similar manner, a color shift should occur in light waves when the light source and observer are in relative motion, though this was only a conjecture that had no verification in his day.

Today, we are actually able to observe a color shift in the light arriving from distant galaxies – usually shifted toward a lower frequency of light than expected. Since red light is the lowest frequency of visible light, this observed shift toward lower frequencies in starlight is known as the *Red Shift*. Incidentally, this does not necessarily mean that Red-Shifted light appears red in color. Blue light that is shifted slightly toward the red end of the spectrum will appear *green*, for example – the next color down in the subjective color spectrum generated in the human mind when we see light; generally, it takes an *extreme* Red Shift before light actually appears red in color. The Red Shift seen by astronomers refers to the fact that the signature pattern of light frequencies given off by reference stars in our galaxy is also seen in other galaxies, but this signature pattern from other galaxies is shifted toward the lower end of the spectrum. This shift toward lower frequencies is taken as Doppler's predicted color shift in light when the light source is in motion, leading to the conclusion that the galaxies all around us are speeding away – presumably still traveling from the "Big Bang." However, with the advent of *Expansion Theory*, we can now take a second look at this whole issue.

The first step in the progression to our current beliefs about the Red Shift and the Big Bang is Doppler's assumption that light should experience a color shift – just as sound changes pitch – when the source and observer are in relative motion. This conjecture is now taken as fact since there does appear to be convincing evidence that electromagnetic radiation of all types exhibits this motion-induced frequency shift. The light from distant galaxies exhibits a Red Shift, presumably due to their Big Bang-induced motion, and NASA even uses frequency shifts in radio signals from space probes to monitor their speed and position. In view of this compelling evidence, is it reasonable to question the existence of the Doppler Effect in electromagnetic radiation?

ERROR

Light Is Not Like Sound Waves and Radio Waves

To address this question, recall that the nature of radio waves was described in Chapter 4, while Chapter 5 described a very different nature for light. This led to a clear distinction in Chapter 5 between the electron bands of radio waves and the electron clusters of light, further leading to a redefinition of the electromagnetic spectrum into distinct low-frequency and high-frequency sections. As such, we can now see that radio waves and light are very different manifestations of expanding electrons – in stark contrast to today's belief that they are merely different frequencies of "electromagnetic waves" within a continuous "electromagnetic energy spectrum."

The description of radio waves according to *Expansion Theory* – as alternating bands of compressed electrons expanding out into space – lends credibility to Doppler's assumed parallels between electromagnetic radiation and sound. That is, sound actually *is* a somewhat physically similar manifestation of alternating bands of compressed air molecules propagating through the atmosphere. Therefore, it should not be surprising if an approaching spacecraft slightly squashes the alternating electron compression bands in its radio waves as it continues producing them while advancing upon those it just transmitted. This is what happens to the alternating compression bands in the atmosphere (sound waves) when a moving sound source exhibits a Doppler Shift as it approaches, so it is reasonable to assume that both cases result in frequency shifts for similar reasons. *Expansion Theory* does not dispute the Doppler Effect in radio waves; in fact, it demystifies this concept by showing that radio waves are not ethereal waves of "pure energy," as currently thought, but are compression waves within a sea of matter particles, somewhat like sound.

However, *Expansion Theory* does claim that radio waves are very different in form and function to light. Light, unlike radio waves, is not compression waves in a sea of electrons, but rather, it is composed of very definite, self-contained electron clusters. Even today's Standard Theory reflects this distinction, speaking of photons of light as well as photons of higher frequency radiation such as X-rays and gamma rays, yet not generally referring to "radio-wave photons." Although, in theory, it should be equally valid to consider the much lower-frequency radio waves (with much longer wavelengths) to be huge photons of electromagnetic energy that can be meters in length (according to Standard Theory), such odd-sounding "radio photon" descriptions are discarded in favor of the "wave" description. This suggests that the different natures of radio waves and "light waves," as stated in *Expansion Theory*, are recognized even today – if only indirectly. So then, unlike radio waves whose frequency is determined by the number of electron compression bands passing by per second (i.e. how dense or "squashed" these waves are), the frequency of light is determined by the number of electrons in each of its component electron clusters.

NOTE

     Therefore, the same straightforward reasoning that makes the Doppler Effect a valid concept for both sound waves and radio waves does *not* apply to light.

If we are to embrace the belief that the motion-induced Doppler Effect explains the observed Red Shift in starlight, we would have to explain how relative motion could alter the *number of electrons* in the electron clusters of light since that is what determines its frequency in *Expansion Theory*. In fact, this same burden must be borne by Standard Theory as well. That is, before concluding that the observed Red Shift in starlight indicates a motion-induced Doppler Shift, standard theorists should be required to show how this would *physically* occur, given their understanding of the actual physical nature of light itself. After all, we do have a clear physical explanation for this effect in sound waves, so a similar clear physical explanation must be arrived at for light waves if our beliefs are to be anything other than pure assumption based on circumstantial evidence. Incidentally, even the clear physical explanation just discussed for frequency shifts in radio waves is found only in *Expansion Theory* – no clear physical explanation can be found in current theory.

Yet, as shown in Chapter 5, the actual physical nature of light is very poorly understood today, being full of unexplained mysteries and unresolved paradoxes. Does the Red Shift indicate that "light waves" are changing in frequency, or that "light photons" are changing in energy content (since the energy content of photons corresponds to their frequency in today's quantum-mechanical models of light)? If light has a wave-particle dual nature then surely the Red Shift must be explainable from both perspectives. But how can there be two very different physical mechanisms occurring – one causing a frequency shift in "pure energy waves" and the other causing a quantum-mechanical energy change in "light photons"? We can always invent an abstract explanation to justify any belief, but do we truly accept that two completely different physical processes are at work to create the identical frequency shift in both waves and photons?

It is sometimes suggested that both processes result from a common underlying cause – the expansion of Einstein's four-dimensional "space-time fabric" as the whole universe expands. Yet, this simply introduces more assumptions, more unexplained mystery and abstraction, and more questions. For example, why do we need such mysteries as "stretching space-time" to explain Red-Shifted starlight but *not* to explain Doppler-Shifted radio waves from our spacecraft? Transmissions from the spacecraft that roam our solar system clearly are not frequency shifted due to "stretching space-time," but merely due to their motion. Yet, although both radio waves and starlight are considered to be merely different frequencies of the same electromagnetic wave phenomenon today, we have such enormously differing explanations for their respective "Doppler Shifts." Why is one treated as a simple classical wave, much like sound, while the other is shrouded in all the wave-particle mysteries and paradoxes that come along with *Quantum Theory*, and often even the further four-dimensional space-time mysteries of both *General and Special Relativity Theories*? This is a very curious and telling inconsistency in today's science.

Given these many problems with today's beliefs, along with their far-reaching implications for our understanding of the physics and nature of our universe, it is extremely important that we think very carefully about our interpretations of Red-Shifted starlight. Yet, have we given this effect due consideration or are we simply running with assumptions that were made decades ago? Let's continue to delve even deeper into this important issue, moving from the mysteries and inconsistencies just mentioned, to the serious violations of the laws of physics that are readily found in our current expanding-universe beliefs.

VIOLATION

 An Expanding Universe Violates the Laws of Physics

A much larger problem with today's Red Shift beliefs lies in the conclusions drawn from the fact that this effect becomes more pronounced as we look out at more distant galaxies. The Doppler-Shift interpretation of this would mean that the more distant the galaxy the faster it speeds away from us. Astronomers are now beginning to believe this means there exists a mysterious new "anti-gravity" force that for some reason only appears between galaxies, pushing them apart with a strength that somehow *increases* with ever more distant galaxies. That is, this "anti-gravity force" does not act between the stars within galaxies, nor between the moons and planets of their solar systems, but only within the space *between* galaxies for some unknown reason. This is a force that has no other precedent in our experience or our science, and which clearly violates the *Law of Conservation Of Energy*. There is no known power source for this mysterious new inter-galactic repelling force, and no explanation for how or why such a hidden power source would provide even greater strength with increasing distance. It is even becoming common to conclude that it is Einstein's "space-time fabric" that is mysteriously being stretched faster and faster between ever more distant galaxies, rather than the mere motion of galaxies through space – effectively circumventing our laws of physics altogether.

Other Explanations

Although there is currently a widely held belief in scientific circles that the expanding universe idea is the only viable explanation for the observed Red Shift in starlight, the preceding discussion shows that such a notion is actually far from scientifically viable. But further, there is no shortage of other truly viable explanations, both from the perspective of Standard Theory as well as *Expansion Theory*. For example, there is a widely known effect, called the *Compton Effect*, in which high-frequency electromagnetic radiation has been shown to exhibit a Red Shift in the lab when passing through a variety of different materials. Given this knowledge, it is significant that the universe is known to be filled with gas, dust, and all forms of radiation, which distant starlight must pass through for billions of light years in order to reach us. Surely if a simple lab experiment shows a measurable Red Shift effect, we might expect that light passing through billions of light years of material may also be similarly affected – and even more so with increasing distance, as observed by astronomers.

In fact, such effects are so common in nature that this principle is easily demonstrated even with a thin sheet of plastic and a common remote control system found in any home TV or stereo system. It is a simple matter to hold a piece of transparent plastic in front of the remote control transmitter, often creating measurable frequency shifts in the infrared light beam it produces. Such frequency shifts are very common when high-frequency electromagnetic radiation passes through materials, and would certainly have to be given serious consideration in any scientific analysis of Red-Shifted starlight – especially when it has crossed unimaginable intergalactic distances to reach us. In fact, even Edwin Hubble, who first suggested the Red Shift/expanding universe connection in 1929, was quick to point out that he was uncertain whether Red Shifts truly indicated velocities, and that they may instead indicate a new effect not yet discovered or understood.

Further Red-Shift Explanations – from _Expansion Theory_

On a related note, the description of light in _Expansion Theory_ allows for very simple physical explanations for frequency shifts in starlight passing through vast intergalactic distances. Since the frequency of a light beam is determined by the number of electrons in its component electron clusters, a frequency shift simply indicates that electrons were added (Red Shift) or removed (Blue Shift) in transit. It is easy to imagine processes that could either cause electrons to accumulate or to be stripped off of the electron clusters in starlight as they zip through light years of gas, dust, and radiation (i.e. various other dense clouds and clusters of electrons). In fact, this new suggestion according to _Expansion Theory_ may well be the mechanism that lies behind today's observations of the Compton Effect and other similar phenomena. Also, the density of material and radiation in the early universe might have been greater, further explaining the observation that Red Shifts seem to increase at an ever more rapid rate when we essentially look back in time at more distant (early) galaxies. Once again, it is suggested that this is a far more viable alternative to the current belief that more distant galaxies are somehow accelerating away ever faster due to an unexplained intergalactic "anti-gravity" force following an equally mysterious "Big Bang."

There are also many additional explanations for Red-Shifted starlight once _Expansion Theory_ is considered. For example, it is conceivable that starlight could be Red-Shifted or Blue-Shifted merely because of size differences in stars. The new understanding of light and gravity shows that a larger star would have a greater amount of expansion and thus a greater effective surface gravity (regardless of its mass), which could well affect the formation of the electron clusters that eventually expand off as starlight. The tremendous resulting gravity of a larger star would create a large, crushing expansion force (or "downward gravitational pull") against the electron clusters it produces. This could affect the number of electrons in the clusters it produces, perhaps delaying their escape until more expansion pressure builds and thus producing larger electron clusters across the board (a Red Shift). Indeed, even in Standard Theory it is currently believed that gravity does somehow cause *Gravitational Red Shifts* in light, though this effect is poorly understood today. So, another possible explanation for the Red Shift data is that we are not looking at galaxies that are speeding away ever-faster with distance, but simply galaxies that are composed of ever-larger stars earlier in the evolution of our universe.

In fact, *Expansion Theory* suggests that such Red Shifts due to sizable outward expansion of the light source may even occur on the grand scale of overall galaxies. Since the gas discs of early solar systems and galaxies are believed to have become more condensed as they matured and formed planets and stars, this describes a physical shrinking of the overall galaxy. However, although this compacting effect would seem to imply an *inward-pulling* acceleration, we must remember that this compacting is actually due to the *outward expansion of matter*. That is, just as a planet does not condense from a disk of gas and particles due to an "inward-pulling gravitational force," but rather, due to the ongoing outward expansion of the particles in the disk, so it is with condensing galaxies. The particles composing the early disk, whether it is a planetary disk or a galactic one, increase the disk's density by continually expanding and filling the space around them, which also causes an outward accelerating expansion force much like that responsible for the gravity of a mature planet or star.

Therefore, the "shrinking" condensation of early galaxies would actually have been a forceful, *outwardly expanding* condensation. This outward acceleration of the overall galactic disk, as it also either emitted or re-emitted/reflected its starlight, could well have caused a Red Shift to the overall galaxy, much as a Gravitational Red Shift results from the outward acceleration of any large star. Again, this effect would likely occur more strongly for the younger (and more distant) galaxies that we observe, when they were most likely in this stage of physically-connected, condensing gas disks. And, once again, an increasing Red Shift with distance is what we do observe.

Also, a closely related effect within early, condensing galaxies is that they would likely have exhibited tremendous outward explosions from their central regions as matter became densely compressed and crushed at their compressed centers. In fact, it is currently believed that just such explosive "Black Holes" of millions of solar masses exist at galactic centers, which would imply a tremendous outward explosion of expanding electron clouds and clusters as immense amounts of atomic matter are crushed. This outward pressure would give an enormous outward acceleration (and possibly a Red Shift) to galactic centers, which again may well occur to a greater degree in the more distant galaxies of the early universe. Such an outward-exploding dynamic may even explain the central bulge of many early, oval-shaped galaxies or the spherical halo of older stars that is often seen surrounding the center of spiral galaxies. Such true physical outward acceleration could well cause or contribute to the observed Red Shifts of starlight. This new understanding is now possible once expanding matter is understood, and once light itself is physically understood in terms of electron clusters whose production could well be affected by such mechanical stress.

So, we can see that, not only are there many problems with current Red-Shift beliefs, but there are also many other possible explanations that could well explain observations without "space-time" expanding ever-faster from a "Big Bang" due to a mysterious "intergalactic anti-gravity force." According to *Expansion Theory*, there is no particular reason to consider our universe to be expanding outward at an ever-increasing rate, or even for it to be expanding outward at all. Our universe may well be an essentially static universe, where any relative motion that may exist between galaxies would likely be the result of an expansionary growth toward each other or a forceful push from a large nearby explosion event such as a Supernova or "Black Hole." Yet, one further phenomenon exists that is sometimes taken as evidence of an expanding universe – a characteristic of the night sky that has become known as *Olbers' Paradox*.

Olbers' Paradox

The more we learn about the heavens, the more odd it might seem that the night sky is dark. After all, with countless billions of stars now known to be out in space it may seem that there should be stars in every conceivable portion of the sky, lighting up the night sky. The fact that this does not occur has become known as *Olbers' Paradox*, after

Heinrich Olbers who popularized this question as a paradox in the nineteenth century. A number of answers to this question have been proposed over the centuries, though none is considered universally convincing or satisfactory. Even today, the dark night sky is often considered a challenging scientific puzzle that leads some astronomers to suggest that the answer lies, at least in part, in the mysterious, ever-increasing expansion of our universe following the "Big Bang" creation event. It is not uncommon for scientists to claim that this ever-increasing outward acceleration either Red-Shifts distant starlight below a visible wavelength, or spreads the light energy out until it is too weak to see, or puts many stars at such a distance that their light still hasn't reached us. Yet, none of these common expanding-universe-based answers to Olbers' Paradox hold up under scrutiny; clear problems can readily be found with each of these proposed solutions to this apparent paradox.

For example, the suggestion that the very fabric of space – Einstein's "space-time" – is expanding too fast for the light from distant stars to reach us is at odds with Einstein's own characterization of light. Einstein claimed that light always travels at the same constant speed for all observers regardless of relative speed, so it follows that, if the light from a distant star would have reached us in a *non*-expanding universe, then it would also be reaching us now even in an *expanding* universe.

Further, the notion that this distant starlight is Red-Shifted by the universe's expansion such that it is below visible wavelengths, or is too weak to be seen, is also readily dispelled. The Hubble Space Telescope has demonstrated that perfectly visible starlight can be seen to be reaching us from every conceivable region of space – no matter how tiny or distant. It is also an even simpler matter to set up a ground-based photographer's camera to point at the night sky with a long exposure on a motorized platform that rotates the camera in step with the Earth's rotation. This ensures that even the weakest starlight registers on the film, while also ensuring that there is no streaking or blurring due to the Earth's rotation. The result of such an exposure does indeed show a bright white night sky literally filled with starlight.

All of these points show that Olbers' Paradox is not actually a paradox at all, and certainly does not constitute evidence for the ever-accelerating universe theory, as commonly claimed today. This apparent "paradox" is nothing other than a simple oversight that overlooks a straightforward optical-contrast effect familiar to any photographer. If there is a competing light source, such as a full moon, the dimmer light of stars will be overpowered in contrast and will not be seen. But on a moonless night, our eyes (or a photographic film) are not overpowered by the moon's brightness and are able to detect the light of many stars. And if we move away from the city, far more stars can be seen since there is even less contrasting or competing light. At this point, the accumulated light from the thousands of visible stars in our Milky Way galaxy becomes a limiting factor, outshining the weaker starlight beyond our galaxy. And even if our galaxy didn't outshine the other stars, there is a limit to how faint a light the human eye can detect. It is simply a fact of our biology that we will always see a dark night sky at least to some degree.

Further to this point, we know that nocturnal animals such as owls can see at night quite well when it is only dimly lit by starlight – even though we experience almost total darkness. Therefore, if the starlight from above is reflected so brightly to an owl that it can easily see by it, then the night sky would certainly appear quite bright and filled with starlight if the owl turned its eyes skyward toward this light source overhead. No doubt an owl is capable of seeing thousands more stars than we will ever see with the unaided eye, as well as a far brighter night sky than we will ever experience. To an owl there would very likely be no "paradox" of a dark night sky.

But there is no reason to end this line of thought with the owl. As just mentioned, the ultimate eye on the universe – the Hubble Space Telescope – has shown us that, with the proper eye design, it is possible to see the light from thousands of stars in *every conceivable patch of sky* no matter how tiny and distant the field of view. This would certainly rival even the owl's night vision, and would likely result in an experience of a bright night sky that resembles the almost glaringly bright long-exposure photographs of the night sky mentioned earlier. There is ample evidence that the universe is indeed spread out before us in plain view across the entire night sky, and is quite visible with the right equipment or techniques. The dark night sky is neither a paradox nor evidence for the mysterious, ever-accelerating anti-gravity force that is beginning to gain scientific acceptance today as part of the Big Bang/expanding universe theory.

What is Time?

No discussion of the big questions in science would be complete without addressing the issue of *time*. Although time is typically measured by the regular advance of a ticking clock, it is also an integral part of many scientific theories today, a central player in many science-fiction stories, and a topic that has captured the imagination of poets and philosophers for centuries. Although much has been said about time, it is still a source of much mystery, speculation, and controversy for many physicists and philosophers. However, as the preceding chapters have shown, many of the current mysteries in our physical world are readily demystified when viewed from the perspective of *Expansion Theory* – a perspective that can help provide clarity to the issue of time as well.

Time is a concept that we are all very familiar with, yet there are a wide variety of philosophical, scientific, and fanciful beliefs surrounding it. Einstein claimed that time is one of four dimensions composing the very fabric of our physical universe. He also believed that the flow of time is completely relative, progressing at a different rate on every moon, planet and star simply because they travel at different relative speeds. If this were true, however, every moving object in the universe would effectively have different rates for all known physical, chemical, and biological processes in nature, making our universe a very unpredictable and chaotic place indeed. It is also becoming common to hear ideas of time travel, where a machine or a strange natural phenomenon allows us to travel backward or forward in time – and not only in tales of science-fiction, but increasingly even as serious science. In fact, the distinction between science and science-fiction is virtually non-existent when it comes to such ideas as "wormholes in space-time" created by Black Holes that lead to a different time or even different universes altogether. Many scientists even claim that an infinity of parallel universes is somehow created at every moment as all the possible outcomes for every atom in our universe spawn an infinity of different realities in parallel universes each moment. Beliefs such as these are often considered to have serious scientific merit by followers of the theories of *Quantum Mechanics, General Relativity,* and *Special Relativity.*

At the root of much confusion about time is the fact that it is often given a life of its own in our thinking, as if it were actually a separate entity or dimension with its own self-contained nature and identity. It is often said that time might somehow be sped-up, slowed-down, or even directed backward, causing everything in our universe to follow suit as if time were an external master puppeteer that could *itself* be manipulated from within the very universe it controls. While such philosophies and beliefs can be worthwhile and enjoyable to contemplate, the simple expansion principle that runs throughout the physical universe in *Expansion Theory* must explain the entire operation of our universe – including time. Just as the mysterious notion of a separate, ethereal "energy" phenomenon has a far simpler explanation in the straightforward physical dynamics of expanding electrons, so it must be with time. This means that time has a significantly more simple and less fanciful explanation than is found in most of our science fiction, and even much of our science today.

The most familiar and straightforward example of time can be found in the operation of a simple analog clock. We often think of clocks as devices that measure the separate entity of *time* itself, tracking it as it moves steadily forward. In this view, if we were somehow able to cause time to flow backward, we would expect to see our clocks run backward as well, as they dutifully tracked the flow of time. However, in actuality, a clock is not a special device that somehow "tracks the flow of time," but is merely an arbitrary mechanism that is deliberately engineered to proceed at regular intervals, and that is ultimately driven by *expanding electrons* – not by a separate "time" entity.

In fact, if our clocks *were* driven "forward" by "time," then this would make today's concept of energy obsolete. We are all well aware that our clocks are driven by "energy" (which we can now see as expansion) in one form or another – and would cease to function without such power sources – yet, the notion that "time" somehow drives our timepieces "forward" still persists. The same is equally true of all human inventions, all life forms, and all natural phenomena. *Without* expanding subatomic and atomic matter there would be no animation anywhere in nature or in human inventions – regardless of any notion of "time," and *with* expanding matter we have the cause of all known animation in the universe – again, regardless of the notion of "time." Therefore, the only requirement for the operation of our universe is *expansion*, while the notion of time is clearly a completely superfluous abstract human invention – a misunderstanding of the singular principle of expansion underlying our universe, as is the notion of energy. *Expansion Theory* finally frees us from the need to represent our universe in terms of these two mysterious and still poorly understood phenomena (energy and time), replacing these interim terms with a clear physical understanding of the single, unifying principle of expanding matter.

Further, not only does this puppeteer called "time" not exist in *Expansion Theory*, but it most definitely does not operate at a rate that mysteriously varies with the relative speed of each individual object in our universe, as Einstein claimed. Even our core measurement of the progression of time – the 24-hour day – is merely a result of the shadow cast by our planet as it rotates at an arbitrary rate within the expansion-driven mechanism of an arbitrary solar system. If the spin of our planet were reversed, for example, would that mean that time now runs in reverse? Of course, the answer is "no" – the progression of the sun across the sky would merely reverse direction, moving from West to East instead, but no fundamental physical, chemical, or biological processes would literally progress "backward in time."

And *all* timepieces are mechanisms that are just as arbitrary as this example, since there *is* no separate entity called "time" driving events forward or backward. In fact, there *is* no forward or backward unfolding of events at all; this concept is also a fanciful invention of the human imagination. There is only the ongoing, unchanging expansion of matter and the various forms of straightforward mechanics that follow from it throughout our universe – as it always has been and always will be. Time is simply a useful abstraction that we superimpose upon our observations for the purpose of organization and convenience. In this sense, it *could* be said that time exists in a dimension outside of the events occurring around us, but this separate dimension is not the mysterious dimension that it is believed to be today – which somehow drives events "forward" – but merely the dimension of the human imagination.

ERROR

X   The "Arrow of Time"

A major reason for this belief in the forward progression of time is due to the phrase "the arrow of time," which has become a commonly used term in science today. This phrase refers to the observation that events in our universe tend to lose energy, wind down, and wear out overall, as if everything were driven to follow such paths of decline by some sort of unidirectional "arrow of time." For example, a hot cup of coffee will naturally cool down but will not spontaneously heat up. A battery-powered device will continually drain its battery until it ceases to function, but its battery will never spontaneously recharge. And all devices eventually wear out or break down with use (and even with disuse eventually due to rust and other processes of molecular degradation), but they never repair themselves. This belief in the general "winding-down" of our universe is formally stated in a currently held law of nature known as the *Second Law of Thermodynamics*. This law is usually stated formally as follows:

LAW

The Second Law of Thermodynamics

Processes within closed systems always tend toward maximum entropy (randomness) and minimum enthalpy (heat content).

The first part of this law states that any isolated system or process that we choose to observe will tend to become more random and disordered over time, but not the reverse. For example, the ordered structure of a sugar cube will spontaneously dissolve into a random dispersal of sugar molecules in water, but the reverse process will not occur – the ordered structure of the original sugar cube will not spontaneously re-form from the randomly-distributed molecules in the sugar water. The second part of this law states that energy naturally disperses in an isolated system, but does not generally collect or concentrate itself spontaneously within that system. The clearest example of this is the fact that hot objects naturally cool down (their heat energy disperses), but they do not spontaneously heat up; energy must be deliberately focused upon objects or added to the closed system from an external source for this to occur. The *Second Law of Thermodynamics* is considered to be one of the classic cornerstones of physics, along with the *Law of Conservation Of Energy*.

However, a closer look shows that this "winding-down" generalization is not evidence of an external "arrow of time" that always flows in a "forward" direction defined by disorder and decay, but is a sizable over-generalization resulting from a narrow, selective view of the dynamics of our universe. For example, if a string is suspended in a glass of sugar water in a dry environment, ordered sugar crystals will spontaneously form on the string as the water evaporates into the surrounding air. And further, objects actually *do* heat up in our universe just as often as they cool down since prior heating is obviously a requirement for every object that cools. Similarly, batteries *do* charge up in our universe since prior charging is a requirement before any battery can drain. And, while bones may break and stars burn out, bones also heal and new stars are born – constantly and spontaneously.

In some of these examples, deliberate effort is expended by humans to increase the energy or order of a system, and in some cases not, but we must remember that they are *all* still examples of the *spontaneous dynamics of expanding matter in our universe*. Even human-driven heating, charging, or repair are still spontaneous expansion-driven events in the grand scheme of things since we owe the very existence and operation of our bodies and minds to the ongoing dynamics of expanding matter. To characterize the heating of a cup of coffee in a microwave oven as an artificial event is to place humans and our inventions outside of the mechanics of our universe – as if they are not produced and continually supported by expanding matter. But, of course, humans and our actions and inventions are just as much a natural product of our universe and its function as are shining stars and orbiting planets – how could it be otherwise?

Therefore, the widely held belief in a "forward arrow of time" directing our universe to wind down is not the foregone conclusion that it is often considered to be. In actuality, every particle of matter in our universe is continually active and expanding, driving continual cycles of order and disorder, heating and cooling, birth and death. Even if it is determined that these cycles ultimately cause a greater overall degree of disorder and "energy" dispersal in the universe as they proceed, this would still not be evidence of an external, forward-flowing puppeteer called "time." It would simply be recognition of general principles in nature that arise from the only manner that the dynamics of expanding matter can unfold in our universe. The *Second Law of Thermodynamics* is a useful statement of the fact that isolated (or closed) systems generally behave according to certain overall rules that we have identified – a necessary realization if we are to carry out reliable experiments or design useful machines. But such observations merely identify a reliable principle that we can take advantage of within the larger overall dynamics of our universe of expanding matter; they are not, however, evidence of an external, forward-flowing "time" entity.

Is Time Travel Possible?

The preceding discussion also speaks to the notion of *time travel*. When we look at a calendar, we can identify a particular day in the future or the past, but *where* does this day exist? In what universe does this day physically exist so that we can leave the present and travel to it? Of course, there would have to be an infinity of universes that physically exist to house every imaginable fraction of a second stretching backward and forward throughout an infinity of time, all awaiting our arrival. Although many scientists consider this to be a serious possibility today, according to *Expansion Theory* our particular universe has only one physical incarnation at any moment, and that incarnation is what we call *the present*. The past has no physical existence anywhere since there is simply no place for it to exist once it has passed. The events of a moment ago occurred right here where we are presently and involved the same objects that now surround us; therefore, the only place that the past could possibly still simultaneously exist is in the neural connections of our minds in the form of a *memory*. When we envision the past, it does literally exist in a physical form of sorts, but only in the space within our heads in the form of representative neural connections – there is simply no other physical location for the events of the past. Similarly, the only place that the future – and its infinity of possible outcomes – exists is also in our minds in our faculty of *imagination*. We can look at a future day on the calendar and *imagine* many different ways that it might unfold, but its singular physical outcome will not literally exist anywhere until that day arrives – after which, it will only ever be a memory.

The human mind can conceive of "time" as an external driving entity that might even run backward, reversing all known events like a movie running in reverse; but, while such time-travel concepts make for enjoyable science-fiction, installing them into our science is a very different matter. Such blurring of the lines between science and science-fiction has reached unprecedented levels today – again, largely due to the widespread acceptance of mysterious theories such as *General Relativity, Special Relativity,* and *Quantum Mechanics* into our science – often making it difficult to discern truth from fiction. We have only our scientific beliefs to guide us in our quest for knowledge and understanding, and if these guiding beliefs become corrupted, then our efforts can become equally misguided. If we are not careful, our science can become derailed by our legacy of well-intentioned, but ultimately misguided, theories and beliefs – particularly when this legacy continues unchecked despite many clear logical flaws and even violations of our currently-accepted laws of physics, as is the state of our science today.

Yet, despite the preceding discussions showing that our current conceptualization of "time" as an external driving entity is merely an abstract invention, there is a deeper level to this discussion. *Expansion Theory* shows that an entirely different type of time *does* exist external to our universe of expansion-driven dynamics – one that could be called *primordial time.*

Primordial Time

Recall the earlier discussion where our universe of expanding electrons could be seen as countless spherical ripples in an ocean of primordial particles that exists entirely outside the physics of our universe – literally creating and supporting our universe and its laws of physics. These ripples would likely have a characteristic speed in the ocean of primordial particles, resulting in the universal expansion rate, X_S, upon all electrons. Since it is the expansion of the electron that underlies the very existence of all matter and energy in our universe, the expansion *rate* of this particle determines the rate at which all events unfold in our universe. And, since this expansion rate is determined by the speed of ripples in the ocean of *primordial* particles that compose and define our universe, we could consider the *pace* of all the expansion-driven events in our universe to arise from the dimension of *primordial time*.

Primordial time is a very different concept from the passage of "time" as measured by a clock. A clock is merely an arbitrary mechanism made of atoms and designed to progress at regular, arbitrary intervals of "time." Primordial time, however, is the dimension that pre-dates and supports our universe of matter and "energy." A clock could never measure primordial time, since the clock's very existence and operation is based on atoms and electricity, which themselves arise from expanding electrons whose very existence and nature depends on the underlying passage of primordial time. A clock whose existence and operation is built upon the activity of expanding electrons cannot measure the primordial time dimension that pre-dates and drives even these electrons. Unlike "time," primordial time is not defined by the unfolding of events in our universe, but rather, it is the dimension that supports the very existence and expansion of electrons so that events in our universe *can* occur. We apply our abstract notion of passing "time" to events *within* our universe, while primordial time truly does pass *outside* our universe. In fact, it could be said that "time" as we define it is *itself* created and supported by primordial time.

Since the pace of primordial time (i.e. the speed of ripples in the ocean of primordial particles) determines the universal expansion rate of all electrons, X_S, it then determines the dynamics of interactions between electrons when they meet. Rapidly expanding electrons, for example, would presumably bounce off each other more vigorously than slowly expanding ones. This would affect how readily electrons settle into groups to form protons and neutrons, how readily these expanding protons and neutrons settle to form stable atomic nuclei, and the resulting nature of atomic structure and chemical bonds. Further, the nature of all forms of "energy" is also determined by the pace of primordial time, since gravity, electricity, magnetism, and electromagnetic radiation are all manifestations of expanding electrons.

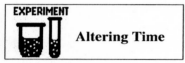

EXPERIMENT

Altering Time

Although it was shown earlier that the concept of altering time is a misguided abstraction, what about altering *primordial time*? If we could somehow alter the pace of primordial time, what effect might this have on our universe? As a starting point for answering such a question, lets first consider a simple analogy. In the broadest sense, a primordial dimension supporting the dynamics of all matter and "energy" in our universe is somewhat like a movie projector that supports the progression of a movie. As the primordial particles move along at their natural pace, they support the expansion of the spherical ripples that we call electrons which, in turn, compose and animate all matter and "energy" in our universe. Likewise, as a movie projector rolls a film along at its normal pace it supports the natural unfolding of all the characters and events in the movie. Using this simple analogy, if the ocean of primordial particles suddenly froze to a complete standstill, it would seem to follow that our universe would also freeze – just like freeze-framing a movie. In this case, not only would *primordial time* stop, but also "time" as we know it would freeze as well since the activity of all matter and "energy" in our universe would also come to a standstill.

Since this movie-projector analogy appears to work nicely for normal forward speed as well as freeze-framing, it may seem reasonable to assume that the concepts of fast-forward, slow-motion, and rewind also transfer equally well to the physical world. In that case, increasing the speed of primordial time would cause our universe to progress faster, slowing the primordial dimension would slow everything in our universe, and reversing it would cause our universe to run in reverse. This is beginning to sound like some of the more fanciful beliefs that are held about regular "time" today, but is this movie analogy really valid?

Before following the movie analogy any further, it is perhaps prudent to temper the discussion with some possible pitfalls right from the start. First, as pointed out earlier, our universe is not simply composed of electrons that happen to be expanding (rather than charged), but rather, our universe only exists *because* electrons are expanding patterns in the primordial ocean, and we cannot assume that it would exist in any form otherwise. Electrons would not *be* the electrons that compose our universe if they were not continuously expanding entities since it is this expansion that defines the nature – and the very existence – of these singular fundamental particles composing our universe. We can imagine electrons as tiny inert spheres without expansion if we wish, but this image never has and never will have any relation to reality. Therefore, if primordial time were to suddenly stop, our universe might not simply "freeze-frame," but instead, may well *cease to exist altogether.* All matter and energy may well irreversibly lose its form and function since the electrons upon which everything is based may cease to have any meaning once they stop expanding. Could our universe still return from such a state of non-existence if primordial time were simply unfrozen again? That depends on whether the primordial particles resume the same motion they had before being frozen, or whether this process forever alters their original motion. If we could somehow flash-freeze a body of water, for example, we would not expect the frozen waves and ripples to simply continue on as before if they were suddenly unfrozen since they would have lost their original momentum.

Secondly, what would it mean to run primordial time in *reverse*? It was shown that our current notion of "time" could not run in reverse since "time" is simply a made-up concept superimposed on the ongoing dynamics of the expanding matter composing our universe. Since our universe and its physical laws are entirely defined by these expanding-matter dynamics, it can only continue to exist if these dynamics continue as they are now; and so, it is meaningless to try to conceptualize "time" – which is entirely based on these ongoing dynamics – as running backward.

Likewise, the ripples in the primordial ocean that we know as electrons arise from the dynamics of this ocean of primordial particles. In order for every particle in this ocean to cease and reverse its direction, there would have to be yet another dimension of "pre-primordial time" that controls this whole ocean of particles in the same manner that a movie projector controls a movie. Otherwise, it would be next to impossible for some process within the same dimension as primordial time to cause every particle to suddenly reverse direction. That would be the equivalent of finding some process in our world that can precisely reverse the direction of every water molecule in a lake in order to reverse the advance of ripples or waves within it. Only a world that unfolds like frames in a movie, and that is driven by some external movie-projector-like process, could reasonably be expected to reverse in this way. The primordial dimension would have to be such a world driven by a further projector-like dimension beyond it in order for its ocean of primordial particles to run in reverse. And even if this feat were achieved, it would mean that electrons were *shrinking* entities instead of expanding ones. No doubt this would drastically change all laws of physics and all the qualities of matter and "energy" in the universe.

So, we can see that the seemingly straightforward notions of both *freezing* time and *reversing* time – found in much of today's science-fiction and, increasingly, even in our science – actually present enormous difficulties and have deep, far-reaching implications. This leaves only the question of time running *faster* or *slower*. While such ideas were shown earlier to be misguided when applied to our abstract notion of regular time as shown on our clocks, they are not so unreasonable when applied to primordial time; in fact, some astronomers may even have found evidence of this occurring, as we will see shortly. The reason that the concept of variations in the forward progression of primordial time differs from the problematic concepts of it freezing or reversing lies in the fact that primordial time can effectively change in *pace* quite naturally without any intervention at all. That is, while reversing primordial time required some extraordinarily complex physics to reverse all dynamics within the primordial ocean, and freezing it required all primordial particles to stop in their tracks, an alteration in the speed of its regular progression turns out to be quite natural, if not expected.

To see why this would be the case, we simply need to recall the earlier discussion of ripples on the surface of a pond. If a pebble is tossed into the pond, a circular ripple will radiate outward from the location where it landed. This circle is initially very tiny and grows extremely rapidly in size at first, then effectively slows its growth as it spreads out ever larger across the pond. The natural speed of ripples in water remains the same throughout this process, of course, but a tiny circle doubles in size far quicker than a large one given this same ripple speed (see Fig. 6-1 earlier). Therefore, although the pace of a circular ripple's expansion effectively slows down as it grows across a pond, neither the laws of physics nor even the speed of the water molecules have changed to cause this effect.

Likewise, it would be reasonable to expect the expanding electrons in the primordial ocean to follow this same principle – expanding extremely rapidly to begin with and eventually slowing as they grew. This would be a straightforward consequence of their nature as expanding patterns in a sea of primordial particles. Therefore, we should expect that primordial time has effectively slowed due to this effect as our universe has aged from its original birth through to today. This means that "energy" would have been more energetic and matter would have been more unstable and less complex and ordered than today, as all the component electrons expanded far more rapidly early on. The speed of light would have been greater, as would the speed and strength of electricity and magnetism. Also, only the simplest atoms could have formed, if indeed any could at all, since protons and neutrons would have been in a highly agitated and unstable state as their component electrons expanded rapidly against one another. Interestingly, astronomers looking back at radiation from the earliest times in our universe are beginning to believe they have found evidence that the speed of light, along with other natural constants, may have varied over the eons for some unknown reason. *Expansion Theory* suggests a likely reason for such observations, if indeed they stand up under scrutiny.

The implications that follow from this notion of the slowing of primordial time further show why altering the speed of regular "time" is a rather problematic concept. As primordial time slows down, regular time within the universe does not slow down accordingly, with clocks simply moving slower and the speed of all regular events following suit. Instead, in all probability, the very laws of nature would change. The speed of light, the speed and strength of electricity and magnetism, the stability and characteristics of atoms, and even the strength of gravity (i.e. the atomic expansion rate) would all likely change. This would result in a very different universe, with significant alterations in our known laws of physics today. It is quite possible that such differences in physics existed in our early universe, and any successful alteration of the pace of primordial time would no doubt cause similar changes to the physics of our universe rather than altering the pace of regular "time" itself. The simplistic abstract notion of "time dilation" merely due to variations in relative speed, as put forth in Einstein's *Special Relativity Theory*, neglects to consider the many deep physical implications of such a concept.

The Theory Of Everything – Has It Finally Arrived?

Our Two Theories of Everything

The state of our science today is commonly thought of as dominated by Standard Theory, with various other theories being proposed from time to time in an effort to arrive at a deeper understanding, and with the ultimate goal being the almost mythical Theory Of Everything. But, in actuality, this current state of science today *is* our "theory of everything" – for the moment. That is, Standard Theory is our best attempt at describing our universe and everything in it, and in fact, it is our *only* attempt so far. The additional theories that have arisen from time to time have not been *replacements* for Standard Theory, nor have they been truly alternate theories, but rather, they are attempts to refine or extend Standard Theory.

Quantum Mechanics, for example, is not a new "theory of everything" to replace the description of our universe found in Standard Theory, but merely a refinement and extension of the areas of subatomic particles and energy *within* Standard Theory. Neither is *Special Relativity* a new "theory of everything," but rather, an extension of how we think of matter, energy, and time within our current "theory of everything" – Standard Theory. Alternate or modified theories of gravity, such as *General Relativity*, also amount to a modification of one component of our all-encompassing Standard Theory. Therefore, not only does this mean that Standard Theory represents the present-day "theory of everything," but it also means that a full, alternate candidate to take the place of Standard Theory has never arisen. All seemingly alternate theories so far actually fall *within* Standard Theory in one way or another, leaving Standard Theory intact as our only known "theory of everything" in the history of science. Essentially, it could be said that Standard Theory is an *energy-based theory of everything* since this attempt at a unifying theory is tied together by a belief in this rather mysterious, ethereal phenomenon that runs throughout it. As such, any truly alternate "theory of everything" would have to be based entirely on a totally unique unifying principle that is so far unknown to science.

As has been shown, the new theory of expanding matter is just such a new scientific principle. *Expansion Theory* is the only theory that does not fit under the umbrella of our current energy-based Standard Theory. It is not a theory of gravity that simply extends or replaces Newtonian gravity *within* Standard Theory, but presents an entirely new theory of gravity based on a principle of expanding matter that belongs in an *entirely separate body of work*. It is not a theory of energy that refines our quantum-mechanical ideas within Standard Theory, but is an entirely new theory of energy that also clearly belongs to the *same entirely separate body of work* since it follows from the same principle of expanding matter. *Expansion Theory* is not a theory that provides yet another model of atomic structure to fit into Standard Theory, but is a new theory of the atom based, once again, on the *same principle of expanding matter* that defines this new, entirely separate body of work. We can now clearly see that Standard Theory is founded upon the mysterious concept of ethereal *energy*, while *Expansion Theory* is based on the solid physical principle of *expanding matter*. *Expansion Theory* is neither a *refinement* nor an *extension* of Standard Theory, but a complete *replacement* of it. It provides the only completely external framework – entirely outside of Standard Theory – to ever fully describe all of physics. In other words, it is the *other* "theory of everything."

We now have Standard Theory and *Expansion Theory* as two completely separate and comprehensive descriptions for the whole of physics – for the first time in history. Therefore, we now have a choice for the first time, rather than a Standard-Theory monopoly. From this new framework we can stand safely inside *Expansion Theory* and take an objective and critical look at Standard Theory without calling into question the only known explanation of our universe. We can now conceptually step inside the Standard-Theory universe and see how it feels, then literally step *entirely outside of it* and into the *Expansion-Theory* universe. How comfortable are we in a Standard-Theory universe where the laws of physics are repeatedly broken by gravity and light bulbs and fridge magnets? We can now compare this with an *Expansion-Theory* universe, where both matter and energy are clearly explained and even physically drawn in diagrams for all to see for the first time ever, and where the mysteries, paradoxes, and law violations of Standard Theory are all resolved.

Clearly, Standard Theory and *Expansion Theory* are the only two "theories of everything" that mankind has ever known. The question now is, which, if either of them, is the correct one? It is safe to say that, while Standard Theory provides us with many useful working models, it also breaks the fundamental laws of physics that form its very foundation at every turn. Not only is it full of unexplainable mysteries and paradoxes that disqualify it as the revolution in understanding that we seek, but it is not even a self-consistent body of work, contradicting itself time and time again. It is definitely a useful and necessary interim "theory of everything" – and the only one known today – but it cannot truly be considered *the* Theory Of Everything that we seek.

So then, is *Expansion Theory* now the *second* interim "theory of everything" to arrive on the scene, or is it truly the final theory – *the* Theory Of Everything? A straightforward way to determine this is to consider whether it does what the Theory Of Everything is expected to do, once discovered. Does it resolve the mysteries and paradoxes that we struggle with today? Does it provide a truly deep, clear, and comprehensive physical explanation for all known phenomena – even for those that lack such an explanation today? Does it allow us to answer further questions about these phenomena that remain unanswered today? Does it show where our natural constants come from? Does it unify the separate areas of physics, showing that all matter, energy, and forces can be explained via one simple, unifying principle that underlies all of nature? Does it finally throw a light switch to illuminate our understanding of our universe, where previously we only had disconnected glimpses in the dark? These are the features of the sought-after Theory Of Everything, and are the very reasons why we are searching for it so intently. Therefore, by definition, any alternate "theory of everything" that achieves these goals must be considered to be a serious contender for *the* Theory Of Everything. These are all qualities that can be readily found in the chapters of this book, and are the reasons why such a theory is considered the "Holy Grail" of physics.

If a fatal flaw *is* found in *Expansion Theory*, we will then be left with *two* flawed "theories of everything" at that point since a great many everyday occurrences already show Standard Theory to be fatally flawed. As has been shown throughout this book, gravity disproves Standard Theory, fridge magnets disprove it, electric charge disproves it, and light bulbs disprove it, to name just a few of the more obvious examples. Standard Theory is already a fatally flawed "theory of everything"; the only remaining question at this point is what type of "theory of everything" is *Expansion Theory*? Does its significance stop at being the only other parallel theory of the universe to ever arise, or does it extend to being the literal Theory Of Everything for which we are searching?

About the Author:

Mark McCutcheon is a 38-year-old, Canadian-born Electrical Engineer and science enthusiast who has always remained keenly aware that there are many mysteries and unanswered questions in our inherited science legacy – an awareness that has culminated in _The Final Theory_.

Printed in the United States
24140LVS00004B/46-48